CLINICAL BIOMECHANICS

CLINICAL BIOMECHANICS
A Case History Approach

Edited by

Jonathan Black, Ph.D.

Professor
Department of Orthopaedic Surgery
Associate Director
McKay Laboratory of Orthopaedic Surgery Research
University of Pennsylvania
Philadelphia, Pennsylvania

and

John H. Dumbleton, Ph.D.

Director
Research and Development
Howmedica International
Limerick, Ireland

With 28 Contributors

Churchill Livingstone
New York, Edinburgh, London, and Melbourne 1981

© Churchill Livingstone Inc. 1981

Distributed in the United Kingdom by Churchill Livingstone,
Robert Stevenson House, 1–3 Baxter's Place, Leith Walk,
Edinburgh EH1 3AF and by associated companies, branches
and representatives throughout the world.

First published 1981
Printed in USA

ISBN 0–443–08022–4

7 6 5 4 3 2 1

Library of Congress Cataloging in Publication Data

Main entry under title:

Clinical biomechanics.

  Bibliography: p.
  Includes index.
  1. Orthopedia—Case studies. 2. Human mechanics—
Case studies. I. Black, Jonathan, 1939–
II. Dumbleton, John H. [DNLM: 1. Biomechanics.
2. Orthopedics. WE103 C6405]
RD733.8.C54   617.3′092′6   80–23094
ISBN 0–443–08022–4

Contributors

Lewis D. Anderson, M.D.
Professor and Chairman
Department of Orthopaedic Surgery
University of South Alabama
Mobile, Alabama

Jurgen U. Baumann, M.D.
Director
Neuro-orthopaedics and Gait Analysis
  Laboratory
Division of Orthopaedic Surgery
Department of Surgery
University of Basel
Basel, Switzerland

Jonathan Black, Ph.D.
Professor
Department of Orthopaedic Surgery
Associate Director
McKay Laboratory of Orthopaedic
  Surgery Research
University of Pennsylvania
Philadelphia, Pennsylvania

Arnold S. Broudy, M.D.
Instructor
Department of Orthopaedic Surgery
University of Pittsburgh
School of Medicine
Pittsburgh, Pennsylvania

Edmund Y.S. Chao, Ph.D.
Director
Biomechanics Research
Department of Orthopaedics
Mayo Clinic
Rochester, Minnesota

Stanley M.K. Chung, M.D.
Division of Orthopedic Surgery
Kapiolani Children's Medical Center
Honolulu, Hawaii

A.U. Daniels, Ph.D.
Research Associate Professor of
  Surgery
Co-Director
Orthopedic Bioengineering Laboratory
Division of Orthopedic Surgery
Department of Surgery
University of Utah College of
  Medicine
Salt Lake City, Utah

John H. Dumbleton, Ph.D.
Director
Research and Development
Howmedica International
Limerick
Ireland

Harold K. Dunn, M.D.
Professor
Department of Surgery
Co-Director
Orthopedic Bioengineering Laboratory
Division of Orthopedic Surgery
Department of Surgery
University of Utah College of
  Medicine
Salt Lake City, Utah

Ian F. Goldie, M.D.
Associate Professor
Department of Orthopaedic Surgery
University of Göteborg
Göteborg, Sweden

Donald R. Gore, M.D.
Department of Orthopedic Surgery
Medical College of Wisconsin
Milwaukee, Wisconsin

Norbert G. Gschwend, M.D.
Professor
Klinik Wilhelm Schultess
Switzerland

Theodore T. Hirata, Ph.D.
Division of Orthopedic Surgery
Kapiolani Children's Medical Center
Honolulu, Hawaii

Karl H. Jungbluth, Ph.D., M.D.
Director
Department of Accident Surgery
Clinics of Surgery of the University
    Hamburg
West Germany

Donald B. Kettelkamp, M.D.
Professor
Department of Orthopaedic Surgery
Indiana University School of Medicine
Indianapolis, Indiana

Harold E. Kleinert, M.D.
Clinical Professor
Department of Surgery
University of Louisville
School of Medicine
Louisville, Kentucky

James B. Koeneman, Ph.D.
Director
Biomedical Products
Lord Corporation
Erie, Pennsylvania

A.J. Clive Lee, Ph.D.
Senior Lecturer
Department of Engineering Science
University of Exeter
Exeter, Devon
England

Robin S.M. Ling, M.D., F.R.C.S.
Consultant Orthopaedic Surgeon
Princess Elizabeth Orthopaedic
    Hospital
Exeter, Devon
England

Paul A. Lotke, M.D.
Associate Professor
Department of Orthopaedic Surgery
University of Pennsylvania
Philadelphia, Pennsylvania

Emmet M. Lunceford, Jr., M.D.
Director
The Moore Clinic
Columbia, South Carolina

Thomas W. McNeill, M.D.
Assistant Professor
Department of Orthopedic Surgery
Rush Presbyterian St. Lukes Medical
    Center
Chicago, Illinois

Larry S. Matthews, M.D.
Associate Professor
Section of Orthopaedics
The University of Michigan
University Hospital
Ann Arbor, Michigan

Bernard F. Morrey, M.D.
Consultant
Department of Orthopedics
Mayo Clinic
Rochester, Minnesota

M. Patricia Murray, Ph.D.
Chief
Kinesiology Research Laboratory
Veterans Administration Center
Wood, Wisconsin

Hans-Dieter Sauer, M.D.
Department of Accident Surgery
University Clinics of Surgery
Hamburg, West Germany

David A. Sonstegard, Ph.D.
Technical Manager
Orthopaedic Inplant Laboratory
3M Company
St. Paul, Minnesota

David H. Sutherland, M.D.
Director
Gait Analysis Laboratory
Chief
Orthopedic Surgery
Children's Hospital and Health Center
Associate Professor
Division of Orthopaedic Surgery and
  Rehabilitation
University of California
San Diego, California

Allan M. Weinstein, Ph.D.
Professor
Biomedical Engineering
Co-Director
Biomaterials Laboratory
Tulane University
New Orleans, Louisiana

Urs P. Wyss, Ph.D.
Engineer
Sulzer Bros. Ltd.
Zurich, Switzerland

Foreword

Biomechanics as a discipline developed in the 1960s and is chiefly connected with the application of engineering principles, especially principles of mechanics, to orthopaedics. The participation by engineers in orthopaedics was part of the larger change toward interdisciplinary and multidisciplinary work that was especially manifested in the biomedical field as a whole. The same period was also one of growth for orthopaedic surgery, as it was demonstrated that replacement of joints could bring dramatic relief of pain and restoration of joint function to arthritis patients. The performance of a joint replacement depends upon the conditions of operation, such as load and speed, implant design, and materials selection. Obviously there are many areas where direct application of engineering practice, knowledge, and experience could have direct benefit to implant performance. The utility of engineering concepts and techniques is, however, wider in scope and such principles may be applied in numerous non-implant areas of orthopaedic surgery and practice.

At the same time that engineers recognized opportunities and challenges in orthopaedics, orthopaedic surgeons realized that a knowledge of engineering principles would be beneficial in the solution of certain problems. While on the one hand, an engineer seeking to work in the orthopaedic area must develop a knowledge and appreciation for medicine in general and orthopaedics in particular, on the other hand, the orthopaedic surgeon must develop a knowledge and appreciation for engineering in general and mechanics and materials science in particular. To date, neither group has achieved a resounding success, but it is fair to say that engineers have made a considerable effort. However, few orthopaedic surgeons can today claim a corresponding knowledge of engineering practice.

The situation that prevails today is reflected in the attitudes of the different professionals towards the term *biomechanics*. To some extent, biomechanics is all things to all people, and published work runs the gamut from three-dimensional finite element analysis of stresses in long bones to simple descriptive accounts of the role of muscles, ligaments, and tendons during body segment movement. Somewhere in between lies a region in which there is sound orthopaedic engineering that communicates directly with the orthopaedist and contributes directly to the solution of his problems.

We hope that this book lies in that difficult-to-attain middle ground. The principal purpose of this book is to assist the orthopaedic surgeon in broadening his appreciation of engineering principles and in furthering their application in the clinical arena. The secondary goal is to provide, in relatively condensed form, accounts of orthopaedic problems with mechanical aspects of etiology and treatment. This latter purpose should make this work of interest to medical students, surgical residents, engineers, and perhaps to a more general audience who concern themselves with biomedical problems.

Unfortunately, to paraphrase Winston Churchill, a common language separates the engineer and the surgeon. The engineer is accustomed to arguing by analysis, primarily of a mathematical nature. The surgeon argues by similarity and comparison, thus making minimal use of mathematical representations except for statistical verifications of hypotheses. Biomechanics, in the sense of the application of engineering principles to medical problems, tends to progress from a mathematical base that is not accessible to the majority of medical practitioners. We cannot expect, nor would it be appropriate to demand, that the surgeon reach the level of mathematical expertise that his engineering colleague uses routinely. Thus the attempt has been made to take a sound, analytical approach that eschews mathematical complexity except where absolutely essential.

In addition to aiming for an appropriate mathematical level, the authors of this book have departed from the traditional engineering approach for a more medical approach: that of presenting the subject matter through study of groups of clinical cases. While this is of immediate advantage to the non-engineering reader in that it represents more familiar ground, the disadvantage is that the tidy, engineering approach of analysis cannot be used. But clinical medicine is not tidy, patients are individuals, and for this book to succeed, it must bring engineering principles to bear on their problems.

Each chapter presented deals with a different clinical entity or problem but may not require different biomechanical principles. Each chapter divides, more or less, into three parts. The first describes the problem or disease state along with a survey of methods of treatment and complications. The middle part presents a series of cases treated by one method and illustrates at least one case where the outcome did not meet expectations. The final part provides a view of the biomechanical aspects of the disease and its treatment and discusses the mechanical aspects of treatment inadequacies. The final chapter is an outline of biomechanical principles, most previously treated, and may be used as a ready reference for the reader.

A group of distinguished American and European clinicians were invited to contribute the balance of the chapters in this work. Many were initially enthusiastic, some reluctant. All asked, at one time or another, "What do you mean by the biomechanics (of this situation)?" In each case we answered, "What do you as a clinician, see as the important biomechanical aspects?" From this dialogue, the drafts and revisions proceeded to the finished work seen here.

J.B.
J.H.D.

Acknowledgments

The authors wish to thank the many contributors to this book for their labors. The development of the concepts covered here required much blood, sweat, and tears on the part of all.

In addition to the authors of these chapters, many others have contributed as critics, illustrators, typists, photographers, and proofreaders. While all are to be thanked, we wish to take note especially of A. Siegal who prepared the vast majority of the master photographs of x-rays and illustrations and Ms. A. Rainer whose illustrations grace many of the chapters.

Contents

1

Direct Repair and Dynamic Splinting of Flexor Tendon Lacerations

H. E. Kleinert, M.D.; A. S. Broudy, M.D.

INTRODUCTION

The repair of flexor tendon lacerations has traditionally represented one of the principal challenges of hand surgery. An optimal result with normal motion and strength requires four mechanical elements: (a) a healed tendon junction sufficiently strong to permit loading in tension without rupture; (b) the absence of incapacitating adhesions which restrict excursion; (c) the essential components of the fibroosseous tunnel or pulley system to prevent bowstringing; and (d) supple finger joints which do not compromise the capacity of the muscle-tendon units to move the phalanges through a normal arc of motion.

ANATOMIC CONSIDERATIONS

Coordinated flexion of the fingers requires the synchronous action of the extrinsic flexor tendons and the intrinsic muscles.[20] Paralysis of the lumbricals and interossei destroys the normal flexion pattern. The long flexors flex the distal and proximal interphalangeal joints prior to initiation of flexion of the metacarpal-phalangeal joints. In effect this causes the fingers to close upon themselves before they move toward the palm. The intrinsic muscles flex the metacarpal-phalangeal joints simultaneously with flexion of the interphalangeal joints. This permits the grasping of large cylindrical objects.

The mechanical efficiency of the flexor digitorum profundus and sublimis tendons depends on the gliding or excursion of each tendon, and or the presence of an effective pulley system.[10, 15] Wherever the tendons traverse a concavity (as in a flexed wrist or flexed metacarpal-phalangeal or interphalangeal joint), a pulley is essential to prevent bowstringing of the tendon. At the wrist, the transverse carpal ligament performs this pulley function for the flexor tendons.

In the fingers, the flexor digitorum sublimis and profundus pass through the fibroosseous tunnel. The floor of the tunnel comprises the periosteum overlying the phalanges and the volar plates of the metacarpal phalangeal and proximal interphalangeal joints. The roof of the tunnel consists of three major, dense fibrous bands known as annular pulleys.[6, 12] The most proximal takes origin from the junction of the volar plate and deep transverse intermetacarpal ligament. The second annular

1

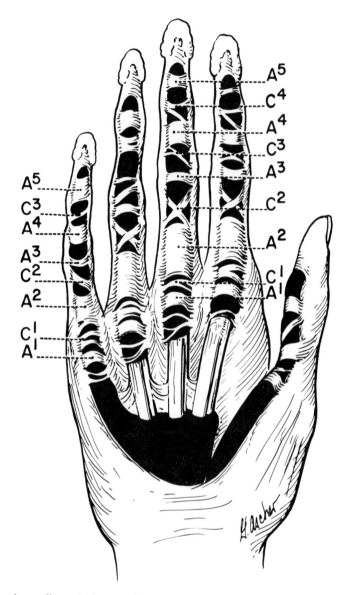

A^5
C^4
A^4
C^3
A^3
C^2

A^2

A^5
C^3
A^4

A^3
C^2

A^2

C^1
A^1

C^1
A^1

Fig. 1 – 1. Annular pulleys: A_1 is over the metacarpal phalangeal joint; A_2 is over the base of the proximal phalanx. A_3, a smaller annular pulley, is over the proximal interphalangeal joint; and A_4 is located over the middle phalanx. Some specimens appear to have an A_5 near the profundus insertion. Cruciform pulleys: C_1 is between A_1 and A_2; C_2 is over the distal proximal phalanx; C_3 is at the proximal end of the middel phalanx; and C_4 at the distal end of the middle phalanx.

pulley is located at the level of the proximal half of the proximal phalanx. A third annular pulley lies at the midportion of the middle phalanx. Adjoining the annular pulleys and overlying each joint is a thinner, flexible cruciform pulley. (Figs. 1 – 1, 1 – 2, 1 – 3)

The flexor digitorum sublimis and flexor digitorum profundus lie in intimate contact as they pass through the fibroosseous tunnel.[12] The flexor digitorum sublimis

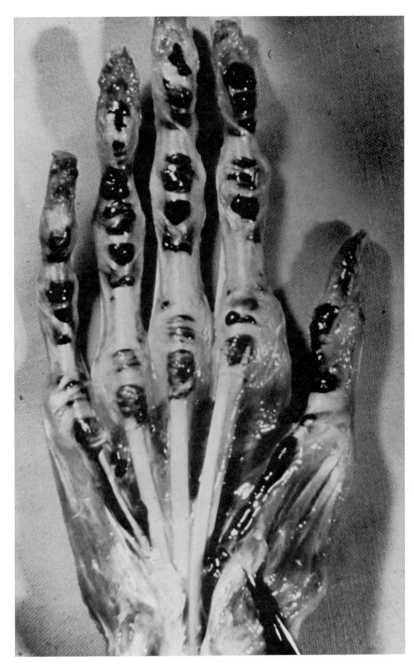

Fig. 1–2. An injected specimen delineates the fibroosseous pulley system, which may be slightly variable. There may be more than four annular and three cruciate pulleys; however, there is consistently a minimum of four annular and three cruciate pulleys.

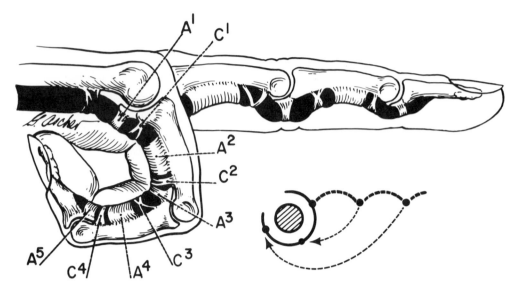

Fig. 1–3. A lateral view of the fibroosseous pulley system. Note diagram: as the finger flexes, each phalanx in succession approximates a part of a circle, easily accommodating the fingers to various-sized objects.

divides distal to the first annular pulley, winding from a volar to a dorsal position relative to the flexor digitorum profundus tendon. The two slips of the sublimis tendon merge dorsal to the profundus, and decussate to form Camper's chiasm before passing distally to insert along the middle phalanx. The presence of adhesions, or the bunching of tissues at the tendon repair site can impede the smooth gliding of the individual tendons upon each other or within the confining fibroosseous canal.

TREATMENT OPTIONS

Flexor tendon lacerations are categorized by grouping into five zones.[10] The treatment options and prognosis are related to the anatomic characteristics of each zone. (Fig. 1–4) Zone I extends distally from the insertion of the flexor digitorum sublimis to the insertion of the flexor digitorum profundus. Zone II begins in the region of the distal palmar crease at the first annular pulley. This zone encompasses the fibroosseous tunnel or pulley system. Zone III is bounded proximally by the distal edge of the flexor retinaculum or transverse carpal ligament. Zone IV is commonly referred to as the area of the carpal tunnel. All lacerations proximal to the carpal tunnel are considered to be in zone V.

The treatment options for a lacerated flexor tendon vary in each zone. In zone I only one tendon (the flexor digitorum profundus) and no critical pulleys are present.[13] Treatment may include: (a) primary repair; (b) delayed primary repair; (c) secondary repair (when the vincula remains intact); (d) tendon advancement (up to 1 cm); (e) tendon graft (select a motivated individual); (f) tenodesis (static, active); (g) capsulodesis; and (h) arthrodesis. Where primary repair is feasible, the advantages are evident. Active motion of the distal interphalangeal joint is maintained and disability time is minimized.

In zones III and V the tendons are not enclosed in a confined space. It is our

FLEXOR TENDON "ZONES"

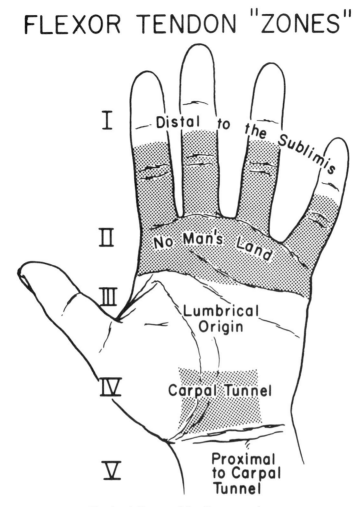

Fig. 1–4. Zones of the flexor tendons.

practice to perform primary repair of both flexor tendons. If too much tendon substance is missing to allow end-to-end approximation of the profundus tendon, a side-to-side approximation with an adjacent profundus tendon may be created. Other options include tendon graft or tendon transfer.

In zones II and IV the two flexor tendons traverse confined spaces in proximity to fibrous pulleys. Historically, the dismal results of primary repair of severed flexor tendons, particularly in zone II, stimulated a search for alternative solutions.[12] To avoid tendon repair in zone II, Bunnell promoted the use of tendon autograft.[5] The improved results he demonstrated have been reproduced by others including Pulvertaft,[17] Boyes,[1, 3] and Tubiana.[22] Tendon grafting, however, was not a panacea. The rehabilitation period was frequently excessive. The rupture of grafts and the development of adhesions necessitated additional reconstructive surgery.

In an effort to restore normal anatomy, diminish rehabilitation time, and improve the ultimate functional outcome, several workers have promoted a return to the technique of primary repair.[10, 12, 15, 23] This recent trend has been enhanced by several developments. The availability of fine, nonreactive yet strong synthetic suture material and the use of loupe magnification have improved technique. Accurate tissue apposition, with the elimination of raw surfaces and bulky repair sites, allows smooth gliding. Another major innovation has been the routine use of immediate, controlled mobilization.[15] Immediate motion with protection of the healing tendon junction serves to maintain the suppleness of the joints and to minimize the formation of heavy, firm adhesions between the healing tendon and the surrounding tissues.

TECHNIQUE OF DIRECT REPAIR AND DYNAMIC SPLINTING

Tendon repairs are routinely performed under axillary block anesthesia and tourniquet control. The preoperative cleansing is supplemented by copious irrigation with Ringer's lactate and 0.5 percent neomycin sulfate solution. The skin lacerations are extended proximally or distally with midlateral or Bruner zig-zag incisions to obtain adequate exposure. The cut end of the distal tendon segment should be visible without applying traction. The proximal segment may retract if the vincular attachments are severed. Flexion of the wrist and finger joints with milking of the forearm and hand may produce the cut tendon end. The introduction of delicate tendon forceps into the fibroosseous tunnel may facilitate retrieval of the proximal tendon segment, obviating the need for additional surgical exposure.

Once the tendon ends are exposed, the proximal tendon is transfixed in an advanced position by a small Keith needle. This relieves the tension at the repair site during the repair.

The tendon sheath may be opened by creating a flap or a window between the critical pulleys. If a flap is reflected, it can be reattached with nonreactive sutures. Tendon repair is performed under loupe magnification ($2\times$ to $4.5\times$). An intratendinous suture (modified Kessler, Mason, Allen, or modified Bunnell) is used to provide strength. The cross limbs are placed about 1 cm from the cut tendon end. Disruption of the outer surface of the tendon is avoided by grasping the cut surface only with delicate forceps.

Debridement of the cut tendon ends is performed when necessary. The intratendinous suture is tied without tension to avoid bunching of the tendon. The tendon margins are then approximated with a running, fine caliber suture of the epitenon. (Fig. 1–5)

With lacerations of both the profundus and sublimis tendons, it is our practice to repair both tendons. As the sublimis approaches its insertion the tendon slips flatten out. A simplified technique with figure-eight sutures is employed, plus a running suture at this level.

Meticulous closure of the tendon sheath may prevent abutment of the tendon repair site against the margins of the sheath defect. If the entire sheath cannot be restored, the critical pulleys must be maintained.

Following tourniquet release, wound irrigation, and the securing of hemostasis, the skin is closed with interrupted 6-0 nylon sutures. A nylon figure-eight suture loop is passed through the nail. A dorsolateral plaster splint is fashioned to maintain the wrist in 45 degrees of flexion and the metacarpal-phalangeal and proximal interphalangeal joints in about 10 degrees flexion. (Fig. 1–6) A rubber band connects the nail loop with a safety pin fastened at the wrist. The tension in the rubber band maintains the finger in flexion. Three days postoperatively the patient is allowed to initiate active extension of the injured digit. The dorsum of the finger should contact

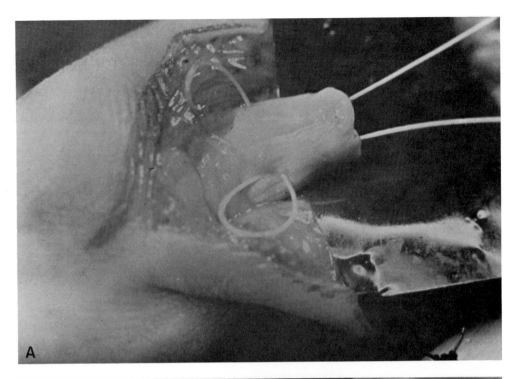

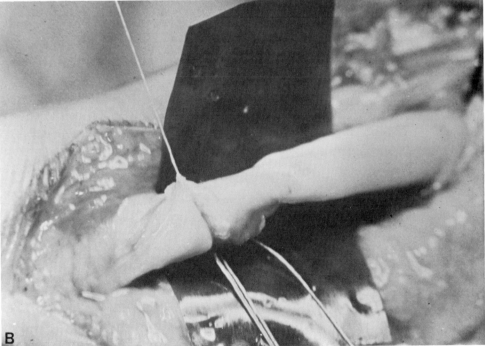

Fig. 1 – 5. (A and B) Approximation of tendon ends using the modified Kessler suture.

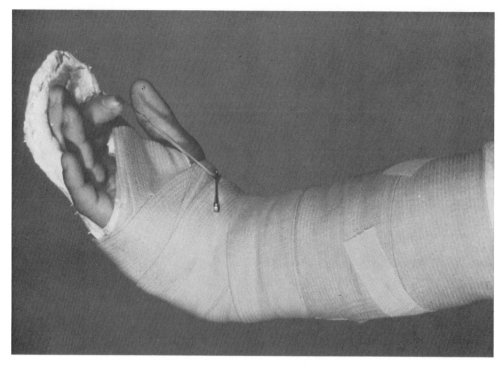

Fig. 1 – 6. Dynamic splint with rubber-band traction.

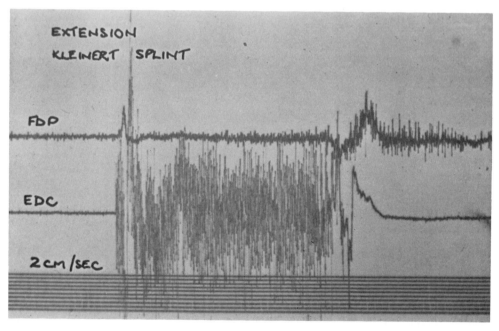

Fig. 1 – 7. EMG demonstrating extensor activity but minimal flexor activity with finger motion in the rubber band splint. (From Lister, G. D., Kleinert, H. E., Kutz, J. E., and Atasoy, E.: Primary flexor tendon repair followed by immediate controlled mobilization. J. Hand Surg., *2:* 441 – 451 1977, (by permission of the publisher.)

the splint. The patient then ceases voluntary extension and allows the tension in the stretched rubber band to return the digit to the flexed position. Electromyographic (EMG) studies have demonstrated that during active extension, the long extrinsic flexors are relaxed. (Fig. 1–7) Even if the patient does show active flexion by EMG, there is only a small burst of electrical activity since the rubber band immediately returns the finger to a flexed position.

Active flexion and passive extension are not allowed. At 24 days the splint is removed and active flexion is started (if both tendons are repaired). If only the profundus is repaired the rubber band is attached to a wrist cuff, protecting the repair for an additional 2 weeks, for two reasons: (a) delayed healing—removal of the sublimis may interfere with profundus circulation; and (b) one repair is half as strong as two. Power flexion and passive extension are avoided until 8 weeks postoperatively. Thereafter splinting is utilized to increase the arc of motion. When necessary, tendolysis is deferred until 6 months postoperatively.

CASE PRESENTATIONS

Case 1. K. T., a 17-year-old male, lacerated his right index finger when he fell on glass. Examination revealed no active flexion of the interphalangeal joints. The radial side of the finger was anesthetic. There was a transverse laceration across the entire midvolar aspect of the proximal phalanx. (Fig. 1–8A)

The wound was explored under axillary block. The tendon ends were easily re-

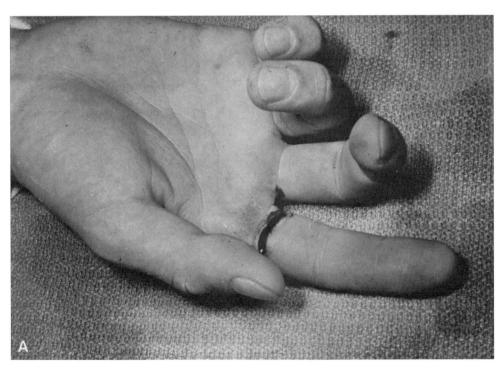

Fig. 1–8. (A) "The straight finger points the way to the lacerated flexor tendon," (Lister, G. D.) that is, loss of the physiologic position of flexion from lack of muscle tone pull. *(Fig. continues on pp. 10, 11.)*

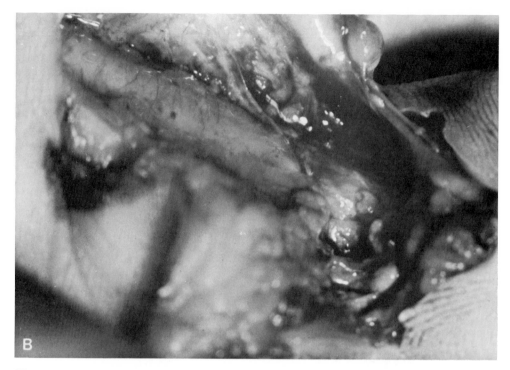

Fig. 1 – 8. (B) Exposed, lacerated fibroosseous canal with no visible tendon ends.

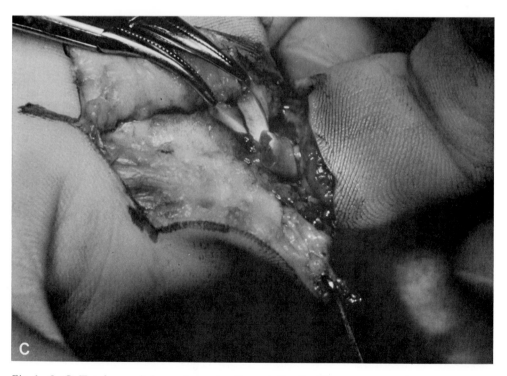

Fig. 1 – 8. (C) Tendon ends located proximally and distally. They are grasped only by their cut ends.

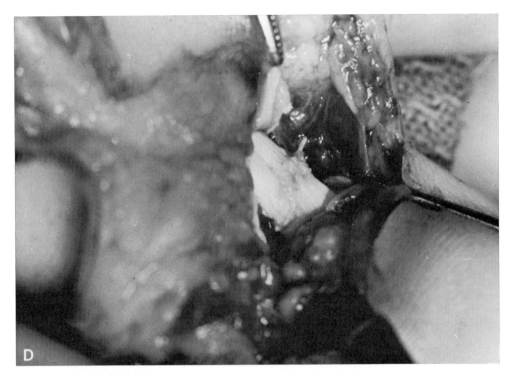

Fig. 1 – 8. (D) Repaired tendons prior to closure of pulley laceration.

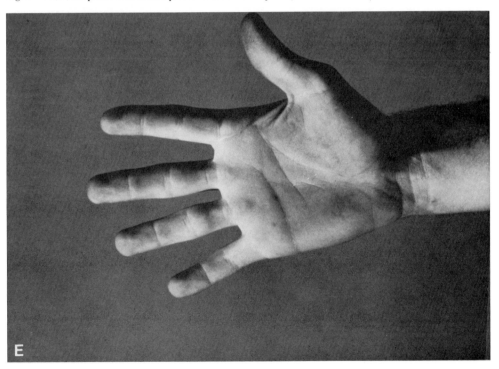

Fig. 1 – 8. (E, F, and G) Complete flexion and extension of index finger postoperatively. *(Fig. continues on p. 12.)*

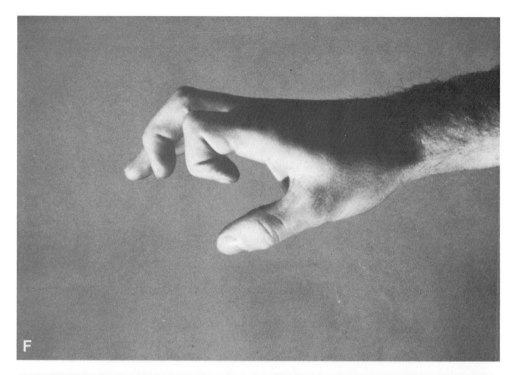

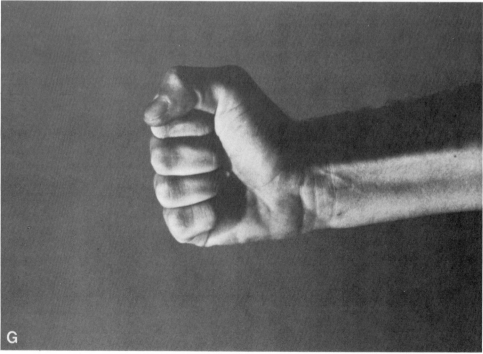

Fig. 1 – 8. (Cont.) (F and G)

trieved distally. The proximal ends were not retrieved by "milking." The proximal incision was extended, and the tendon ends were found just outside and proximal to the A_1 pulley. The A_2 pulley was lacerated. Both tendons were repaired using a modified Kessler stitch, and the A_2 pulley was repaired.[14] The radial digital nerve was repaired under magnification. A volar flexion splint with dynamic rubber-band traction was applied. After 3 weeks the splint and sutures were removed. The patient was instructed regarding extension and flexion exercises with avoidance of resistance and stress on the repaired tendons. Examination on his last visit, 11 months postoperatively, revealed complete active flexion and extension. (Fig. 1–8B to G)

Case 2. Z. F., a 44-year-old female, sustained a crushing laceration and fracture of the right middle finger. Following removal of the splint she was unable to flex the interphalangeal joints. (Fig. 1–9A)

Initial examination revealed no active flexion, and limited passive flexion, of the interphalangeal joints. After a period of passive exercises and dynamic splinting, passive flexion was obtained.

Under wrist-block anesthesia, and through a volar zig-zag incision, a flexor tendolysis was performed and nearly normal flexion was demonstrated at operation. The following day active range of motion exercises were initiated. Seven months postoperatively the patient has nearly normal flexion and extension. (Fig. 1–9B to F)

Case 3. M. F., a 21-year-old male, gave a history of having an excision of a flexor sheath ganglion of the right ring finger. Reexploration for scar tissue left him with a sensory deficit on the ulnar aspect of the finger. Cortisone injected for fullness at the base of the finger was followed by tenosynovitis and bowstringing of the flexor tendons. (Fig. 1–10A)

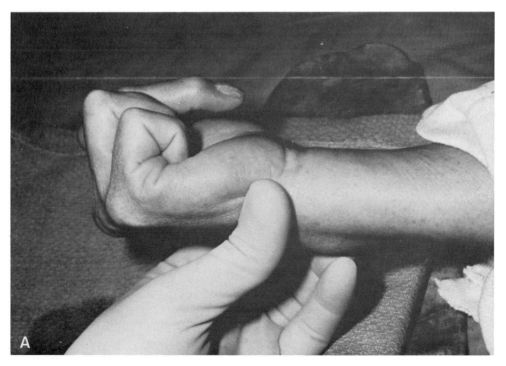

Fig. 1–9. (A) Limited flexion of middle finger following crush injury with fracture. *(Fig. continues on pp. 14, 15.)*

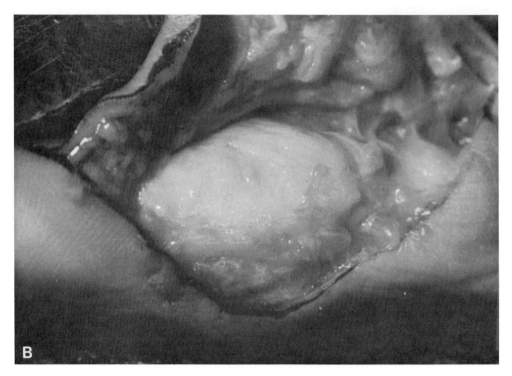

Fig. 1 – 9. (B) Scar in the pulley mechanism.

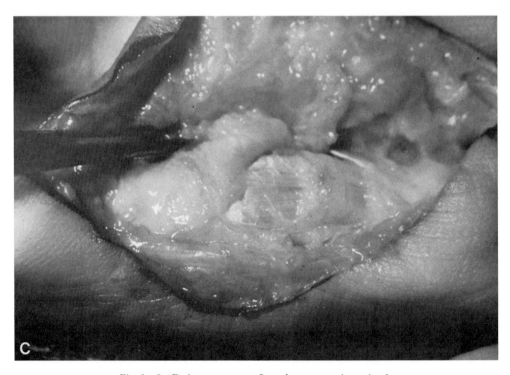

Fig. 1 – 9. (C) Appearance of tendon as scar is excised.

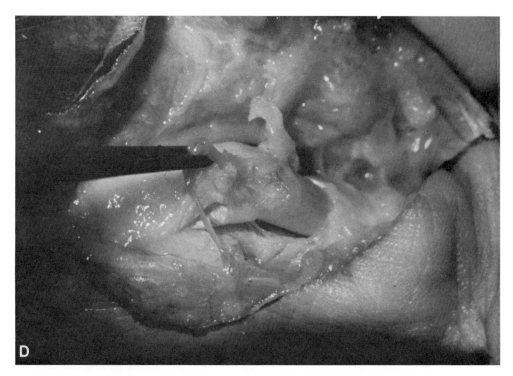

Fig. 1–9. (D) Appearance of tendon as scar is excised (cont.).

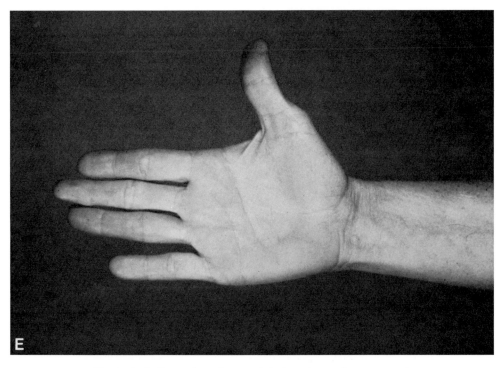

Fig. 1–9 (E) Extension after tendolysis. *(Fig. continues on p. 16.)*

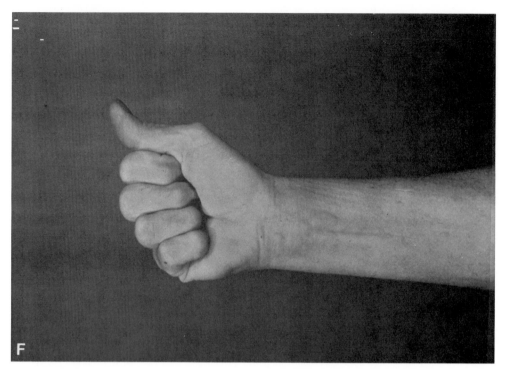

Fig. 1–9. (F) Flexion after tendolysis.

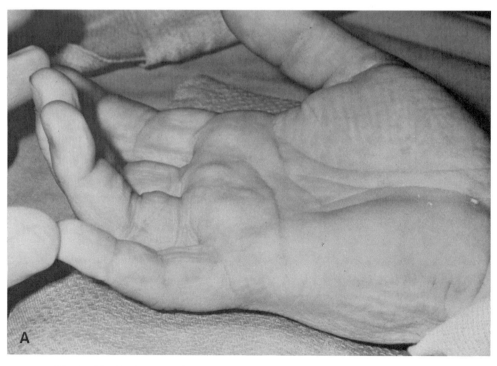

Fig. 1–10. (A) Preoperative bowstringing of flexor tendons of ring finger.

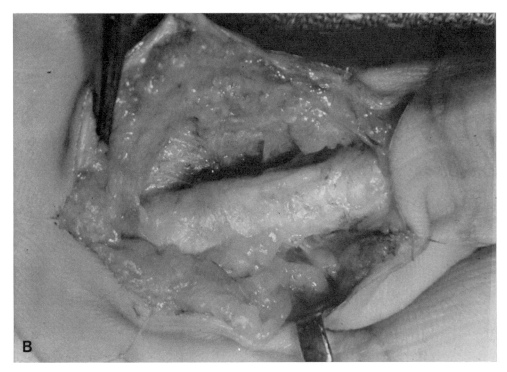

Fig. 1 – 10 (B) Exploration revealed absence of A$_2$ pulley.

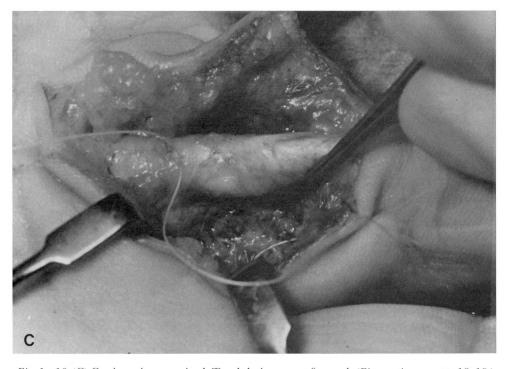

Fig. 1 – 10. (C) Copious tissue excised. Tendolysis was performed. *(Fig. continues on pp. 18, 19.)*

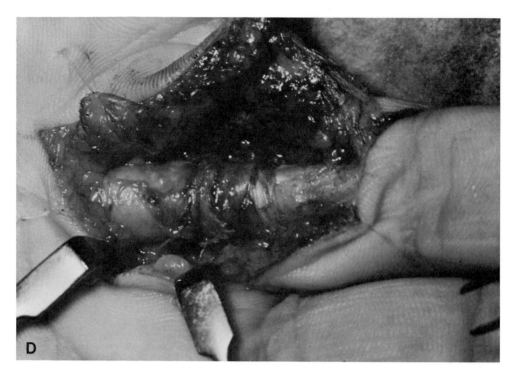

Fig. 1 – 10. (D) Pulley reconstructed using half the palmaris longus, 12 cm in length.

Exploration through the previous scar revealed absence of the A_2 pulley. (Fig. 1 – 10B) A copious quantity of synovium and granulation tissue was excised from the fibroosseous sheath. (Fig. 1 – 10C) The pathology report was of severe chronic synovitis with foreign body granuloma formation. All cultures were negative.

Pulley reconstruction was carried out with the Weilby method. (Fig. 1 – 10D and E) Half of the palmaris longus tendon, 12 cm in full length, was woven through holes in the scarred, fibrous periosteal remnants of the A_2 pulley. Postoperatively the patient was started on active motion exercises.

At 3 months postoperative he had full flexion and extension, no pain, and no evidence of bowstringing. (Fig. 10F and G).

RESULTS

In analyzing the results in a large series of flexor tendon repairs, the criteria of assessment must be scrutinized. In a series of 60 patients with a mean followup of 5.3 months, 40 repairs (67 percent) were excellent. 8 (13 percent) were good, 5 (8 percent) were fair and 7 (12 percent) were poor. The criteria used are listed in Table 1 – 1. Poor results were attributed to tendon rupture or the development of adhesions. Pediodic reviews consistently show about the same 80-percent-plus of good to excellent results.

MECHANICAL ANALYSIS

The versatility of finger motion results from the dynamic balance of muscles and tendons[4, 20]; there exists an equilibrium of active and static forces. The term visco-

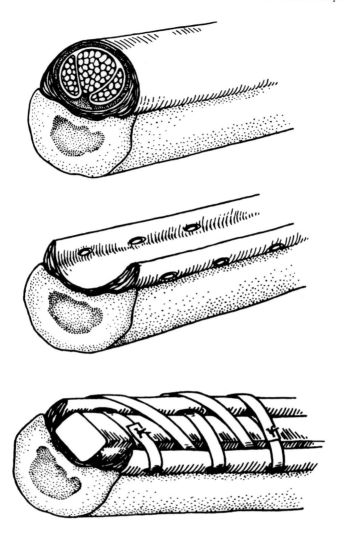

E

Fig. 1 – 10. (E) Weilby method of pulley reconstruction. A tendon graft is woven through perforations made in the fibroosseous rim. (From Kleinert, et al., Chirurgie des tendons de la main, Monographies du G.E.M., 1976, Fig. 88, p. 100.) *(Fig. continues on p. 20.)*

elastic force has been used to describe the sum of the forces that resist lengthening of a muscle tendon unit such as the flexor digitorum profundus. Components contributing to the viscoelastic force include: (a) muscle cells; (b) perimyseum; (c) fascia; (d) the tendon sheath; (e) the joint capsule; and (f) joint inertia.

Whenever the viscoelastic forces on the volar side of the finger exceed those on the dorsal side, the finger will flex. With persistence of active contraction of the flexor muscles and relaxation of the extensor muscles, finger flexion proceeds to the limit of joint flexion determined by the anatomy of the articular surfaces and the tension in the joint capsule.

The position of each finger joint is related to the position of more proximal joints. For example, extension of the wrist increases the viscoelastic forces on the volar side of the finger, producing finger flexion even without active contraction of the flexor

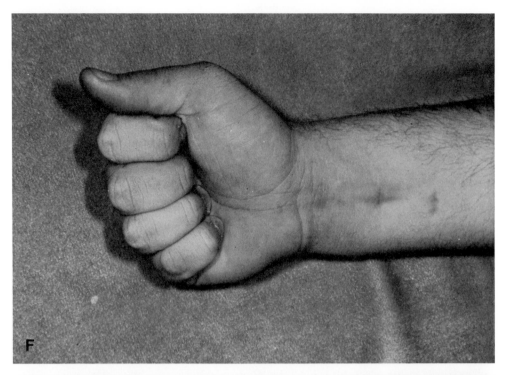

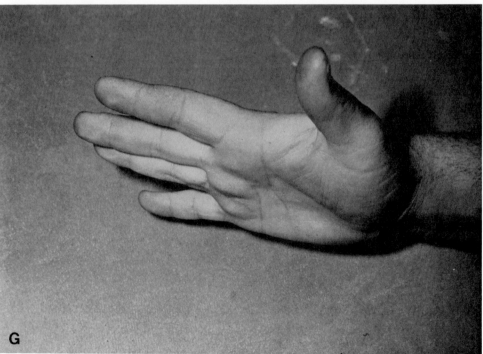

Fig. 1 – 10. (F and G) Three months postoperatively. Full flexion and extension.

TABLE 1-1. CRITERIA FOR
ASSESSMENT OF RESULTS OF
IMMEDIATE REPAIR AND
CONTROLLED MOBILIZATION IN
SEVERED FLEXOR TENDONS

Result	Flexion Deficit*	Extension Deficit**
Excellent	Grade I	Grade I
Good	Grade II	Grade II
Fair	Grade III	Grade III
Poor	Either worse than Grade III	

*Minimal distance between digital pulp and the distal palmar crease on active voluntary flexion; grade I: Pulp within 1 cm of distal palmar crease; grade II: 1.0-1.5 cm of distal palmar crease; grade III: 1.6-3 cm of distal palmar crease.
**Sum of angles that each joint lacked of full voluntary extension; grade I: Less than 15 degrees; grade II: 16-30 degrees; grade III: 31-50 degrees.
(From Lister, G. D., Kleinert, H. E., Kutz, J. E., and Atasoy, E. Primary flexor tendon repair followed by immediate controlled mobilization. J. Hand Surg. 2:441-451, 1977.)

muscles. Similarly, wrist flexion produces extension of the interphalangeal joints (tenodesis effect). In other words, joint motion is produced by viscoelastic forces as well as by voluntary active muscle contraction.

When the flexor digitorum profundus and sublimis contract, the respective tendons move proximally, exerting a force on the phalanx to which they are attached. The excursion of a tendon is defined as the distance moved by the tendon which gives rise to a given movement of a joint.

The excursion of the profundus tendons varies with each finger.[8] At the wrist the measured excursion varies from 3 cm for the index finger tendon to 4.5 cm for the ring finger tendon, with intermediate values in the other two digits. When measured at the metacarpal-phalangeal joint, the excursion is 0.5 cm smaller. The excursion of the sublimis tendons is from 0.5 to 0.75 cm smaller than for the corresponding profundus tendons. The available amplitude of a muscle is the distance moved between the fully relaxed and fully contracted position. The amplitude required to move a joint through its full range of motion may be less than the available amplitude.[4]

When a tendon crosses multiple joints, the available amplitude is usually of the same order as the sum of the required amplitudes for all of the joints. For example, with the wrist in neutral position, the flexor digitorum profundus is capable of bringing the finger into full flexion. If the wrist is fully flexed, however, full power flexion of the finger is not possible. The available amplitude of the flexor digitorum profundus is not sufficient. This explains why the wrist flexion position is preferred postoperatively for protecting the repaired tendon.

The analysis of joint motion may be simplified by considering the proximal bone at a given joint to remain stationary and the distal bone to move around it. The axis of an interphalangeal joint for finger flexion is defined as that point which does not move in relation to either phalanx. With the finger extended, the flexor tendon lies close to the axis. In flexion the tendon moves farther from the axis. This creates a longer moment arm and provides greater mechanical advantage. For simplicity the tendon can be considered to be pulling on the end of a lever perpendicular to the tendon extending from the joint axis.

The joints of the fingers are somewhat unique, owing to the presence of the pulleys.[4] These condensations of fibrous tissue create slings which hold the tendons adjacent to the plane of the skeleton. Thus the tendon lies at a constant perpendicular distance from the axis, independent of the position of the joint. The pulley, in effect, reduces the mechanical advantage of the tendon by shortening the moment arm. A given tendon excursion therefore provides more joint movement with a smaller amount of effective strength. In the hand, force is sacrificed in favor of precision and range of motion.

The contribution of the pulleys becomes evident when the pulley system is disrupted. In the absence of all pulleys, the flexor tendons bowstring across the flexed joints in the manner of fishing line tautened along a flexed rod having few eyelets. The finger contour bulges, and finger motion is lost. In a study performed on fresh cadaver specimens, the following measurements were made when maximum tension was applied to the long flexor tendons[6]: (a) absence of all pulleys resulted in failure of the fingertip to touch the palm by 25 to 30 mm; (b) the second and fourth annular pulleys were required for normal tendon function (that is, full flexion); (c) the presence of A_2 alone brought the fingertip to within 12 to 15 mm of the palm; A_2 can be narrowed from 18 to 20 mm to 5 mm without a change in function; and (d) the presence of A_4 alone brought the fingertip to within 20 to 25 mm of the palm.

When, with uniform tension in a tendon, the tendon changes direction across a pulley, there is an equal and opposite reaction force in the pulley; the system is in equilibrium in that all tension factors add up to zero.

Alternatively, one can analyze the mechanical situation by resolving the tension forces into perpendicular components. In the Y plane the pulley provides an equal and opposite reactive force. In the X plane the pulley does not alter or react to the tension. There is thus the resultant force T_x moving the tip of the phalanx proximally along the X axis.

1. Boyes, J. H.: Immediate versus delayed repair of the digital flexor tendon. Ann. West. Med. Surg., *1*:145, 1947.

2. Boyes, J. H.: Flexor tendon grafts in the finger and thumb: An evaluation of end results. J. Bone Joint Surg., *32A*:489, 1950.

3. Boyes, J. H.: Evaluation of results of digital flexor tendon grafts. Am. J. Surg., *89*:1116, 1955.

4. Brand, P. W.: Biomechanics of tendon transfer. Orthop. Clin. North Am., *5*:205–230, 1974.

5. Bunnell. S.: Repair of tendons in fingers and description of two new instruments. Surg. Gynecol. Obstet., *26*:103, 1918.

6. Doyle, J. R., and Blythe, W.: The finger flexor tendon sheath and pulleys: Anatomy and reconstruction. pp. 81–87. AAOS Symposium on Tendon Surgery in the Hand. C. V. Mosby Co., St. Louis, 1975.

7. Green, W. L., and Neibauer, J. J.: Results of primary and secondary flexor tendon repairs in no-man's land. J. Bone Joint Surg., *56A*:1216–1229, 1974.

8. Kaplan, E. B.: Functional and Surgical Anatomy of the Hand. J. B. Lippincott Co., Philadelphia, 1953.

9. Ketchum, L. D.: Primary tendon healing: A review. J. Hand Surg., *2*:428–435, 1977.

10. Kleinert, H. E., Kutz, J. E., Atasoy, E., and Stormo, A.: Primary repair of flexor tendons. Orthop. Clin. North Am., *4*:865–876, 1973.

11. Kleinert, H. E., and Meares, A.: In quest of the solution to severed flexor tendons. Clin. Orthop., *104*:23 – 29, 1974.

12. Kleinert, H. E., Kutz, J. E., and Cohen, M. J.: Primary repair of Zone 2 flexor tendon lacerations. pp. 91 – 104. AAOS Symposium on Tendon Surgery in the Hand. C. V. Mosby Co., St. Louis, 1975.

13. Kleinert, H. E., Forshew, F. C., and Cohen, M. J.: Repair of Zone 1 flexor tendon injuries. pp. 115 – 122. AAOS Symposium on Tendon Surgery in the Hand. C. V. Mosby Co., St. Louis, 1975.

14. Kleinert, H. E., and Bennett, M. B.: Digital pulley reconstruction employing the always present rim of the previous pulley. J. Hand Surg., *3(3)*:297 – 298, 1978.

15. Lister, G. D., Kleinert, H. E., Kutz, J. E., and Atasoy, E.: Primary flexor tendon repair followed by immediate controlled mobilization. J. Hand Surg., *2*:441 – 451, 1977.

16. Matthews, P., and Richards, H.: Factors in the adherence of flexor tendons after repair. J. Bone Joint Surg., *58B*:230 – 236, 1976.

17. Pulvertaft, R. G.: Tendon grafts for flexor tendon injuries in the fingers and thumb. J. Bone Joint Surg., *38B*:175, 1956.

18. Schneider, L. H., and Hunter, J. M.: Flexor tenolysis. pp. 156 – 162. AAOS Symposium on Tendon Surgery in the Hand. C. V. Mosby Co., St. Louis, 1975.

19. Schneider, L. H., Hunter, J. M., and Norris, T. R.: Delayed flexor tendon repair in no-man's land. J. Hand Surg., *2*:452 – 455, 1977.

20. Smith, R. J.: Balance and kinetics of the fingers under normal and pathological conditions. Clin. Orthop., *104*:92 – 111, 1974.

21. Smith, R. J.: Surgical treatment of the clawhand. pp. 181 – 203. AAOS Symposium on Tendon Surgery in the Hand. C. V. Mosby Co., St. Louis, 1975.

22. Tubiana, R.: Graft of the flexor tendons of the fingers and the thumb. Technique and results. Rev. Chir. Orthop., *46*:191, 1960.

23. Verdan, C.: Primary repair of flexor tendons. J. Bone Joint Surg., *42A*:647, 1960.

24. Verdan, C.: Half a century of flexor tendon surgery. J. Bone Joint Surg., *54A*:472, 1972.

25. Whitaker, J. H., Strickland, J. W., and Ellis, R. K.: The role of flexor tenolysis in the palm and digits. J. Hand Surg., *2*:462 – 470, 1977.

2
Recurrent Anterior Dislocation of the Shoulder

B. F. Morrey, M.D.; E. Y. S. Chao, Ph.D.

CLINICAL AND ANATOMIC CONSIDERATIONS

Clinical Setting

Most instances of recurrent anterior dislocation of the shoulder occur in young males, with equal frequency on the left and right sides.[48, 49, 62] Recurrence is particularly frequent in individuals in the second and third decades of life[44, 50, 62]; over 90 percent of cases of multiple recurrence occur before age 30, while more than 90 percent of cases of single dislocation occur after age 30.[46] Direct trauma to the posterior aspect of the shoulder, or from a backward fall on the flexed elbow and extended shoulder, was described by Bankart[5] as the common mechanism. However, a fall on the outstretched arm with the elbow bending, causing the humerus to extend and the humeral head to be directed anteriorly, is probably a more common mechanism of injury.[4] Indirect trauma to the abducted arm, forcing it into external rotation, is likewise frequently implicated, especially with recurrent episodes of dislocation.[49] Recurrence can occur with minimal subsequent force, depending on the nature of the initial trauma, the preinjury state of the shoulder, and the pathologic lesion caused by the initial injury.[46, 49, 59, 69] In older individuals the initial dislocation usually causes a posterior capsule rupture, a tear of the rotator cuff, or a fracture of the greater tuberosity, and recurrence is unlikely.[47] In younger individuals the initial dislocation often causes a tear or stripping of the capsule from the labrum or anterior aspect of the glenoid, and recurrent dislocation is common.[47, 60, 62]

Some feel that 3 weeks of immobilization can decrease the incidence of recurrence,[62] and Watson-Jones[73] even claimed not to have seen a dislocation following 4 weeks of immobilization. However, many feel that if anterior injury is sustained, recurrences are frequent, especially in young individuals, and that this is only moderately influenced by the mode or duration of immobilization following the initial episode.

Patients with a recurrent dislocation of the shoulder will sometimes give a family history of the condition[26, 49]; occasionally (8 percent) the condition is bilateral, and in about 2 percent of cases[48, 62] the shoulder will have dislocated posteriorly as well. These three features implicate an inherent laxity of the static stabilizer – the capsular ligaments – or an asynchronous function of the dynamic stabilizer – the subscapularis

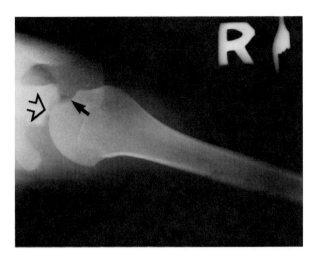

Fig. 2–1. Axillary view showing defect in humeral head (closed arrow) as well as crushed anterior glenoid (open arrow).

of the shoulder as important etiologic considerations. A biomechanical study of the shoulder motion and forces will help to understand the mechanism of injury and the method of treatment.

Roentgenographically, patients with recurrent anterior dislocation of the shoulder frequently reveal a defect of the posterolateral aspect of the humeral head (Fig. 2–1). This lesion was first described by Caird[12] in 1887, and was popularized by Hill and Sacks.[30] The incidence of this bony defect has been variously reported as between 30 percent[9, 48] and 100 percent.[16, 51, 55] The glenoid may be fractured at the time of injury,[2, 27, 36] or may become eroded from multiple recurrences (Fig. 2–1).[48] The incidence of erosion has been reported as "unusual"[1] to "frequent" (73 percent).[63] As we shall see, presence of these bony lesions assumes some degree of significance in the planning of optimum surgical treatment. Such bony defects can also be explained biomechanically. These implications will be discussed in greater detail below.

Functional and Anatomic Considerations

MOTION AND ARTICULAR ANATOMY

The shoulder complex allows greater motion than any other joint in the body. In general, most studies[25, 34, 57] indicate that approximately 60 percent of abduction of the upper extremity occurs at the glenohumeral articulation, with the remainder occurring from scapulothoracic rotation. Superior translation of the humeral head on the glenoid is minimal,[57] and anterior and posterior translation has been discussed,[65, 69] but is difficult to measure *in vivo* since routine roentgenograms do not show this relationship well.

The humeral head exhibits about 30 degrees of retroversion with respect to the plane of elbow flexion[34, 65]; this is balanced by the 30 degrees anterior angulation of the scapula on the thorax. The glenoid is retroverted about 7 degrees,[3, 66] which might offer some protection to the shoulder against anterior dislocation. The width of the glenoid is subject to great individual variation, but averages about 2 cm, which is about 57 percent as wide as the humeral head.[66] Bost and Inman[7] report that only one-third of the humeral head is in contact with the glenoid at any one time. It is therefore obvious that there is no inherent stability from joint articular congruity.

THE GLENOHUMERAL LIGAMENT COMPLEX

The shoulder capsule is thin and redundant. As such, it contributes little or nothing to the stability of the joint. The essential static constraints of the shoulder are the glenohumeral ligaments first defined by Flood in 1830.[24] This ligament complex consists of a superior, middle, and inferior portion, which form a triangular configuration based at the lesser tuberosity (Fig. 2–2).

The superior glenohumeral ligament arises from the tubercle of the glenoid near the origin of the long biceps tendon, and runs inferiorly and laterally to insert at the fovea capitus of the humerus near the proximal tip of the lesser tuberosity.[17] It has little function in preventing anterior shoulder dislocation.[3, 51, 70]

The middle glenohumeral ligament originates from the superior aspect of the glenoid and labrum, and extends laterally and inferiorly to blend with the subscapularis tendon about 2 cm medial to its insertion into the lesser tuberosity.[17, 18, 53] This ligament is a substantial structure, measuring up to 2 cm in width and 4 mm in thick-

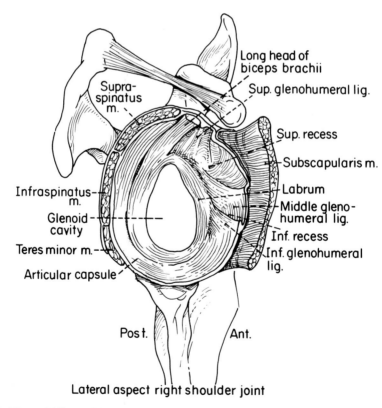

Lateral aspect right shoulder joint

Fig. 2–2. The middle and inferior glenohumeral ligaments are the important static constraints, and the subscapularis is the primary dynamic stabilizer of the shoulder preventing anterior dislocation.

ness.[17] The middle glenohumeral ligament becomes taught when the arm is externally rotated or dorsally flexed with slight abduction.[70]

The inferior glenohumeral ligament originates from the anterior inferior aspect of the glenoid labrum, and attaches to the surgical neck and inferior aspect of the lesser tuberosity.[51] It reinforces the capsule between the subscapularis and long head of the triceps,[20, 53, 70] and is often of considerable thickness, ranging up to 4 mm.[17] The inferior glenohumeral ligament stabilizes the shoulder during external rotation and abduction.[70]

In recurrent anterior dislocation of the shoulder, or in the absence of the middle glenohumeral ligament, the subscapularis bursa is found to communicate with the inferior recess of the joint.[18, 51, 59] A large bursa is evidence of lax or inadequate anterior static stabilizing ligaments, which allow the humeral head to move forward and dislocate.

THE GLENOID LABRUM

The glenoid labrum is a fibrocartilaginous tissue connecting the capsule with the bone of the anterior aspect of the glenoid. Bankart[6] and others[1, 20] have considered lesions of the labrum as essential for recurrent anterior dislocation of the shoulder. However, the labrum seems to represent simply "a fold of capsular tissue,"[51, 69] and in reality only serves as an attachment for the middle and inferior glenohumeral ligaments.[18, 69]

THE SUBSCAPULARIS MUSCLE

The subscapularis originates from the subscapular fossa and inserts into the lesser tuberosity of the humerus in intimate association with the anterior capsule.[18] This muscle is the major dynamic stabilizer to the anterior aspect of the shoulder and is active during abduction and external rotation.[34, 66, 70] Some[19, 22, 51, 70] consider the dynamic action of subscapularis more important than the static effect of capsule and ligaments in preventing anterior shoulder dislocation, and it has obtained a prominent role in the surgical procedures designed to correct the condition.

PATHOLOGIC ANATOMY

There is no single pathologic lesion common to all recurrent dislocations. A number of lesions have been emphasized, and can be summarized in the following categories: (a) lesions of articular size, shape, or orientation; (b) bony defects; (c) capsular, ligamentous, or labral disruption; and (d) subscapularis inadequacy.

Articular Size, Shape, and Orientation

A dysplastic glenoid articulation is not common, but was noted in about 3 percent of 223 surgically treated cases.[48] An anterior, rather than a posterior glenoid tilt was claimed to be present in 80 percent in one series,[15] but few investigators consider the role of anterior glenoid tilt as a common factor predisposing to recurrences. Likewise, the role of humeral retroversion and shoulder dislocation has been discussed,[64] but the data are limited and not quantitated, so no conclusions can be drawn. Few clinicians consider bony or articular features to comprise the essential lesions causing recurrent dislocation of the shoulder.

Bony Defects

Humeral Head

The initial or recurrent dislocation of the shoulder frequently produces a compression fracture of the posterior lateral aspect of the humeral head,[1, 45, 51, 63, 69] with a frequency reported as between 30 percent[20, 48] and 100 percent.[16, 49, 55] Some [9, 19, 55] believe that this compression fracture may predispose to future recurrences. Cadaver experiments reveal that the defect itself, however, does not decrease stability unless the anterior structures are cut.[68]

Glenoid

A crushed or eroded anterior glenoid margin is an effect of an initial or subsequent dislocations.[20, 26, 49, 63] Specific fractures of the glenoid have been reported with the initial shoulder dislocation.[2, 27, 36] Lesions of attrition have been recognized with frequencies of between 2 percent[16] and 44 percent.[63] Experiments removing a thin portion of the anterior glenoid, otherwise leaving the anterior structures intact, decreased the force necessary to dislocate the shoulder by about 10 percent.[68]

The Capsule and Ligaments

McLaughlin[45] states that the primary dislocation imparts an initial injury to the glenohumeral ligaments, predisposing to recurrence. Injury to the static stabilizers of the anterior glenoid consist of detachment of the anterior glenoid, a lax anterior capsule, or stripping of the capsule from the periosteum. The classic Bankart lesion is important only in that it reflects ligamentous or capsular disruption.

In 20 to 33 percent of patients,[19, 45, 60] the labrum is intact, but there is present a large tear or recess that allows the humeral head to slide into the subscapularis bursa, usually between the superior and middle glenohumeral ligaments. A posterior lateral humeral defect is invariably[51] associated with this lesion. Capsular and labral disruption, with medial periosteal stripping of the capsule,[10] has been reported with frequencies ranging from 14 to 40 percent.[19, 45, 51]

Subscapularis Inadequacy

Definite laxity of the subscapularis was observed in all of DePalma's 38 cases,[19] in all of Moseley and Overgaard's 25 cases,[51] and in "nearly every one" of Symenoides 45 patients.[68] Rowe,[63] however, was able to demonstrate this feature only rarely in a careful analysis of 124 surgical repairs. Experimentally, it has been shown[68] that repeated dislocations cause the subscapularis to stretch about 1.5 cm.

In about 10 percent of cases,[19, 45, 69] the subscapularis slides over the humeral head, allowing inferior shoulder dislocation. Inferior tearing or laxity of the subscapularis, coupled with a lax inferior capsule, predisposes to this mechanism of dislocation.

Posterior cuff disruption has been demonstrated in about 15 percent of initial dislocations, but in only 1 to 5 percent of cases having recurrences.[46,62] Posterior injury usually occurs in older individuals, which may account for the infrequent development of recurrences in this age group.[46,62]

SURGICAL OPTIONS

There is no effective nonoperative treatment for recurrent anterior dislocation of the shoulder following trauma. Over 100 surgical procedures have been described to prevent recurrence. In the last 30 years, three basic approaches have proven successful: capsular repair, subscapularis shortening, or bone block to the anterior glenoid (Table 2–1). For completeness it should be mentioned that other approaches, reported as highly successful in the properly selected patient, have been: humeral osteotomy,[12, 74] glenoid osteotomy,[64] latissimus dorsi transfer,[65] and infraspinatus transfer.[14]

Capsular Repair

As described by Perthes[56] in 1906, and popularized by Bankart,[56] capsular repairs are effective in eliminating recurrent dislocations. Although difficult surgical procedures, they are attractive because, if properly done, almost normal shoulder motion is obtained. The surgical approach most frequently transects the subscapularis tendon[20, 63, 69]; however, longitudinal splitting of this muscle has been reported.[8]

The actual capsular repair may be through holes in the anterior glenoid rim,[6, 63] through use of a metal prosthesis at the anterior glenoid,[49] with staples,[8, 20, 53] or with screws.[52] If the capsule is torn in its substance, it may simply be sutured or plicated.[69] It has been claimed that capsular procedures are effective because they create anterior scarring and because the subscapularis must be shortened when repaired.[19] The highly successful results reported by Boyd and Hunt, using a procedure which does not shorten the subscapularis, tends to negate this argument. In addition, the high (11 percent) recurrence rate reported by us,[48] and from the combined military experience,[61] suggests that simply operating on the shoulder and producing hematoma and soft tissue scarring is inadequate to reliably prevent recurrences. The recurrence rate in over 500 procedures primarily involving capsular repairs averages 3.2 percent (Table 2–1).

Transfer or Shortening of the Subscapularis Muscle

Transfer or shortening of the subscapularis are effective because they eliminate external rotation. The Putti-Platt procedure is primarily directed at shortening the subscapularis by doublebreasting the 2.5 cm of its lateral tendinous insertion under the medial muscular portion. The original description, by Osmond-Clark,[54] described suturing the lateral cuff to the "soft tissues" of the anterior scapular neck.

TABLE 2–1. RESULTS OF VARIOUS SURGICAL TECHNIQUES FOR RECURRENT ANTERIOR DISLOCATION OF THE SHOULDER

Primary Pathologic Lesion	Procedure*	Author	Recurrence %	No. Patients
Bony Procedures				
Deficient anterior glenoid	Bone block (Eden-Hybbenitte)	Palmer and Widen (55)	6.7	60
		Hindmarsh and Lindberg (31)	1.7	114
Soft Tissue Procedures				
Labrum avulsion	Bankart	Adams (1)	5.5	18
		Morrey and Janes (48)	0	16
		Rowe (63)	3.5	145
Capsuler/ligamentous	DuToit	DuToit and Roux (20)	4.6	150
		Boyd-Hunt (8)	4.1	49
	Capsular repair/ plication	Townley (69)	0	26
Lax subscapularis or capsule/ligament	Subscapularis shortening	Adams (1)	5.4	37
		Brav (9)	7.3	41
	Putti-Platt	Osmond-Clarke (54)	1.4	140
		Morrey-Janes (48)	14	132
		Quigley (58)	5	79
		Lipscomb (39)	0	93
	Magnuson-Stack	Vare (71)	0	30
		DePalma et al (19)	2.7	75
		Bryan et al (11)	7.5	53
	Capsular repair and subscapularis shortening	Watson-Jones (73)	2	52
		Weber (74)	1.6	62
		Lambdin et al (37)	5	50
"For all conditions— traumatic, congenital, paralytic"	Latissmus transfer	Saha (65)	9	45
Bony and Soft Tissue Procedures				
Labrum tear, humeral head defect, deficient anterior glenoid	Bristow	Helfet (29)	3	30
		McMurray (55)	2.7	73
		Halley (28)	3	31
		Lombardo et al (40)	2	51

*Includes modifications of procedure

Simply doublebreasting the tendon, with no attempt to attach the lateral stump to the anterior glenoid, is a popular modification of the original technique.[9, 68] The average recurrence rate following 670 Putti-Platt procedures has been 4 percent. Some have combined doublebreasted shortening of the subscapularis and repair of the capsule.[50, 73, 74] The recurrence rate in 239 combined procedures has been 3.8 percent (Table 2–1).

The same effect of shortening or tightening the subscapularis muscle is accomplished by the Magnuson-Stack procedure,[42] which transfers the subscapularis from the lesser to the greater tuberosity. A later modification[41] suggests transferring the tendon distally as well. Regardless of the anatomic defect, the procedure seems to be effective, with an average recurrence rate of 3.8 percent among 254 procedures. The loss of motion, however, is greater than that associated with capsular repairs.

Anterior Bone Block

First described in 1918 by Eden[21] and later by Hybbinette,[32] anterior bone block is a popular procedure in Europe.[31, 55] A bone graft, measuring 1 cm wide and 2.5 cm long, is taken from the iliac crest and placed at the anterior aspect of the glenoid. This provides approximately one additional centimeter of contact for the humerus at the anterior aspect of the glenoid. A failure rate of about 4 percent has been reported in 174 patients.[31, 55] We believe this procedure is best reserved for patients with anterior glenoid fracture, for whom the caracoid is inadequate to fill the existing defect and radiographic evidence of degenerative arthritis is common following the procedure.

Combined Soft Tissue Procedure with Bone Block

In 1958, Helfet[29] described the Bristow procedure, which reinforces the anterior-inferior aspect of the joint with the short head of the biceps and coracobrachialis muscles. The modified and popular version of the procedure attaches the coracoid process to the anterior glenoid by a screw, with varying amounts of subscapular shortening.[43] This procedure is felt to provide a dynamic sling to the humeral head, as well as offering an anterior buttress to prevent recurrent dislocation. This is a relatively new procedure, and in 201 reported cases there has been a recurrence rate of 2 percent.

In view of the similar, 95 percent success rate in all of the forgoing procedures, it may seem that there is little from which to choose among them. The ideal procedure is one that is easily performed and that will provide as near normal motion as possible. The most easily performed procedures shorten or transfer the insertion of the subscapularis. The procedure that provides the most normal motion is the capsular repair, but this is more difficult technically. In our experience the Bristow procedure, while not as difficult as capsular repair, does limit motion more than the latter, but is excellent for patients with surgical failures.

We feel that the surgical procedure should be individualized according to the pathologic anatomy determined by clinical, radiographic, and surgical findings. Application of biomechanical principles helps to clarify the pathogenesis of the process and the rationale of the surgical correction.

CASE PRESENTATIONS

Case 1. A 25-year-old female sustained her first dislocation while diving to catch a softball at age 23. Reduction was performed by a physician and she was held in a sling and swath for 3 weeks. She has since had multiple recurrent anterior dislocations. Roentgenograms revealed no significant posterior lateral humeral head defect, and the anterior glenoid was intact (Fig. 2–3). At surgery the patient was noted to have an intact anterior glenoid but a stretched, redundant anterior capsule, and consequently a Magnuson-Stack procedure was performed in order to limit external rotation (Fig. 2–4). Postoperatively, she has had no recurrence but has an approximately 40 degree loss of external rotation.

Case 2. A 20-year-old boy sustained a dislocation of his left shoulder when he fell on the outstretched hand while playing handball at the age of 18. In the ensuing 2

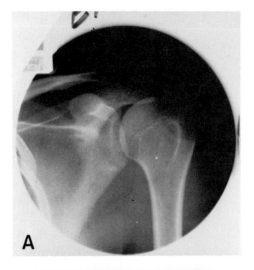

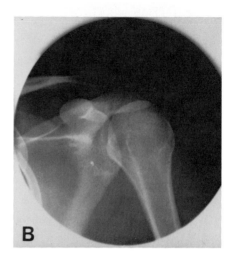

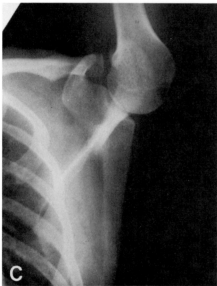

Fig. 2–3. Normal external (A), internal (B), and Stryker (C) views of the shoulder.

years he has had innumerable dislocations, including dislocations of the opposite shoulder. The internal rotation view of the shoulder was essentially normal (Fig. 2–5), although the West Point view did show a small Hill-Sachs lesion of the posteriolateral humeral head (Fig. 2–6). At surgery the humeral head defect was palpated and a small glenoid erosion was noted. The labrum was intact, but the capsule was quite redundant. The patient has had no recurrences following a Magnuson-Stack procedure (Fig. 2–7), but does have a 20 degree lack of full abduction, and an approximately 40 degree loss of external rotation.

Case 3. A 24-year-old man fell while waterskiing and sustained dislocation of the right shoulder. The dislocation was reduced under sedation anesthesia and immobilized in a velpeau dressing for 3 weeks. However, 2 months later, the patient again sustained a dislocation when involved in a motorcycle accident. The frequency of dislocations increased over a 2-year period. X-rays revealed a significant Hill-Sachs

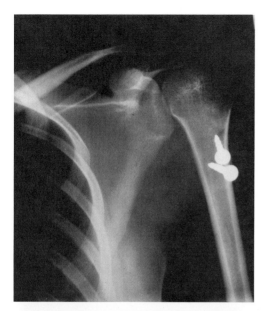

Fig. 2–4. The subscapularis muscle has been advanced laterally and distally and fixed to the humerus.

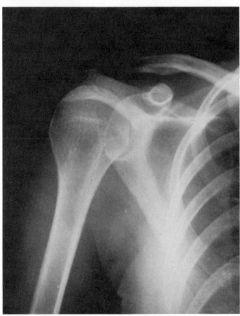

Fig. 2–5. There is no demonstrable Hill-Sachs lesion on this internal rotation roentgenogram.

lesion (Fig. 2–8). At surgery an eroded anterior glenoid was found, but the labrum was still intact. The anterior capsule was "ballooned" and the Hill-Sachs lesion was observed. The capsule was plicated and a Putti-Platt procedure was performed. The patient has sustained no recurrences over the ensuing 3 years. A year following surgery he lacks 15 degrees of external rotation, but otherwise has normal range of motion.

Case 4. A 17-year-old high school athlete sustained an anterior dislocation of his left shoulder when he crashed against an outfield fence going for a fly ball. He had multiple recurrences over a 2-year period. Roentgenograms showed a Hill-Sachs le-

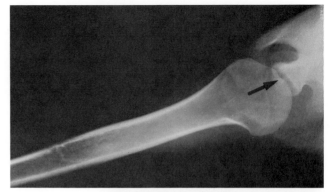

Fig. 2–6. West Point view, demonstrating small humeral head defect.

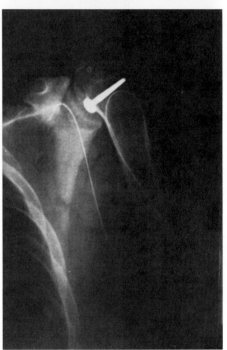

Fig. 2–7. Postoperative anterior-posterior view with shoulder in external rotation. The subscapularis muscle has been advanced laterally and fixed to the humerus.

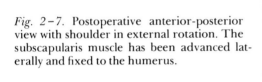

sion (Fig. 2–9). At surgery an eroded anterior glenoid was likewise noted, but the labrum was intact. A Putti-Platt procedure was performed and the patient did well for approximately 1.5 years when, while playing collegiate hockey, he fell on the outstretched arm and sustained a recurrence. He continued to be active in athletics, sustaining innumerable recurrences of the left shoulder dislocation. A second procedure was performed 7 years later, and at the time of surgery the previous repair was found to be disrupted and had stretched out. A Bristow procedure was performed (Fig. 2–10), and the patient has had no recurrences over the last 2 years.

Comment

The foregoing cases demonstrate the various lesions accompanying and causing recurrent anterior dislocation of the shoulder. In the first case there was no bony le-

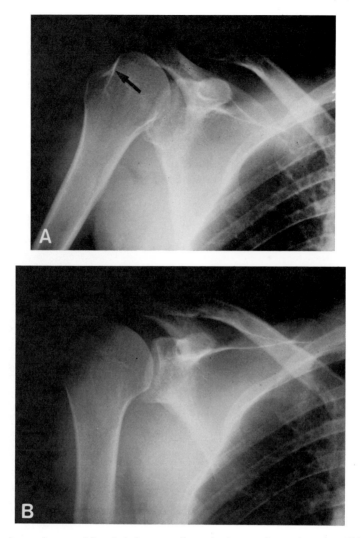

Fig. 2–8. Moderate humeral head defect noted on the internal rotation view (A), but not on the external rotation view (B).

sion, but there was a stretched, redundant capsule. In this instance, since the patient was not particularly athletic, advancing the subscapularis limited external rotation, tightened the capsule, and was a satisfactory solution.

In the second instance there was a minimal erosion of the anterior glenoid, as well as a small Hill-Sachs lesion. The patient wanted to remain active and a capsular procedure would have been a good choice in this individual. He has done well with subscapularis transfer, although he does have moderate limitation of abduction and external rotation.

In the third instance the patient had only minimal radiographic changes of the glenoid and humerus. Shortening the subscapularis and double-breasting the capsule, the Putti-Platt procedure was an effective means of stabilizing the shoulder. Again, however, the patient has mild limitation of external rotation.

In the fourth instance the patient was very active, with a mild erosion of the ante-

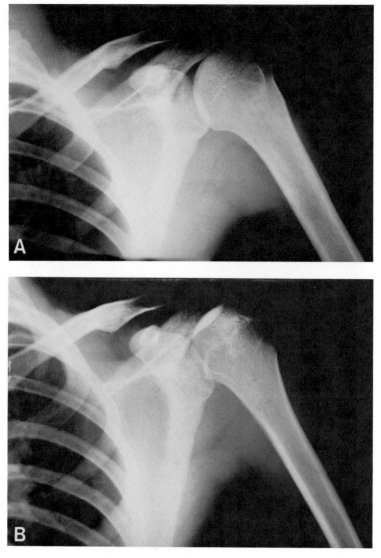

Fig. 2 – 9. Normal external anterior-posterior roentgenogram (A); suggestion of humeral head flattening on the internal rotation view (B).

rior glenoid as well as a moderately large Hill-Sachs lesion. In this instance a soft tissue procedure alone was inadequate because, though external rotation was initially limited, the repair stretched out and the patient was able to rotate sufficiently to allow the humeral head defect to engage the eroded anterior glenoid, causing instability and recurrent dislocations. The treatment of choice in such instances, we feel, is the Bristow procedure. An anterior buttress is provided by the coracoid bone graft, supplemented by overlapping the subscapularis tendon, thus limiting external rotation by a variable amount.

The above cases demonstrate that recurrent anterior dislocations of the shoulder can be prevented by any procedure that limits the amount of external rotation of the shoulder, thereby avoiding the unstable configuration that results in the dislocation.

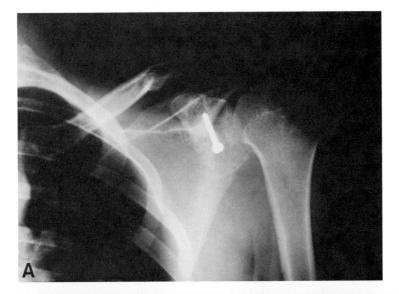

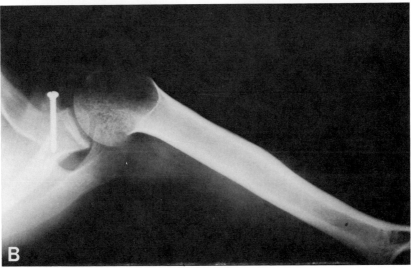

Fig. 2–10. Anterior-posterior (A) and axillary view (B) of left shoulder showing the coracoid transferred to the anterior aspect of the glenoid.

From a biomechanical point of view these procedures can be discussed on the basis of the unique pathologic characteristics of, and optional surgical procedure directed at, each lesion.

We will now analyze the biomechanics of recurrent anterior dislocation of the shoulder, and discuss the biomechanical aspects of the various surgical corrective procedures.

BIOMECHANICS

Very little is actually known about the forces at the shoulder joint. Inman et al.[34] and Walker[72] calculated forces equal to body weight at the shoulder joint when the arm is at 90 degree abduction, but both studies were simple two-dimensional analyses. Three-dimensional force representation is necessary to determine the forces tending to dislocate the humeral head anteriorlym Unfortunately, such data is not available in the literature at the present time.

Center of Rotation of the Glenohumeral Joint

The humerus rotates about an axis that passes through the approximate center of the humeral head.[57] There can be found on this axis a point which represents the center of rotation of the humeral head with respect to the glenoid. The center of rotation of a rigid body in plane or three-dimensional motion can be used as the origin in a force analysis in which all the resultant forces and moments are concentrated at that point. This concept allows the calculation of forces as though they were all acting about the center of the humeral head.

In order to analyze the contribution of ligamentous force in maintaining joint stability, the joint resultant forces (the resultant moment at the glenohumeral joint is zero, since it can be assumed as a ball-and-socket joint) must be distributed through the capsular and bone-articulating surface structure. An equipollent force analysis can be adopted for this purpose. In the present case the anterior shear resultant force must be balanced by the capsular tension, since the joint-articulating surface is unable to provide such restraint. Under dynamic conditions, the intrinsic and extrinsic muscles of the shoulder can also contribute to resist the anterior dislocation force.

Kinematics of the Glenohumeral Joint

The entire shoulder complex is a combination of four individual joints.[35] Gross examination of arm motion with respect to the trunk must recognize the contribution of all related joints of the shoulder. Since anterior dislocation of the shoulder occurs only at the glenohumeral joint, its motion must be isolated from the rest of the shoulder motion. However, the motion of the minor joints of the shoulder (scapulothoracic joint, acromioclavicular joint, and sternoclavicular joint) is strongly constrainted by the ligamentous structure, and they can thus be related to the motion of the major joint of the shoulder (glenohumeral).[38] Operative functional evaluation of patients, and the assessment of capsular tension at different shoulder positions, provides a full understanding of these subtle movements involved in the system. The finite angular motion of anatomic joints is not a vector quantity unless the motion is infinitesmal; therefore it must be path-dependent.

The shoulder is the most mobile joint in the body. Its range of motion is not only large, but can also occur in an infinite variety of combinations. A unique definition must be used to define shoulder motion in order to accurately and consistently define the relationship of the humerus and the scapula. Failure to recognize these fundamental principles can lead to unnecessary complication in defining shoulder motion, such as the well publicized Codman's paradox. Shoulder-joint motion can be defined by the standard, finite angular orientation definition used to describe rigid body relative motion in classical mechanics.[38] By a careful selection of reference axes of rotation, the sequence dependent property of shoulder motion can be eliminated. Based on this definition, an objective functional evaluation of shoulder motion can be established so as to assess the effectiveness of each reconstruction procedure for anterior dislocation of the shoulder joint.

Force Equilibrium of the Shoulder

The intrinsic muscles — the subscapularis, supraspinatus, infraspinatus, and teres minor — are active during abduction and external rotation,[3, 67, 72] and are involved in the pathogenesis of anterior dislocation of the shoulder joint. To satisfy the requirements of the three-dimensional equilibrium solution, the teres minor and infraspinatus muscles were considered as a single unit, and the effect of the deltoid was excluded. The equilibrium analysis was based on three assumptions: (a) each contributing muscle acts with a force in proportion to its cross-sectional area, as calculated by Fick,[23] and equal to 6.2 kg/cm² as calculated by Ikai et al.[33]; (b) that each muscle is equally active; (c) that the active muscle contracts along a straight line connecting the center of its insertion to the center of its origin.

The three-dimensional equilibrium configuration for the arm abducted to 90 degrees with respect to the trunk and in external rotation is demonstrated in Figure 2–11. The unloaded extremity exerts a compressive force (F_x) of 70 kg directed toward the joint articulating surface, an anterior shear force (F_z) of 12 kg, and an inferior shear force (F_y) of 14 kg. These are the orthogonal components of the glenohumeral joint resultant force produced by the active muscles resisting the gravitational force of the arm. The resultant (R) of the three forces is directed 12 degrees anteriorly.

When the arm is loaded so as to produce anterior dislocation of the shoulder, the subscapularis is the primary intrinsic muscle responsible for preventing anterior dis-

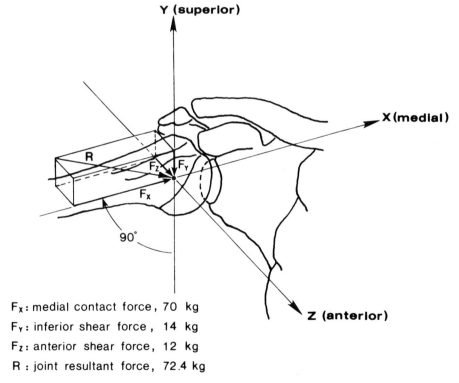

F_x: medial contact force, 70 kg
F_y: inferior shear force, 14 kg
F_z: anterior shear force, 12 kg
R : joint resultant force, 72.4 kg

Fig. 2–11. Equilibrium of the unloaded shoulder with the arm at 90 degrees abduction and 90 degrees external rotation. Note that there are inferior and anterior shearing forces. The anterior shear force is the key factor contributing to the anterior dislocation of the shoulder.

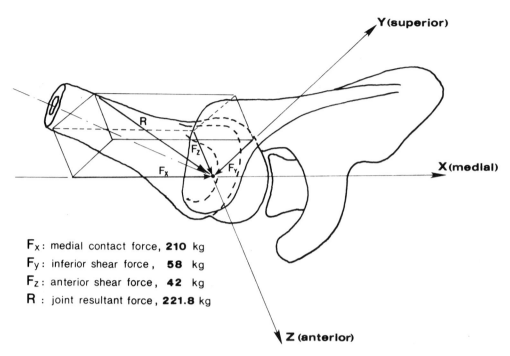

F_X: medial contact force, **210** kg
F_Y: inferior shear force , **58** kg
F_Z: anterior shear force , **42** kg
R : joint resultant force , **221.8** kg

Fig. 2–12. With the shoulder abducted 90 degrees, externally rotated 90 degrees, and extended 30 degrees, and with the intrinsic muscles maximally contracting, there is produced significant anterior shear force, which may exceed the tensile strength of the capsule structure. This is believed to be the most frequent cause of anterior dislocation.

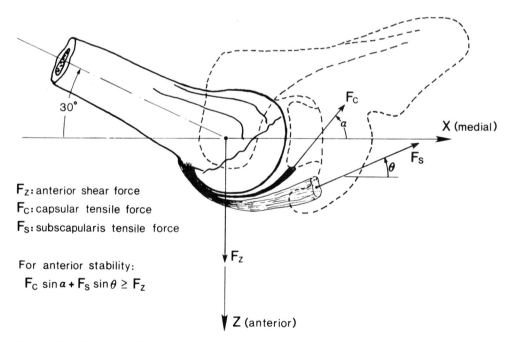

F_Z: anterior shear force
F_C: capsular tensile force
F_S: subscapularis tensile force

For anterior stability:
$$F_C \sin a + F_S \sin \theta \geq F_Z$$

Fig. 2–13. With externally applied forces, the static anterior capsular mechanism is disrupted and the humeral head dislocates. In order to prevent such a lesion, the vertical component of F_c (capsular force) and F_s (subscapularis tension) must be greater than or equal to the anterior shear force, F_z.

placement. The remainder of the intrinsics, however, contract to stabilize the humeral head in the glenoid, and this results in an anterior shear force on the humeral head with respect to the glenoid. If the shoulder is extended backward approximately 30 degrees while at 90 degrees of abduction, and is loaded so that all of the intrinsic muscles are at maximal contraction, the anterior shear force caused by the intrinsic muscles is increased to almost 42 kg (Fig. 2–12). In order to avoid anterior dislocation, this shear force must be balanced by the capsuloligamentous structure, since the articulating surface is too shallow to provide the required constraint in this direction.

Force Inequilibrium Analysis

Under passive loading conditions, the primary stabilizers of the shoulder are the capsule and glenohumeral ligament complex. During external rotation, the subscapularis muscle also provides anterior stability. Since the capsular force, F_c, is directed at an angle α to the medial line (Fig. 2–13), it must be able to generate significant tensile strength in order to resist the anteriorly directed joint shear force, F_z. Because of its common insertion site with the capsular ligament, the subscapularis muscle also contributes to the anterior stability of the shoulder joint. Again, since the subscapularis muscle force, F_s, is obliquely oriented at an angle θ with respect to the mediolateral line (axis x), only a portion of its tension can be utilized to resist the anterior shear force. Under both dynamic and static conditions, the components of the capsular force F_c and the subscapularis force F_s must be greater than or equal

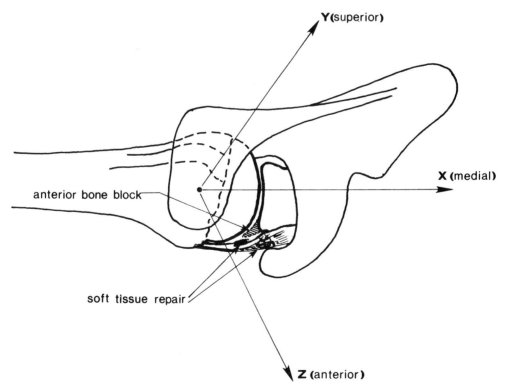

Fig. 2–14. The shoulder is stabilized by repairing the capsule and shortening the subscapularis muscle. Occasionally, anterior bone block is also used to increase the anterior constraining force by increasing the articulating surface contact area.

to the anterior shear force F_z in order to maintain anterior stability. The tensile strength of the capsule and ligament complex averages approximately 50 kg.[60] Hence, with a load causing the shoulder to be forced in abduction, external rotation, and extension, an imbalance of forces can exceed these static and dynamic constraints, causing dislocation (Fig. 2 – 13).

Once the anterior capsular ligaments are torn, less force is required for subsequent dislocations of the shoulder. With the same injury the subscapularis tendon may sometimes be torn or stretched up to 1.5 cm in length,[68] further reducing the constraining force available to prevent recurrent anterior dislocation of the humerus on the glenoid under both static and dynamic conditions. The constraining effect of the capsule and subscapularis can be significant as the tensile strength of this complex has been calculated to be about 120 kg.[60] Under normal physiologic conditions, the subscapularis is capable of generating approximately 85 kg/cm² force. On the basis of the length-tension relationship, less force can be generated if the subscapularis has been stretched. Therefore, surgery may be indicated to shorten the subscapularis, as well as to repair the capsule, and can theoretically increase the anterior stability of the shoulder by restoring the normal tensile strength of the capsule and muscle (Fig. 2 – 14). By shortening the stretched subscapularis to restore its physiologic resting length, greater force of contraction is likewise obtained.

In addition to the dynamic and static constraints of the shoulder joint, bony defects are also important in the mechanism of shoulder dislocation. If, as the arm ex-

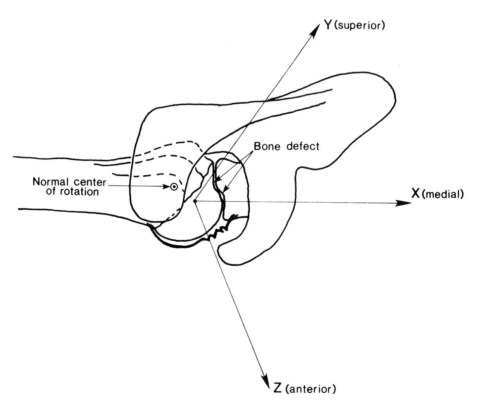

Fig. 2 – 15. A defect in the humeral head allows the center of rotation to be displaced anteriorly and medially, since the anterior constraining structures are not taut and dislocation ensues.

ternally rotates and before the capsular structures are taut, a large defect is present in the posterolateral humeral head, the defect will contact the anterior aspect of the glenoid, which may also be indented owing to articular surface erosion. The large joint-surface compressive force (70 kg) can then displace the glenoid into the defect, causing the center of rotation to shift anteriorly and medially and the humeral head to sublux anteriorly against the lax capsular mechanism (Fig. 2–15). The anterior bone block extends the range of motion before this unstable pattern is realized. From a biomechanical point of view, the bone block can increase the force necessary for subluxation by increasing the range of motion through which the shoulder is stable (Fig. 2–14).

SUMMARY

As the plethora of successful surgical procedures attest, there is no single, essential lesion to account for recurrent shoulder dislocation. Static (capsular ligaments) and dynamic stabilizers (the subscapularis muscle), as well as bony congruity (the humeral head and anterior glenoid) are important to shoulder stability; and, in a given instance, any or all of these factors may be implicated in the recurrence of dislocations. As illustrated, reinforcing the anterior static as well as dynamic stabilizers can correct the problem. If there is significant bony involvement, it is usually in the presence of an associated soft tissue defect, and procedures directed at these structures are normally successful in preventing recurrence. The anterior component of the resultant force produced by the intrinsic muscles during maximal contrasture is approximately equal to the static tensile strength of the capsule and ligaments. The dynamic stabilizer of the shoulder, the subscapularis muscle, is double breasted or advanced, returning the stretched muscle fibers more nearly to their physiologic length, and thus is a more effective shoulder stabilizer. Finally, the bony lesions both of the humeral head and anterior glenoid are secondary to the recurrent dislocations and their presence does not preclude a good result with either the static or dynamic soft tissue repairs mentioned above. If the bony defects are significant, bone graft procedures are effective either alone or in combination with static or dynamic soft tissue procedures.

REFERENCES

1. Adams, J. C.: Recurrent dislocation of the shoulder. J. Bone Joint Surg., *30B*:26, 1948.
2. Aston, J. W., and Gregory, C. F.: Dislocation of shoulder with significant fracture of the glenoid. J. Bone Joint Surg., *55A*:1531, 1973.
3. Basmajian, J. V., and Bazant, F. J.: Factors preventing downward dislocation of the adducted shoulder joint. J. Bone Joint Surg. *41A:* 1182, 1959.
4. Bateman, J. E.: The Shoulder and Neck. W. B. Saunders Co., Philadelphia, 1972. p. 390.
5. Bankart A. S. B.: Recurrent or habitual dislocation of shoulder joint. Br. Med. J., *2:* 1131, 1923.
6. Bankart, A. S. B.: The pathology and treatment of recurrent dislocation of the shoulder joint. Br. J. Surg. 26:23, 1938.
7. Bost, F. C. and Inmann V. T. G.: The pathological changes in recurrent dislocation of the shoulder. J. Bone Joint Surg., *24:*595, 1942.
8. Boyd, H. B., and Hunt, H.: Recurrent dislocation of the shoulder. J. Bone Joint Surg., *47A*:1514, 1965.

9. Brav, E. A.: Recurrent dislocation of the shoulder. Ten years experience with the Putti-Platt reconstruction procedure. Am. J. Surg., *100*:423, 1960.

10. Broca, A., and Hartmann, H.: Contribution a l'etude dis luxations de l'epaule. Bull. Soc. Anat. (Paris) *5me, Serie 4*: 312, 1890.

11. Bryan, R. J., DiMichele, J. D., Toul, G., and Cary, O. R.: Anterior recurrent dislocation of the shoulder. Cl. Ortho. 63:177, 1969.

12. Caird, R. M.: The shoulder joint in relation to certain dislocations and fractures. Edinburgh Med. J. 32:708, 1887.

13. Chaudhuri, G. K., Sengupta, A., and Saha, A. K.: Rotation osteotomy of the shaft of the humerus for recurrent dislocation of the shoulder: Anterior and posterior. Acta Orthop. Scand., *45*:193, 1974.

14. Connally, J.: X-ray defects in recurrent shoulder dislocation. J. Bone Joint Surg., *51A*: 1235, 1969.

15. Das, S. P., Ray, G. S., and Saha, A. K.: Observation of the tilt of the glenoid cavity of the scapula. J. Anat. Soc. India *15*:114, 1966.

16. DeAnquin, C. E.: Recurrent dislocation of the shoulder: Roentgenographic study. J. Bone Joint Surg., *47A*:1085, 1969.

17. DeLorme, D.: die Hemmungs bander des Schultergelenks and ihre Deductung fur die Schulter Luxationen. Arch. fur Klin. Chirurg., *92*:79, 1910.

18. DePalma, A. F., Callery, G., and Bennett, G. A.: Variational anatomy and degenerative lesions of the shoulder bone. AAOS Ins Course Lect *16*:255, 1949.

19. DePalma, A. F., Coker, A. J., and Probhaker, M.: The role of the subscapularis in recurrent anterior dislocation of the shoulder Clin. Orthop., *54*:35, 1969.

20. DuToit, G. T., and Roux, D.: Recurrent dislocation of the shoulder: A 24-year study of the Johannesburg Stapling operation. J. Bone Joint Surg., *38A*:1, 1956.

21. Eden, R.: Zur operation der habitullen schulter luxation mitterlung lines nuren verhaberens bie abriss am inneren pfannenroude. Deutsch Ftschr. Chir *144*:269, 1918.

22. Fahey, J.: Anatomy of the shoulder. Regional Orthopedic Surgery and Fundamental Orthopaedic Problems, No. 11, 1947.

23. Fick, R.: Spezille Gelenk and Nuskelmuchanik, Mechanik Deskniegelenkes. Jena, G. Fischer, *111*:521, 1911.

24. Flood, V.: Discovery of a new ligament of the shoulder joint. Lancet, *671*, 1829–1830.

25. Freeman, L., and Munro, R.: Abduction of the arm in the scapular plane. Scapular and glenohumeral movement. J. Bone Joint Surg., *48A*:1503, 1966.

26. Gallie, W. E., and LeMesurier, A. B.: Recurring dislocation of the shoulder. J. Bone Joint Surg., *30B*:9, 1948.

27. Hall, R. H., Issac, F., and Booth, C. R.: Dislocation of the shoulder with special reference to accompanying small fracture. J. Bone Joint Surg., *41A*:489, 1959.

28. Halley, D. K.: A review of the Bristow operation for recurrent anterior shoulder dislocation in athletes. Clin. Orthop., *106*:175, 1975.

29. Helfet, A. J.: Coracoid transplantation for recurring dislocation of the shoulder. J. Bone Joint Surg., *40B*:198, 1958.

30. Hill, H. S., and Sacks, M.D.: The grooved defect of the humeral head. A frequently unrecognized complication of dislocations of the shoulder joint. Radiology, *35*:690, 1940.

31. Hindmarsh, J., and Lindberg, A.: Eden Hybbinette. Operation for recurrent dislocation of the sternoscapular joint. Acta Orthop. Scand., *38*:459, 1967.

32. Hybbinette, S.: De la transplantation d'un fragment osseux pour remedier aux laxations recidinants de l'epaule conotations it usultots operatiores. Acta Chir. Scand., *71*:411, 1932.

33. Ikai, M., and Fukunaga, T.: Calculation of muscle strength per unit of cross-sectional area of human muscle. Z Angew Physiol Einschl Argeitsphysiol *26*:26, 1968.

34. Inman, V. T., Saunders, M., and Abbot, L. C.: Observations on the function of the shoulder joint. J. Bone Joint Surg. *26:*1, 1944.

35. Kapandji, I.: The Physiology of Joints, vol. 1. The Williams and Wilkins Co., Baltimore, 1970.

36. Kummel, B. M.: Fracture of the glenoid causing chronic dislocation of the shoulder. Clin. Ortho p., *69:*189, 1970.

37. Lambdin, C. S., Young, S. B., and Unsicker, S. L.: A modified bankart Putti-Platt shoulder capsularrophy. J. Bone Joint Surg., *53A:*1237, 1971.

38. Landon, G. C., Chao, E. Y., and Cofield, R. N.: Three dimensional analysis of angular motion of the shoulder complex. Trans. Orthop. Res. Soc., *3:*297, 1978.

39. Lipscomb, A. B.: Treatment of recurrent anterior dislocation and subluxation of the glenohumeral joint in athletes. Clin. Orthop., *109:*122, 1975.

40. Lombardo, S. J., Kerlan, R. K., Jobe, F. W., Carter, V. S., Blazina, M. E., and Shields, C. L.: Modified Bristow procedure for recurrent dislocation of the shoulder. J. Bone Joint Surg., *58A:*256, 1976.

41. Magnuson, P.: Treatment of recurrent dislocation of the shoulder. Surg. Clin. North Am., *25:*14, 1945.

42. Magnuson, P., and Stack, J.: Recurrent dislocation of shoulder. J.A.M.A., *123:*898, 1943.

43. May, V. R. Jr.: Recurrent dislocation of the shoulder. J. Bone Joint Surg., *52A:*1010, 1970.

44. McLaughlin, H. L., and Carallero, W. U.: Primary anterior dislocation of the shoulder. Am. J. Surg., *80:*615, 1950.

45. McLaughlin, H. L.: Recurrent anterior dislocation of the shoulder. I. Morbid anatomy. Am. J. Surg., *99:*628, 1960.

46. McLaughlin, H. L., and MacLellan, D. I.: Recurrent anterior dislocation of the shoulder. II. A comparative study. J. Trauma, *7:*191, 1967.

47. McMurray, T. B.: Recurrent dislocation of the shoulder. J. Bone Joint Surg., *43B:*402, 1961.

48. Morrey, B. F., and Janes, J. J.: Recurrent anterior dislocation of the shoulder. Long-term follow-up of the Putti-Platt and Bankart procedures. J. Bone Joint Surg., *58A:*252, 1976.

49. Moseley, H. E.: Athletic injuries to the shoulder region. Am. J. Surg., *98:*401, 1959.

50. Moseley, H. E.: Recurrent Dislocation of the Shoulder. McGill University Press, Montreal, 1961.

51. Moseley, H. E., and Overgaard, B.: The anterior capsular mechanism in recurrent anterior dislocation of the shoulder. J. Bone Joint Surg., *44B:*913, 1962.

52. Muller, M. E., Allgower, M., and Willernegger, H.: Manual of Internal Fixation. Springer-Verlag, Berlin, 1970.

53. Ogilvie, H.: Recurrent dislocations of the shoulder. The Johannesburg Staple Driver. Br. Med. J., *1:*362, 1946.

54. Osmond-Clarke, H.: Habitual dislocation of the shoulder. The Putti-Platt operation. J. Bone Joint Surg., *30B:*19, 1948.

55. Palmer, I., and Widen, A.: The bone block method for recurrent dislocation of the shoulder joint. J. Bone Joint Surg., *30B:*53, 1948.

56. Perthes, G.: Uber Operationen bie habitueller Schulterbuation. Dtsch. Z Chir., *85:*199, 1906.

57. Poppen, N. K., and Walker, P. S.: Normal and abnormal motion of the shoulder. J. Bone Joint Surg., *58A:*195, 1976.

58. Quigley, T. B., and Friedman, P. A.: Recurrent dislocation of the shoulder: A preliminary report of personal experience with 7 Bankart and 92 Putti-Platt operations in 99 cases. Am. J. Surg., *128:*595, 1974.

59. Reeves, B.: Arthrography of the shoulder. J. Bone Joint Surg., *48B:*424, 1966.

60. Reeves, B.: Experiments on the tensile strength of the anterior capsular structures of the shoulder region. J. Bone Joint Surg., *50B:*858, 1968.

61. Roberts, R. R.: Tri-service medical survey on shoulder dislocations, Bureau of Medicine and Surgery, U.S. Navy Dept., 1963 (Unpublished).

62. Rowe, C. R.: Prognosis in dislocation of the shoulder. J. Bone Joint Surg., *38A:*957, 1956.

63. Rowe, C. R., Patel, D., and Southmayd, W. E.: The Bankart procedure: A long-term end result study. J. Bone Joint Surg., *60A:*1, 1978.

64. Saha, A. K.: Theory of shoulder mechanism. Charles C. Thomas, Springfield, IL, 1961.

65. Saha, A. K.: Anterior recurrent dislocation of the shoulder. Acta. Orthop. Scand., *38:* 479, 1967.

66. Saha, A. K.: Dynamic stability of the glenohumeral joint. Acta Orthop. Scand., *42:*491, 1971.

67. Schering, L. E., Pauly, J. E.: An electromyographic study of some muscles acting on the upper extremity of man. Anat. Rec., *135:*239, 1959.

68. Symenoides, P. O.: The significance of the subscapularis muscle in the pathogenesis of recurrent anterior dislocation of the shoulder. J. Bone Joint Surg., *54B:*476, 1972.

69. Townley, C. O.: The capsular mechanism in recurrent dislocation of the shoulder. J. Bone Joint Surg., *32A:*370, 1950.

70. Turkel, S. J., Panio, M. W., Marshall, J. L., and Girgis, F. G.: Stabilizing mechanisms of the glenohumeral joint: The relative role of the muscles and ligaments in anterior dislocation. (In press).

71. Vare, V. B.: The treatment of recurrent dislocation of the shoulder. Surg. Clin. North Am., *33:*1703, 1953.

72. Walker, P. S.: Human joints and their artificial replacements. Charles C. Thomas, Springfield, IL, 1977.

73. Watson-Jones, R.: Fractures and Joint Injuries, 4th Ed. The Williams and Wilkins Co., Baltimore, OH. 1957.

74. Weber, B. G.: Operative treatment for recurrent dislocation of the shoulder. Injury. *1:* 107, 1969.

3
Plate/Rod Fixation of Dual Forearm Fractures

L. D. Anderson, M.D.; J. H. Dumbleton, Ph.D.

INTRODUCTION

This chapter will deal exclusively with diaphyseal fractures of both the radius and ulna in adults. Specifically, it will discuss the treatment of fractures of both bones of the forearm by open reduction and internal fixation with plates and screws, or with intramedullary rods. Since almost all fractures of the radius and ulna in children can be successfully treated by closed methods, the chapter will not cover these fractures in children. At times, however, in adolescents with little potential left for growth and remodelling, satisfactory closed reduction may not be obtained and maintained; treatment of these fractures in such patients should be the same as that for fractures in adults.

Time and space do not permit discussion of the unstable fracture dislocations of the forearm bones. These are the Monteggia lesion (fracture of the ulna with dislocation of the head of the radius) and the Galeazzi lesion (fracture of the distal one-third of the radius with dislocation of the distal radio-ulna joint). However, many of the points covered in relation to diaphyseal fractures of both bones of the forearm apply to these lesions also. For a detailed discussion of the Monteggia and Galeazzi lesions the reader is referred to the articles by Bado,[5] Reckling and Cardell,[28] Hughston,[19] Evans,[18] Boyd and Boals,[8] and Anderson.[2, 3]

Most articles about fractures of both bones of the forearm include a discussion of the surgical anatomy of the forearm. The anatomy of the forearm creates certain. problems not encountered in the treatment of fractures of the shaft of the humerus, femur, and tibia. The radius and ulna are roughly parallel but come into actual contact only at the ends of the bones. Proximally the bones are held together by the capsule of the elbow joint and the annular ligament. Distally the bones are linked by the anterior and posterior radio-ulnar ligaments, the fibrocartilaginous triangular disc, and the capsule of the wrist joint.

Superficially, this sounds as if there were only two joints. However this is far from the case, since these two joints function as many joints to allow the various complicated movements which take place at the elbow, forearm, and wrist. Proximally these joints include the ulno-humeral, the radiohumeral and the proximal radio-ulnar joints. Distally the articulations are the distal radio-ulnar and the radio-carpal joints.

The ulna is a relatively straight bone. Its medullary canal is relatively narrow and approximately of the same diameter in the middle and distal thirds of the bone. In the proximal one-third of the bone, however, the medullary canal is quite large and increases in diameter as the elbow is approached. On the other hand, the radius is more complex. In 1959, Sage[31] reported a study of 100 radii dissected from cadavers, and described the normal diameter of the medullary canal as well as the various curves and angles of the bone. The importance of maintaining the normal radial bow in the middle third of the radius was pointed out if good pronation and supination were to be achieved. Patrick,[27] has also pointed out that the normal range of pronation is limited by the radius crossing over the ulna and compressing the deep flexor muscles between the two bones. Therefore, anything that encroaches upon this space, such as fibrous tissue or callus, will limit pronation.

The soft tissues of the forearm are also complex. The interosseous membrane is a dense structure with fibers that run from origins on the radius distally to insert on the ulna. The fibers of the interosseous membrane are tense with the forearm in full supination and relaxed in full pronation. Two groups of muscles directly join the radius and ulna. These are the supinator and the pronator quadratus. The supinator and pronator quadratus each take origin on one of the forearm bones and insert on the other. Thus, in a fracture of both bones, these muscles, in addition to their given function, act to approximate the radius and ulna and compromise the interosseous space.

Two other muscles that arise outside the forearm also affect pronation and supination. The pronator teres arises from the medial epicondyle of the humerus and inserts into the mid-radius. It is a strong pronator. In fractures of both bones of the forearm it not only pronates but also coapts the fragments to narrow the interosseous space. The biceps brachii inserts on the bicipital tuberosity of the radius and not only acts to flex the proximal radius, but is also a very strong supinator.

Sage[31] notes that those forearm muscles that arise from the volar side of the forearm and insert into the radial side of the wrist or hand, such as the flexor carpi radialis, exert a pronating force. Similarly, muscles that arise from the dorsal side of the ulna and the interosseous membrane, and insert on the radial side of the wrist and thumb (the abductor pollicis longus and brevis, and the extensor pollicis longus), exert a supinating force.

Thus, in treating fractures of both the radius and ulna in adults, a very complex situation obtains. If a successful functional result is to be achieved, the following criteria must be met for reduction and maintenance of position of the fractures until firm union has taken place:

1.   The length of each bone must be maintained.
2.   Axial and rotational alignment must be restored.
3.   The radial bow must be maintained and encroachment of callus into the interosseous space must be kept to a minimum.

In addition, the method should not require extended immobilization for such a long period as to interfere with restoration of function

Fractures of the shaft of the radius and ulna are almost always the result of a high-velocity trauma. Most of these fractures result from motor vehicular accidents. The injured person may be an occupant of the vehicle or a pedestrian struck by the vehicle. An increasing percentage of fractures of the radius and ulna in recent years has been the result of motorcycle accidents, and these are frequently open fractures. In these vehicular accidents, events happen so rapidly that the patient often has no clear recollection of the actual cause of injury. In most of these cases some type of direct blow to the forearm probably causes the fracture.

Falls also cause these fractures. These are usually major falls from a height, and not the minor type of fall in the home, which results in Colles's fractures and hip fractures in the elderly. Industrial accidents also cause some radial and ulnar shaft fractures. The exact cause of injury varies greatly, from crushing type injuries with the forearm caught between rollers to heavy objects falling on the extremity.

Another cause of fractures of the shaft of the radius and ulna is a blow with a club such as a baseball bat. This most often produces a Monteggia fracture or a nightstick fracture, but such a blow can also fracture both forearm bones. This mechanism of injury is seen most often by those orthopedists who work in city and county hospitals in the United States. In an article reviewing 102 patients with forearm fractures from England, Smith[33] did not list such blows as causing this injury. Of his 102 patients, 46 were injured in traffic accidents, 37 in falls, and 19 in industrial accidents.

In recent years very few reports in the literature have recommended manipulation followed by external plaster immobilization for shaft fractures of both bones of the forearm. Earlier reports, including the one by Knight and Purvis[22] from the Campbell Clinic in 1949, related very poor results from this form of treatment; there were 71 percent unsatisfactory results in patients with fractures of the radius and ulna treated as closed reduction. It was concluded that improper rotational alignment was a causative factor in a high percentage of poor results. Evans[17] has also stressed the importance of rotational alignment in treating these fractures. Knight and Purvis[22] did state that transverse fractures in the medial and distal thirds of both bones could in some instances be handled satisfactorily by closed reduction if the proper rotational alignment could be obtained. On the other hand, it was recommended that fractures in the upper third of the forearm, and oblique and comminuted fractures at any level, should be treated by open reduction and internal fixation. In 1953 Bradford, Adams and Kilfoyle[9] reported that 74 percent of patients with both radius and ulna fractures, on whom closed reduction was attempted, subsequently required open reduction because adequate reduction could not be achieved initially or because reduction was subsequently lost.

With such a high rate of failure by conservative methods, the natural response of the orthopedic surgeon was to turn to operative treatment for these difficult fractures. However, as pointed out by Bradford, Adams and Kilfoyle[9] surgical treatment was resisted by many orthopedists because of concern about postoperative infection, the poor results of bone plating in the past, and the length of the surgical trauma required to internally fix both bones, which made the operation almost like two procedures. However, problems with sepsis and operative technique, and the shortcomings of implants have largely disappeared, and there is now no reason to avoid reduction of dual forearm fractures by internal fixation.

According to Sage,[31] medullary fixation for fractures dates back to the late nineteenth and early twentieth centuries. After suitable metals for internal fixation were devised, Rush and Rush[30] used Steinmann pins for medullary fixation of a comminuted fracture of the ulna, and subsequently devised intramedullary pins for use in various long bones. With the popularity of medullary nailing for fractures of the femoral shaft that developed after World War II, there followed an increase in the use of medullary nailing for forearm fractures. Many devices were used, including single and multiple Kirschner wires, Steinmann pins, Kirschner nails, Lottes nails, and Rush pins. Smith and Sage,[32] in 1957, surveyed, in 338 patients, the results of 555 fractures of the forearm that had been treated with medullary fixation. The patients came from 17 different centers across the United States. It was found that 38 percent of the fractures treated with Kirschner wires developed nonunion. If these patients were eliminated, the nonunion rate in the remaining 321 fractures was a respectable 14 percent. It was stated that the ulna readily lent itself to medullary fixation because of its relative straightness, but that the radius presented more difficult problems. It was emphasized that the success of medullary nailing depended on the fragments being held in apposition by a medullary rod with sufficient stability to prevent rotary, side-to-side, and angular motion.

After this review, Sage,[31] in 1959, subsequently studied the anatomy of the radius and its medullary canal in cadaver specimens. From this study medullary nails were designed for the radius and ulna that had a triangular cross section so as to grip the endosteal surface of the cortex and prevent rotation and telescoping of the fragments. The nail for the radius has a prebent bow that, after it is driven in place, springs back into shape to maintain the radial bow. It is inserted through the radial styloid. Sage reported excellent results for the first 50 patients with 81 fractures treated with these nails, and only 6.2 percent developed nonunion. It is apparent that the type of rigid fixation thought necessary to achieve stable reduction was achieved. In addition to the results of Sage,[31] other orthopedic surgeons have reported good results for fractures of the shafts of the radius and ulna using different types of medullary nails. Reports include those of Street,[34] Ritchie, Richardson and Thompson[29] and Marek.[25] The nails used by these authors for both the radius and ulna are straight. The radial nail is inserted through Lister's tubercle, which tends to prevent collapsing of the radial bone. The nails used by Street are diamond shaped, and those used by Ritchie and Marek are square. Thus these nails, as well as those described by Sage, have a cross sectional configuration that allows the nail to achieve firm purchase on the endosteal surface of the cortical bone, preventing telescoping and rotational movement.

The nails described by Street,[34] Marek,[25] and Ritchie, Richardson and Thompson[29] as well as those described by Sage,[31] have all been reported to give good results in fractures of both bones of tthe forearm. However, the Sage nails are to be preferred, primarily because there is less chance of loss of the radial bow. The indications for the use of intramedullary nails in the forearm must be carefully considered. These nails should not be used in fractures of the distal third of the radius, nor of the proximal third of the ulna, because the medullary canals of the bones at these levels are too large for the nail to achieve good fixation. Sage recommends that autogenous iliac bone grafts be used on all forearm fractures treated with medullary nails. Most authors also recommend external cast immobilization until there is bridging callus or amalgamation of the bone grafts. This usually occurs at about 10 to 12 weeks after repair.

Early in this century Lane[24] and Lambotte[23] reported the use of plates and screws in diaphyseal fractures. Unfortunately, failures were frequent, owing to metal reaction as well as to the inadequate design of the fixation devices. This led Campbell and Boyd[11] to use autogenous tibial cortical bone grafts fixed to the radius and ulna with bone pegs and later with screws for acute fractures as well as for nonunion. This operation met with some success, but if external immobilization was not prolonged, the grafts developed fatigue fractures and fixation was lost.

Inadequacies in the fixation materials and devices available led to disappointing results. As recently as 1949, Knight and Purvis[22] reported that of 20 patients with fractures of both bones of the forearm treated withplate fixation, 65 percent had unsatisfactory results. The finding was attributed to the use of plates or screws of inadequate length, and to bone absorption at the fracture site consequent upon inadequate fixation and persistant motion at the fracture.

The importance of bone contact and compression in fracture healing was emphasized by Eggers et al.[14-16] in the 1940s, and a slotted plate was advocated to allow muscles acting across the fracture to cause compression of the fracture ends. Probably the slots are filled with fibrous tissue within a few days, and it is doubtful if sliding takes place beyond this time. In any case the Egger plates were much stronger than plates previously used, and the results were much better. Jenkins, Lockhart and Egger[20] reported 7.4 percent nonunion in a series of forearm fractures treated with slotted plates for the ulna and either a slotted plate or Rush pin in the radius. In 1961 Caden[9] reported a series of 40 forearm fractures treated with slotted plates, with an incidence of nonunion of only 7.5 percent.

As far as can be determined, Danis,[12] in 1949, was the first to use plates through which active compression of the fracture could be achieved. Attention was drawn to the fact that in fractures treated with rigid plate immobilization, healing occurred with very little periosteal callus, and this was referred to as primary fracture healing. Venable[34] devised a similar plate in 1951. Bagby and James,[6] in 1958, modified a Collason plate to achieve compression by eccentric placement of the screws. In about 1958, Muller, Allgower and Willenegger[26] devised what is now known as the ASIF compression plate, and recommended a technique for using this plate as well as other methods of open reduction and internal fixation, which was published in 1965.

The author used the ASIF plate in experimental fractures in the femurs of dogs, and found it to be very successful in such fractures. At the Campbell Clinic, the compression plate was first used in patients with fractures and nonunions in 1960. In both experimental fractures in dogs and in clinical fractures, it was found that with the rigid fixation of the fracture achieved by the plate, resorption was seen only if the screws loosened and rigid fixation was lost.

In 1970, Dodge and Cody[13] reported excellent results in fractures of the radius and ulna treated with ASIF compression plates. Union was achieved in all but 2 of the 119 fractures, although there was loss of fixation in 4 fractures. In 3 of these, a four-hole plate had been used, and it was recommended that plates with at least five holes be used.

In 1972, the experience at the Campbell Clinic and the University of Tennessee over a 10-year period was reported with 330 acute diaphyseal fractures of the radius and ulna treated with compression plates.[4] The overall union rate for the ulna was 96.3 percent, and for the radius 97.9 percent. This rate of union is at least as good as percentages reported in other large series in the literature. The average time to union in this series was 7.4 weeks, which also compares favorably with the time in other series. There was no definite evidence that compression stimulated osteogenesis; it is believed that compression achieved at least the following three biomechanical factors of real importance:

1. Compression of the fracture to force the interdigitating spicules of bone together and increase rigidity.
2. Narrowing of the space between the spaces to reduce the gap to be bridged by new bone.
3. Protection of the blood supply through enhanced fracture stability.

In the present series, almost all of the failures could be attributed either to infection or to errors in surgical technique. The technical errors occurred early in the series for the most part, and included such things as using a plate that was too short and placing screws too close to the fracture. The overall result of the errors in technique was that rigid fixation was not achieved. Such errors should be preventable in the future.

Thus it has been shown that modern techniques of intramedullary rod fixation and plate and screw fixation can both give excellent results in dual forearm fractures. Whichever method is used, success demands that accurate reduction and rigid fixation of the fractures be achieved and maintained until bony union has occurred. Success also depends on appropriate postoperative management. Following open reduction and internal fixation, the time of immobilization in plaster must be highly individualized. Most authors who advocate medullary nails recommend a period of postoperative case immobilization ranging from 6 to 12 weeks.

With compression plate fixation of fractures, the need for external cast immobilization depends on several factors, such as the configuration and degree of comminution of the fracture, whether compression was achieved, and the reliability of the patient. At times, with an intelligent, co-operative patient with transverse fractures in which good compression has been obtained, no cast immobilization may be needed. A compression dressing is worn for the first few days and the patient is cautioned not to overuse the extremity until there is good radiographic evidence of union. On the other extreme, for an uncooperative patient with a comminuted fracture and where little compression is obtained, the fracture should be protected until good union has occurred, usually 7 to 8 weeks. When the surgeon is uncertain about the patient's cooperation and the rigidity of fixation, a compromise may be helpful. The extremity can be left out of a cast while under observation in the hospital, and exercises can be carried out under supervision of the therapist. At 10 to 12 days the sutures are removed and a cast applied just prior to discharge from the hospital. During the period of exercises in the hospital, the patient usually regains a good range of motion. The patient rapidly regains this motion when the cast is removed a few weeks later.

CASE PRESENTATIONS

Case 1. S. K. is an 18-year-old white female with congenital insensitivity to pain. This condition first became known in early childhood, when she chewed on the tips of her fingers and toes, a habit that has continued until the present time. She also has had problems with trophic ulcers beneath the metatarsal heads. These have been treated with walking casts and special shoes. She was originally seen after she fell from a shopping cart in which she was being pushed and injured her left forearm. Roentgenograms of the left forearm showed fractures of the radius and ulna in the middle third of the bones, with only minimal displacement. (Fig. 3–1A and B). She was placed in a well molded, long arm cast. She had no complications while in the hospital, but was warned that there was a good chance that the fractures might change position and require open reduction and internal fixation. She was discharged from the hospital one week after injury to be followed as an outpatient.

The patient returned to the outpatient clinic two weeks after injury; roentgenograms made through the cast showed complete displacement of the fractures (Fig. 3–1C and D). She was advised that she should reenter the hospital for open reduction and internal fixation of the fractures with compression plates and screws. The patient was readmitted to the hospital 10 days later, and underwent open reduction

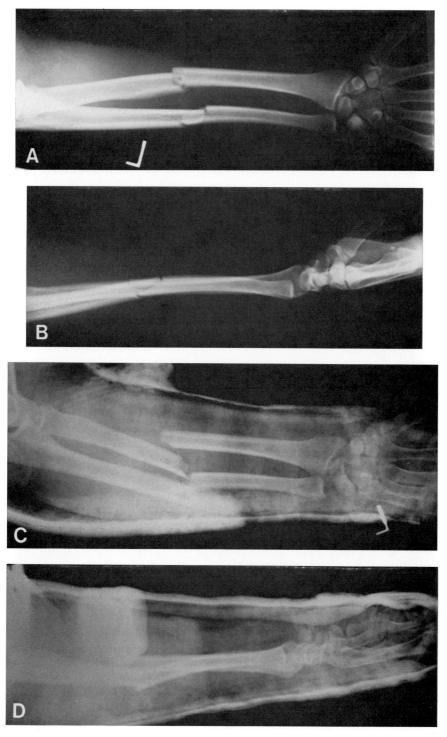

Fig. 3-1. (A and B) Minimally displaced fractures of the left radius and ulna. (C and D) The fractures displaced and angulated after two weeks immobilization in a long arm cast.

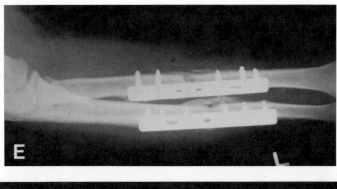

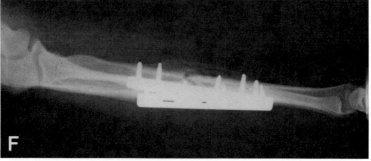

Fig. 3 – 1 (E and F) Four months after internal fixation with compression plates and screws with bone grafting. The fractures are well united. Both plates were applied dorsally.

and plate fixation of the left radius and ulna the following day. The radius was approached through a dorsal Thompson incision, and a seven-hole compression plate was placed on the dorsal aspect of the radius. The plate was fixed to the distal fragments with three screws and to the proximal fragment with two screws. The two screw holes nearest the center of the plate were left empty because of moderately severe comminution of the fracture. The ulna was then approached through an incision along the subcutaneous border. A six-hole plate was used for fixation, and three screws were used in both the proximal and distal fragments. Bone grafts were then obtained from the right iliac crest and placed about fractures. Care was taken to prevent the bone grafts from encroaching on the interosseous space. Roentgenograms made in the operating room showed good position and fixation of the fractures. After closure of the incisions a long arm cast was applied postoperatively. She had no problems and was discharged a week after surgery, to be followed in the outpatient clinic.

She returned on several occasions for examination and roentgenograms of the forearm. Three months later, roentgenograms showed sufficient healing that the patient was left out of a cast. Films made four months after surgery, showed good healing and amalgamation of the bone grafts. (Fig. 3 – 1 E and F). She had full flexion and extension of the elbow and both pronation and supination were greater than 75 percent normal.

This patient's fractures of the radius and ulna were initially only minimally displaced, and closed treatment in a well molded long arm cast was elected as the initial method of treatment. Approximately 2 weeks later, roentgenograms made through the cast showed that position of the fractures had been lost. Open reduction and internal fixation was then chosen as the next form of treatment. The only possible alternative method have been remanipulation of the fractures and continued treatment in a cast. This hardly seemed a viable choice, since this method had already failed. Once open reduction was decided upon, plate and screw fixation was chosen rather than intramedullary rod fixation, because at the small diameter of the medullary canals of the radius and ulna and the fear of splitting the bones if rods were used.

Case 2. M. R., a 56-year-old white female, was admitted to hospital after having been involved in an automobile accident in another city some 40 miles away. She complained of pain in her right forearm and also had lacerations about both knees. Roentgenograms made in the emergency room showed fractures of the right radius and ulna in the distal one third of the forearm. (Fig. 3–2A and B). There was marked comminution of the fracture of the ulna. The lacerations were sutured in the emergency room and a posterior plaster splint was applied prior to admission to the hospital. Following admission, the patient complained of pain in the chest which delayed surgery while studies were done to rule out serious chest pathology. None was found, and 3 days later, open reduction and internal fixation of the fractures of the right radius and ulna was carried out. The radius was approached through an anterior Henry approach and fixation was secured with a seven-hole compression plate. Three screws were used for fixation in the distal radial fragment and two in the proximal fragment. The two holes nearest the center of the plate were left empty because of comminution at this level of the bone. The ulna fracture was then approached through an incision along the subcutaneous border.

There was marked comminution of the ulna fracture. After reaming, a 4 mm Sage nail was inserted onto the distal fragment through the fracture site, and driven in a retrograde fashion out the ulna styloid and soft tissues at the wrist. The fracture was then reduced and the nail driven up the proximal fragment. The comminuted fragments were then fitted accurately into place and fixed with two circumferential wire loops of number 20 wire. (Fig. 3–2C and D) Autogenous bone grafts were obtained from the left iliac crest and packed about the fractures. The incisions were closed and a posterior plaster splint was applied to the extremity. Intraoperative roentgenograms showed good position of the fractures and of the internal fixation devices. Post operatively there were no complications, and the patient was placed in a long arm cast prior to discharge one week after admission.

The patient was followed as an outpatient, and one month after surgery roentgenograms showed sufficient union of the fractures that she was left out of a cast. On later followings, she complained of pain over the tip of the nail at the ulna styloid. Films made on 14 weeks after surgery, showed sufficient union of the fractures and amalgamation of the bone grafts that it was thought safe to remove the nail. (Fig. 3–2E and F). This was accomplished 4 weeks later. When last seen on six months after original injury, the patient had slight limitation of dorsiflexion of the wrist but full elbow motion, as well as full pronation, and supination.

This patient had fractures of the radius and ulna in the distal one third of the forearm. With the degree of comminution present, especially in the ulna, it was thought that closed reduction and cast immobilization stood little chance of success. Open reduction was elected as the treatment of choice. The fracture of the ulna was comminuted over a distance of more than 2 cm, and was too close to the distal end of the ulna to permit placement of at least two screws. For this reason a Sage ulna nail was used with circumferential wire loops to hold the comminuted fragments in

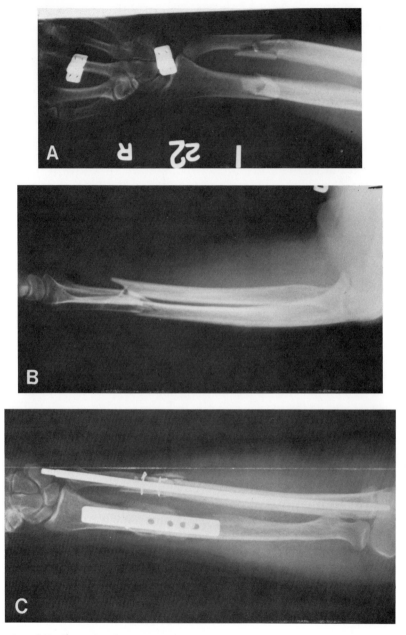

Fig. 3–2. (A and B) Comminuted fractures of the radius and ulna in the distal one third of the forearm. (C and D) Intraoperative films showing the fracture of the radius internally fixed with a compression plate and screws on volar surface. The comminuted ulna fracture fixed with Sage intramedullary nail and two wire loops. Bone grafts were placed about the fracture. (E and F) Roentgenograms 4 months after internal fixation. The fractured bone united.

place. The fracture of the radius was located in the distal third. The diameter of the medullary canal in the distal radial fragment was too great for an intramedullary nail to achieve fixation. For this reason a compression plate was used for internal fixation of the radius.

Case 3. T. P., a 17-year-old white male, was a passenger in an automobile that

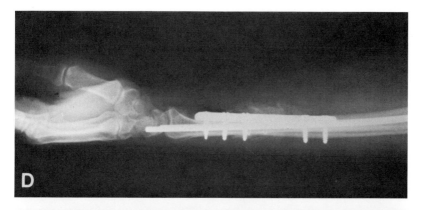

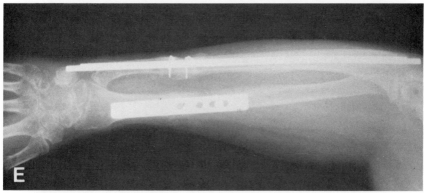

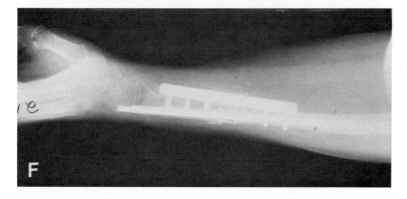

struck a bridge abutment. Following the collision he noted pain and deformity of the left forearm. He was brought to the hospital, where roentgenograms revealed fractures of the radius and ulna in the middle one third of the bones, (Fig. 3–3A). The patient had no other injuries. He was admitted to the hospital and on the third day after injury was taken to the operating room for open reduction and internal fixation of the fractures. The fracture of the ulna was approached through an incision along the subcutaneous border. The fracture of the radius was exposed through a dorsal Thompson approach. The deep branch of the radial nerve was exposed in the substance of the supinator muscle and retracted. Fixation of both the radius and ulna was achieved using seven-hole compression plates. Three screws were used in both the proximal and distal fragments. The center screw hole was left out of both the radial and ulna plate because if one had been used in this position,

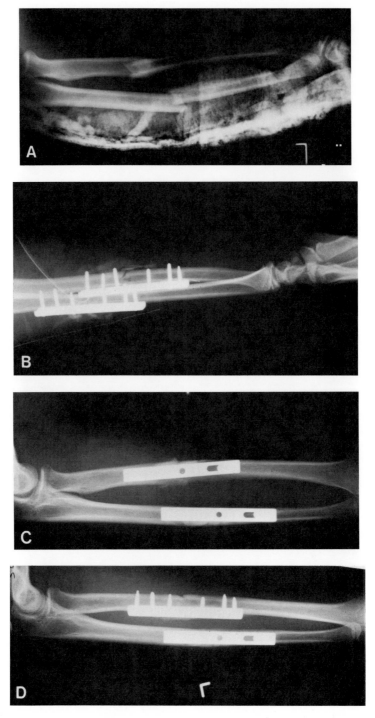

Fig. 3 – 3. (A) Fracture of the left radius and ulna in the middle one third of the forearm. (B) Intraoperative roentgenogram after internal fixation of the fractures with compression plates and screws. Both plates were applied dorsally. (C and D) Films approximately 3 months after surgery. The fractures show good union and maintenance of anatomic alignment.

the screw would have been too close to the fracture. Because of moderate comminution of the fractures, autogenous iliac bone grafts were packed about both the radial and ulna fracture sites. Roentgenograms made in the operating room showed good position and alignment of the fractures. (Fig. 3–3B). At the end of the operation, a plaster splint was applied to the arm. The patients postoperative course was uneventful, and he was discharged a week later to be followed in the outpatient department.

The patient returned to the clinic on several occasions. Six weeks after surgery, he was left out of a cast and encouraged in an active exercise program. Films made 3 months after injury showed solid union of both fractures, and he was noted to have a full range of elbow, forearm, and wrist motion, (Fig. 3–3C and D). When last seen 7½ months after injury the patient was completely asymptomatic and had normal function in the left upper extremity.

This patient had fractures of the radius and ulna in the middle one third of the forearm. Open reduction and internal fixation was chosen as the primary form of treatment. The alternative method, closed reduction and cast immobilization, usually gives poor results with fractures at this level because the midforearm is fleshy, and satisfactory position is very difficult to maintain even if it is achieved initially. At this level in the forearm, and with this degree of comminution, either intramedullary rod or plate and screw fixation for these fractures is appropriate. Compression plate fixation was elected.

Case 4. E. B., a 37-year-old white male, was intoxicated and walking along a highway when he was struck by an automobile. He was brought to the emergency room where he was found to have a closed fracture of the midshaft of the left humerus (Fig. 3–4A), with a complete radial nerve palsy, and closed comminuted fractures of the left radius and ulna in the proximal one third, (Fig. 3–4B). The left upper extremity was initially treated in a hanging arm cast with a radial nerve outrigger splint.

One week after injury, the patient was taken to the operating room for open reduction and internal fixation of the left radius and ulna. The fracture of the radius was very proximal and was approached through a Thompson dorsal incision. The deep branch of the radial nerve was identified and retracted, and a six-hole compression plate was applied to the dorsal aspect of the radius. The comminuted fragments were fitted back into place as well as possible. Three screws were used in the distal fragment and two in the proximal fragment. The centermost screw hole was left empty because of the severe comminution. The ulna fracture was then exposed through an incision along the subcutaneous border of the ulna. A seven-hole compression plate was used for fixation with three screws placed in the distal fragment and two in the proximal fragment. Again, because of marked comminution, the two center screw holes were left empty, (Fig. 3–4C and D). Autogenous iliac bone grafts were packed about both fractures and the incisions closed. A hanging arm cast with radial nerve splint was applied at the end of the procedure as treatment for the humeral shaft fracture.

The incisions healed primarily. When in bed the patient was maintained in the semi-Fowler position so that the hanging arm cast could function. He was discharged on two weeks after surgery, to be followed as an outpatient. He made numerous follow-up visits. Union of the forearm fractures progressed rapidly but the fracture of the shaft of the humerus developed a delayed union which required continued use of the hanging arm cast until 4½ months after his original injury (Fig. 3–4E, F and

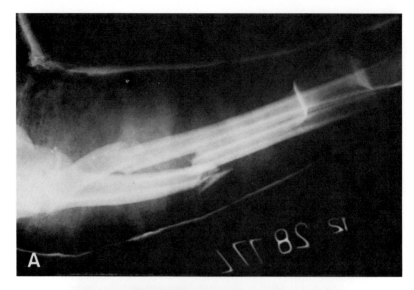

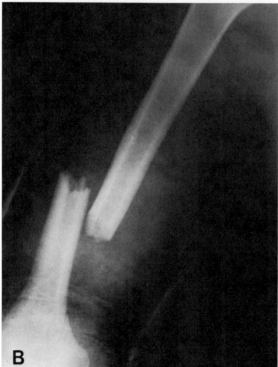

Fig. 3–4. (A) Fractures of the proximal one fourth of the left radius and middle one third of the ulna. Both are comminuted. (B) There was an ipsilateral fracture of the left humerus with complete radial nerve palsy.

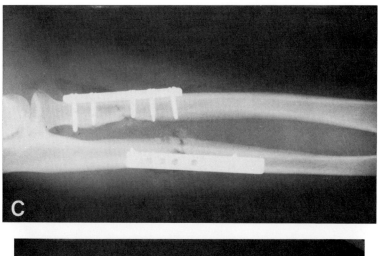

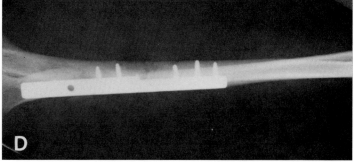

Fig. 3–4. (C and D) Intraoperative films after fixation of both fractures with compression plates. Autogenous iliac bone grafts were used. Both plates were applied to the dorsal aspects of the radius and ulna.

G), when it was finally discontinued and an active exercise program begun for the extremity. At about the same time, signs of return of radial nerve function developed.

On the patient's most recent visit, 9½ months after injury, there was full return of radial nerve function, and all fractures were well healed. Motion in the left elbow and forearm have been slow to return. On this visit he had elbow extension to 40 degrees, flexion to 120 degrees, and pronation and supination of 50 degrees each. He was continued on his exercise program in the hope of regaining additional motion.

This patient's combination of injuries presented a difficult treatment problem. With an ipsilateral fracture of the shaft of the humerus and fractures of the radius and ulna in the proximal forearm, there seemed no viable alternative to open reduction of the forearm fractures. Closed manipulation of the radius and ulna fractures appeared impossible with the concomitant fracture of the humerus. The fracture of the humerus was treated by a closed method—a hanging arm cast. In retrospect this may have been a mistake, as will be discussed in the biomechanics section of this chapter.

Early motion following fixation of forearm fractures is highly desirable in order to obtain good function. Perhaps the result might have been better if internal fixation of the fracture of the humerus had been carried out initially. On the other hand,

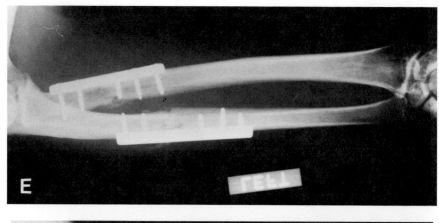

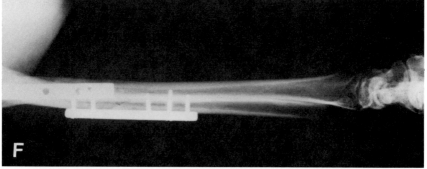

Fig. 3–4. (E and F) Roentgenograms 6 months after surgery. Both fractures are well united in excellent position. Note how far proximally the radial plate reaches. If it were not located dorsally it would impinge on the coronoid process during pronation.

fractures of the humerus such as this usually unite in 8 or 9 weeks, and the conservative treatment was thus thought to have merit.

For internal fixation of the forearm fractures, either a plate and screws or an intramedullary nail would have been suitable for the ulna fracture. The fracture of the radius was too proximally located for a medullary nail to achieve good fixation, and for this reason a plate and screws were used.

BIOMECHANICS

Since the anatomy of the forearm is complex, there has been very little work done on the biomechanics of the system, either regarding the forces applied to the intact bones or to the forces acting on internal fixation devices at the radius and ulna. However, it is clear from the anatomy that there will be several muscle groups acting to displace forearm fractures. In flexion, the brachialis, brachioradialis, and biceps brachii are active, with the latter muscle being the primary flexor of the elbow.

The directions of the muscle pulls are shown in Figure 3–5. According to the "tension band" principle, a plate must be inserted on the tensile side of the bone in order to have the bone bear a considerable share of the load. For the forearm, it is stated that the posterior surface is the one under tension, and this is given as the reason why dorsal bowing and a gap on the dorsal surface occurs.[26] The reasoning be-

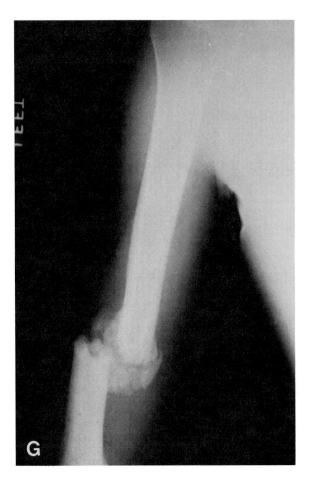

Fig. 3–4. (G) Roentgenograms of the ipsilateral fracture of the humerus. Delayed union only permitted mobilization of the forearm fractures at 6 months, and probably compromised the overall result.

hind this recommendation seems to be based on the action of the brachioradialis muscle which certainly would tend to cause dorsal bowing (Fig. 3–6), but is not the main flexor of the elbow. The action of the other two muscle groups is less clear, but it seems that their actions would tend to close the fracture. In fact, the location of a plate is strongly determined by the degree of fit of the plate to bone rather than by considerations of the tension-band principle which, although clear for the weight-bearing long bones, is not so straightforward for the bones of the forearm.

Forces acting to displace fractures of the forearm bones are also present in the actions of pronation and supination. Figure 3–7 shows the situation. If the fracture occurs in the upper third of the radius, each fragment is acted upon by muscles having a complementary effect, so that supinators act on the upper fragment and pronators on the lower fragment. There will be a rotation of the fragments with respect to one another, and any fixation device would thus be under the action of torsional forces. If the fracture occurs in the middle of the radial shaft there is less displacement, since pronation of the inferior fragment is due to the pronator quadratus and supination of the upper fragment is checked by the pronator teres.

The role of the interosseous membrane is also of interest, since the membrane can act to transfer load from the radius to the ulna. The amount of load transfer is un-

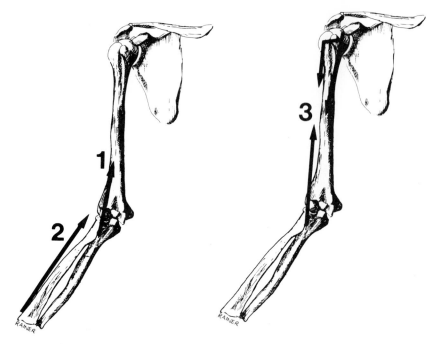

Fig. 3 – 5. Flexor muscles of the elbow, 1: brachialis; 2: brachioradialis; 3: biceps brachii.

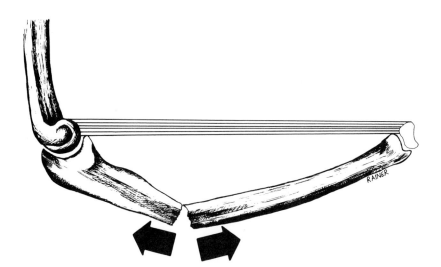

Fig. 3 – 6. The action of the brachioradialis in causing dorsal bowing of the radius.

certain, but Walker[36] has carried out experiments to show that load transfer does take place except in pronation, when the fibers of the membrane are not tight. The stiffness of the membrane is 200 N/mm. One specimen failed at a deflection of 5 mm, giving a membrane strength of 1,200 N. The load transfer properties of the membrane were also demonstrated by loading the radius. With the membrane intact, the deflection of the radius was 0.06 mm at a load of 500 N, but this increased to 0.69 mm after cutting the membrane.

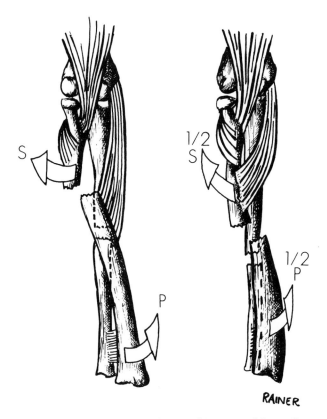

RAINER

Fig. 3–7. Pronation and supination forces acting on fràctures of the radius and ulna (adapted from Kapandji, I. A.: The Physiology of the Joints. Volume 1: Upper Limb. E. & S. Livingstone, Edinburgh, 1970.)

The foregoing studies have resulted in the model for the forearm shown in Figure 3–8; this does not include the muscle forces, but only the wrist attachments, the interosseous membrane, the radio-humeral joint, and the humero-ulnar joint. The attachments are regarded as springs with different stiffnesses. The situation considered is for a load applied at the wrist. The force carried by the membrane may be calculated by regarding the radius as a simply supported beam with a load applied at the midpoint. It is assumed that the deflection of (0.69 − 0.06 mm), or 0.63 mm, is due to the membrane pull. The force, F, to give this deflection, δ, is given by

$$F = \frac{48EI\delta}{\ell^3}$$

where E is the tensile modulus, I the moment of area, and ℓ the length of the radius. Here, I is taken as 2421 mm^4 and ℓ is 250 mm, giving F a value of 46.4 N. Since the membrane pulls at an angle of 30 degrees to the radius, the actual force in the membrane is F · cos 60, or 92.8 N. The membrane would transfer a force of 92.8 · cos 30 or 80.4 N from the radius to the ulna for the case considered, representing 16 percent of the applied force.

The main forces acting on fixation devices have been discussed above. The magnitude of the forces is not known, but some estimates may be made for the action of flexion. Figure 3–9 shows the forearm flexed at 90 degrees with a hand-held weight

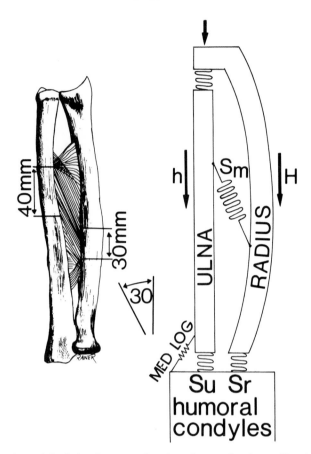

Fig. 3–8. Idealized model of the forearm showing the mechanism of load transfer between the radius and ulna.

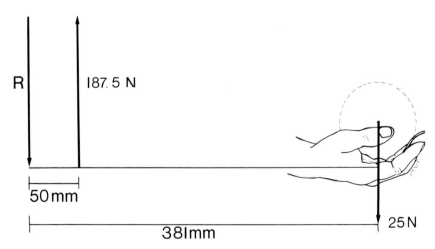

Fig. 3–9. Determination of the force in the biceps muscle with the arm at 90 degrees flexion and a hand-held weight of 25 N.

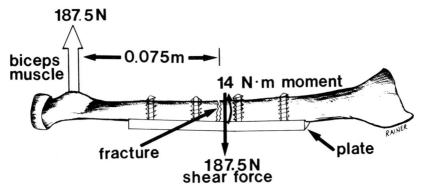

Fig. 3-10. Forces acting on a plate for a midshaft fracture of the radius; only the action of the biceps is considered.

of 25 N. The force in the main flexors is obtained by taking moments about the humero-ulnar joint (at 0), and the flexor force is 187.5 N. Figure 3-10 shows the situation with a plate applied to the radius. In order to withstand the applied muscle force, the plate must develop at the fracture site a shear force of 187.5 N and a moment of 14 N.m. Since failures appear to occur by loss of alignment rather than by plate fracture or bending, it is the fixation of the screws which gives way under these muscle forces.

As recorded at the beginning of the chapter, successful functional results in patients with fractures of both bones of the forearm can be achieved only if the length of each bone is maintained. As demonstrated in Case 1, this is difficult to achieve by closed methods, even in fractures that are only minimally displaced initially. For these reasons, most authors now believe that the best results in fractures of both bones of the forearm are achieved by some form of open reduction and rigid internal fixation.

The choice of plates and screws versus intramedullary nails is at times optional, and depends on the training and experience of the orthopedic surgeon. For example, fractures in the middle half of the radius and distal two thirds of the ulna can usually be treated satisfactorily by either method. However, if fractures in this location are comminuted over a distance of 2.5 cm or more, intramedullary rods are usually better. The fracture can be reduced and circumferential wires can be used to hold the comminuted fragments in position. With this degree of comminution, plates are very difficult to use. This was the reason that an intramedullary nail was chosen for fixation of the ulna fracture in Case 2.

Intramedullary nails should not be used in fractures of the proximal ulna or in fractures of the proximal or distal radius. The medullary canal is too large in these areas for the nail to grip the endosteal surfaces of the cortical bone and provide rigid fixation. Intramedullary nails that fit loosely cannot control rotation of the fragments. Also, telescoping of the fragments on the nail is not prevented. As has been shown in both experimental[1] and clinical fractures,[32] fixation with a loose fitting intramedullary nail produces a high rate of delayed union and nonunion.

Fractures of the radius and ulna in patients with small bones having medullary canals measuring less than 3 mm in diameter are not considered suitable for treatment with intramedullary nails.[31] When, in such patients, the medullary canals of the forearm bones are reamed to the necessary 4 mm to accept the Sage nails, the cortex is thinned and weakened to such an extent that the bones may split or shatter when the nails are driven down the canals. In such patients with small forearm bones, compression plates are better than intramedullary nails. Case 1 is an example of this problem; the medullary canals of the patient's radius and ulna at their narrowest point measured only 3 mm, and for this reason plates and screws were used rather than medullary nails.

To be successful in the treatment of forearm fractures, plates and screws must be sufficiently long and strong to provide rigid fixation. The fracture must be reduced as anatomically as possible. Four-hole plates may be adequate for transverse fractures with no comminution; however, even in this type of fracture it is probably better to use a five- or six-hole plate. All fractures that are oblique or comminuted should be fixed with at least five- or six-hole plates. When there is comminution or obliquity of the fracture the surgeon must be especially careful not to insert a screw too close to the fracture line; to do so may cause the screw to split into the fracture and cause additional comminution. At times such a split from a screw hole into the fracture may cause complete loss of fixation. In comminuted or oblique fractures it is better to choose a longer plate and not drill or use screws in one or two of the holes nearest the fracture. Screws should not be placed closer than 1 cm from the fracture line. The roentgenograms in all four of the cases discussed in this chapter exemplify instances in which screw holes were left empty for this reason.

Fixation with plates and screws can be used at almost all levels for diaphyseal fractures of the radius and ulna, except where there is extensive comminution. In fractures of the ulna, the plate may be placed either on the dorsal or anterior aspect of the bone; placement depends on the fit of the plate to the bone.

The case for the radius differs from that of the ulna. In fractures of the distal one third of the radius, the plate should be applied to the volar surface. In its distal third the radius is much flatter on its anterior surface, and the plate fits more snugly on the bone. In addition, there is much better soft tissue coverage of the plate and screws. If an attempt is made to apply the plate to the dorsal surface of the distal radius, Lister's tubercle precludes a good fit and the plate may interfere with the function of the extensor tendons, especially the outcropping tendons to the thumb. Case 2 is an example of a fracture in the distal third of the radius in which the plate was applied to the volar surface for the reasons just outlined.

On the other hand, for fractures of the proximal two thirds of the radius, the plate should be placed on the dorsal surface. There are two principal reasons for this, one being largely biological and the other mechanical. If a dorsal Thompson approach is used,[7] the supinator muscle can be exposed and the deep branch of the radial nerve identified in the substance of the muscle and carefully retracted. Exposure and retraction of the deep branch of the radial nerve is difficult if not impossible through an anterior approach.

Mechanically, the reason for applying the plate to the dorsal surface of the proximal radius is that the dorsally placed plate does not mechanically block pronation. In very proximal radial fractures, a plate placed on the anterior surface will actually bump into the coronoid process of the ulna. At best, this results in permanent loss of pronation. At worst, the screws may loosen from repeated stresses and a nonunion will develop. In the middle third of the radius a plate on the anterior aspect will encroach on the interosseous space during pronation and tend to limit this important motion. Cases 1 and 3 are examples of fractures in the middle third of the radius in which the plate was applied dorsally to avoid encroachment on the interosseous space. Case 4 is an example of a very proximal radial fracture in which the dorsal application of the plate prevented it from impinging on the coronoid process of the ulna during pronation.

Whether or not primary bone grafting of fractures of the radius and ulna should be carried out at the time of open reduction and internal fixation is a question that is still debatable. Certainly a good case for bone grafting can be made. There are difficult fractures that most agree should be treated by open reduction and internal fixation. As long as the fracture has already been exposed, it adds relatively little in time or risk to obtain and apply autogenous iliac bone grafts. Theoretically, bone grafting should increase the percentage of fractures going on to union and decrease the time of immobilization as well as the time to union. Some authors, such as Sage,[30] recommend bone grafting routinely, and this is especially useful when intramedullary nails are used.

On the other hand, in a study carried out at the University of Tennessee and the Campbell Clinic in Memphis,[4] an attempt was made to select certain patients for bone grafting. In a study of 330 fractures of the radius and ulna treated with compression plates, only those with comminution greater than one third the circumference of the bone were grafted. With these criteria, 90 fractures were bone grafted and 240 were not. Results showed that the overall percentage of fractures that united was 97.7 for those with grafts and 97.0 percent for those without grafts. These data are subject to different interpretations, but it is believed that more difficult comminuted fractures can do as well as simpler uncomminuted fractures if bone grafts are used. All four of the cases discussed in this chapter had comminution of the fractures and were bone grafted.

The reasons for bone grafting are both biological and mechanical. In comminuted fractures the fixation is not usually as rigid as in noncomminuted fractures. Comminuted fractures usually cannot be compressed as much as desirable, and there are usually areas of the cortical surfaces that are not in contact. If union does not occur in such fractures, either the metal will break or the screws will loosen. Bone grafts will fill defects between the cortical ends and increase the speed of union. By speeding the healing process, union of the fracture can take place before mechanical fixation is lost.

REFERENCES

1.　Anderson, L. D.: Compression plate fixation and the effect of different types of internal fixation on fracture healing. J. Bone Joint Surg., *47A*: 191–208, 1965.

2.　Anderson, L. D.: Fractures. In Crenshaw, A. H., ed: Campbell's Operative Orthopaedics, Vol. 1, 5th Ed. C. V. Mosby Co., St. Louis, 1971.

3.　Anderson, L. D.: Fractures of the shafts of the radius and ulna. In Rockwood, C. A., and Green, D. P., Eds., Fractures, Vol. 1. J. B. Lippincott, Philadelphia, 1975.

4.　Anderson, L. D., Sisk, T. D., Tooms, R. E., and Park, W. I.: Compression plate fixation in acute diaphyseal fractures of the radius and ulna. J. Bone Joint Surg., *57A*: 287–297, 1975.

5.　Bado, J. L.: The Monteggia lesion. Clin. Orthop., *50*: 71–76, 1967.

6.　Bagby, G. W., and James, J. M.: The effects of compression on the rate of healing using a special plate. Am. J. Surg., *95*:761–771, 1958.

7.　Boyd, H. B.: Surgical approaches. In Crenshaw, A. H. (ed): Campbell's Operative Orthopaedics, Vol. 1, 5th Ed. C. V. Mosby, St. Louis, 1971.

8. Boyd, H. B., and Boals, J. C.: The Monteggia lesion. A Review of 159 cases. Clin. Orthop., *66*: 94–100, 1969.

9. Bradford, C. H., Adams, R. W., and Kilfoyle, R. M.: Fractures of both bones of the forearm in adults. Surg. Gynecol. Obstet., *96*: 240–244, 1953.

10. Caden, J. G.: Internal fixation of fractures of the forearm. J. Bone Joint Surg., *43A*: 1115–1121, 1961.

11. Campbell, E. C., and Boyd, H. B.: Fixation of onlay bone grafts by means of Vitallium screws in the treatment of unusual fractures. Am. J. Surg., *51*: 748–756, 1941.

12. Danis, R.: Les Coopteurs et les Coopteurs: Uncles Theoric et Pratique de l'Osteosynthese. Masson & Cie, Paris, 1949.

13. Dodge, H. S., and Cody, G. W.: Treatment of fractures of the radius and ulna with compression plates: a retrospective study of one hundred eight patients. J. Bone Joint Surg., *54A*: 1167–1176, 1972.

14. Eggers, G. W. N.: Internal contact splint. J. Bone Joint Surg., *30A*: 40–51, 1948.

15. Eggers, G. W. N., Shindler, T. O., and Pomerat, C. M.: The influence of the contact-compression factor on osteogenesis in surgical fractures. J. Bone Joint Surg., *31A*: 693–716, 1949.

16. Eggers, G. W. N., Ainsworth, W. H., Shindler, T. O., and Pomerat, C. M.: Clinical significance of contact-compression factor in bone surgery. Arch. Surg., *62*: 467–474, 1951.

17. Evans, E. M.: Rotational deformity in the treatment of fractures of both bones of the forearm. J. Bone Joint Surg., *27*: 373–379, 1945.

18. Evans, E. M.: Pronation injuries of forearm with special reference to anterior Monteggia fracture. J. Bone Joint Surg., *31B*: 578–588, 1949.

19. Hughston, J. C.: Fracture of the distal radial shaft. Mistakes in management. J. Bone Joint Surg., *39A*: 249–264, 1957.

20. Jenkins, W. J., Lockhart, L. D., and Eggers, G. W. N.: Fractures of the forearm in adults. Southern Med. J., *53*:669–679, 1960.

21. Kapandji, I. A.,: The Physiology of the Joints. Volume 1: Upper Limb. E. & S. Livingstone, Edinburgh, 1970.

22. Knight, R. A., and Purvis, G. D.: Fractures of both bones of the forearm in adults. J. Bone Joint Surg., *31A*: 755–764, 1949.

23. Lambotte, A.: Chirurgie Operatorie des Fractures. Masson & Cie., Paris, 1913.

24. Lane, W. A.: Treatment of simple fractures by operation. Lancet, *1*: 1489–1493, 1900.

25. Marek, F. M.: Axial fixation of forearm fractures. J. Bone Joint Surg., *43A*: 1099–1114, 1961.

26. Muller, M. E., Allgower, M., and Willenegger, H.: Technique of Internal Fixation of Fractures. Springer-Verlag, New York, 1965.

27. Patrick, J.: A study of supination and pronation, with special reference to the treatment of forearm fractures. J. Bone Joint Surg., *28*:737–748, 1946.

28. Reckling, F. W., and Cardell, L. D.: Unstable fracture-dislocations of the forearm. The Monteggia and Galeazzi lesions. Arch. Surg., *96*:999–1007, 1968.

29. Ritchie, S. J., Richardson, J. P., and Thompson, M. S.: Rigid medullary fixation of forearm fractures. Southern Med. J., *51*:852–856, 1958.

30. Rush, L. V.: Atlas of Rush Pin Technics. Mississippi Doctor, 31: book section, August 1953; March 1954.

31. Sage, F. P.: Medullary fixation of fractures of the forearm. A study of the medullary canal of the radius and a report of fifty fractures of the radius treated with prebent triangular nail. J. Bone Joint Surg., *41A*: 1489–1516, 1959.

32. Smith, H., and Sage, F. P.: Medullary fixation of forearm fractures. J. Bone Joint Surg., *39A*: 91–98, 1957.

33. Smith, J. E. M.: Internal fixation in the treatment of fractures of the shaft of the radius and ulna in adults. J. Bone Joint Surg., *41B*: 122–131, 1959.

34. Street, D. M.: Spectator Letter, 1955. Venable, C. A.: An impacting bone plate to attain close apposition. Ann. Surg., *133*: 808–812, 1951.

35. Venable, C. S., Stuck, W. G., and Beach, A.: The effects on bone of the presence of metals; based upon electrolysis; an experimental study. Ann. Surg., *105*:917–938, 1937.

36. Walker, P. S., Human Joints and Their Artificial Replacements. Charles C. Thomas, Springfield, Illinois, 1977.

4
Intertrochanteric Osteotomy of the Femur

I. F. Goldie, M.D.; J. H. Dumbleton, Ph.D.

INTRODUCTION

Osteotomy of the upper end of the femur is an old procedure. It has been a surgical remedy for a variety of hip disorders. Originally, the object of the procedure was to restore function and correct deformity. Today, the prime indication is pain in the hip joint. The milestones in the development of osteotomy have been laid by both American and European orthopedic surgeons including Barton,[3] Lorenz,[12] Schanz,[27] Pauwels,[22] McMurray,[13] Blount,[4] Nissen,[18, 20] and Muller.[15, 16]. Before the advent of hip replacement surgery, femoral osteotomy was a widely accepted procedure, with an initial enthusiasm in America which, with time, however, faded. In Europe, the procedure has maintained a continuing popularity.

The conditions suitable for osteotomy of the upper end of the femur have been, and still are, with few exceptions: congenital dislocation of the hip, epiphyseolysis, paralysis of hip muscles, leg-length shortening, Perthe's disease, osteoarthrosis, pseudarthrosis of the femoral neck, and coxa vara.

In this presentation a review will be made of intertrochanteric femoral osteotomy in the treatment of osteoarthrosis. First, however, a brief outline of the history of osteotomy will be given.

HISTORY

Doctor John Rhea Barton[3] of the Pennsylvania Hospital in Philadelphia described how ". . . on the 22nd day of November 1827 assisted by Doctors Hewson and Parrish . . . the muscles in contact with the bone around the part of the great trochanter were carefully detached and a passage thereby made, just large enough to admit the insinuation of my forefingers before and behind the bone; the tips of which now met around the lower part of the cervix of the femur, a little above its root. The saw was readily applied and without any difficulty a separation of the bone was effected. The operation, though severe was not of long duration . . . ."

The patient to be operated on was a young sailor who had sustained a fracture of the hip. Healing had resulted in an ankylosis with the hip in adduction and 50 degrees flexion. By Doctor Barton's osteotomy the deformity was corrected, the leg became half an inch shorter, and satisfactory function was maintained for 6 years. Gradually, after this period, the hip became stiffer and ". . . with all the train of evils, abuse of health, intemperance etc . . ." the patient ". . . subsequently dies of phthisis pulmonaris."

In 1894, Kirmission[11] presented his concept on correcting the lordosis and adduction contracture in congenital dislocation of the hip. He suggested that an oblique osteotomy, 4 cm distal to the tip of the great trochanter, would improve the alignment of the femur by bringing it into adduction and hyperextension. The procedure was successful in four patients.

An extension of Kirmisson's concept was presented in 1919 by Lorenz,[12] who believed that the object of a femoral osteotomy should be a reduction of load in the hip joint. He therefore designed the "bifurcation operation." An oblique intertrochanteric osteotomy (in the coronal plane) was performed. The distal femur was medialized, which resulted in the shaft projecting into the acetabulum.

Schanz,[27] who also believed in reducing the load in the hip joint, preferred to locate the osteotomy more distally, at the level of the ischial tubercle. The proximal fragment became adducted and a larger zone of contact between the pelvis and femur was achieved, thus distributing the weight-bearing load over a larger surface.

McMurray[13] has been called the "father of modern femoral osteotomies." In 1935 he presented his modification of the Lorenz "bifurcation operation" to be used in the surgical treatment of osteoarthrosis. McMurray felt that the osteotomy should be intertrochanteric and oblique. The femoral shaft was to be displaced medially and extended below the capsule of the hip joint, — that is, just below the capsule of the hip joint at the edge of the acetabulum. Moreover, a union was to be sought between the portions of the divided femur, since a weak painful hip joint would otherwise be the invariable outcome. Of 15 cases, 12 turned out excellent with complete relief of pain, improved — but still reduced — movement, and less strain to the lumbar and sacroiliac regions. In the remaining 3 cases the results were unsatisfactory. The reason for this was that ". . . the operation incision was too small so that the whole area could not be inspected and the shaft was not placed in the correct position. As a result, body weight continued to pass through the hip joint as before."

The contributions of Pauwels[22] are fundamentally important from a biomechanical point of view. He analyzed the position of the femoral head, and suggested either a valgus- or varus osteotomy. The analysis was of particular importance in contributing to the management of pseudarthrosis of the femoral neck. In such cases he suggested a "Y-shaped" osteotomy with a medial displacement of the femoral shaft, by which the inner cortical region served as a support to the femoral head.

Blount[4] became the strongest supporter in America of the Pauwels osteotomy: "My colleagues have criticized me for being too enthusiastic about the high femoral osteotomy and recent converts to this procedure have accused me of being too rigid in my indications," he wrote.

The rationale for osteotomy up to the 1960s had been based mainly on biomechanical principles. The predominating idea was that hip joint deformity — irrespective of its cause — should be treated by surgical correction, including realignment, angulation, or both.

A new approach was announced by Nissen.[18-20] He advocated the theory, which in time became an accepted practical truth, that there was a biologic response to osteotomy. This was especially true for cases with osteoarthrosis. Following the complete division of bone there appeared to be an arrest of the degeneration of both the femoral head and acetabulum. The evidence for this theory was: (a) clinical, by the recorded improvement with alleviation of pain; and (b) radiographic, by demonstration of the regression of sclerosis and of bone cysts, and by an increase in trabeculation of the cancellous bone and a widened joint (indicating some influence on the degenerated cartilage). The conclusion Nissen drew was: "If osteotomy could be shown constantly to cause regression of primary osteoarthritis at a relatively late stage, it would be logical to employ it years sooner than is at present the custom even in the absence of troublesome deformity in order to obtain the biological effects of freedom from severe pain and arrest of degeneration."

Following osteotomy, immediate postoperative stability was achieved either by plaster casts or by internal fixation with a variety of designs of plates and screws as initiated by Blount in 1945. The contribution of Müller[15] was the introduction of the compression plate as a device for immediate stable fixation of the osteotomy. By use of this plate it became possible to mobilize the patient, although without weight-bearing, on the day after operation. Today, this method is universally employed and has made osteotomy a less complicated procedure than ever before.

As will be seen below, osteotomy still holds a position in the surgical management of hip osteoarthrosis, despite the advent of implant procedures. In other conditions mentioned above, femoral osteotomy is the only procedure possible and must therefore be part of the armamentarium of the orthopedic surgeon.

INTERTROCHANTERIC OSTEOTOMY OF THE FEMUR IN THE TREATMENT OF OSTEOARTHROSIS OF THE HIP JOINT

Osteoarthrosis of the hip joint is a degenerative disease. It causes destruction of the joint cartilage, making the articulating surfaces less congruent. The degree of involvement varies from slight cartilaginous pitting to gross destruction and deformation of cartilage and bone structures. Pain is frequent. Impairment of function occurs. Alignment deformities develop.

Some 50 percent of patients have symptoms that resolve clinically over a period of 12 to 15 years without necessitating active treatment. Eighty percent can be managed conservatively. It is the remaining 20 percent who are likely to need surgical procedures.

Over the past 10 years, partial and total joint replacement surgery has dominated treatment, but there are still instances in which intertrochanteric femoral osteotomy is the procedure of choice. Patients with incapacitating osteoarthrosis who are below age 60 or who suffer from excessive obesity, or both, fall into this latter group.

It should also be emphasized that surgical intervention must rely entirely on evaluation of the patient's situation, the need changing for each individual.

Pain is the main indication for intertrochanteric osteotomy in osteoarthrosis of the hip. Experience has shown that better results are obtained when the primary indication is pain that occurs both upon weight-bearing and during rest. Secondary indications are deformity and changes in the radiographic appearance of the hip. It is surprising to note that good results are obtained in cases in which advanced destruction has occurred. However, the best results seem to be obtained in those cases in which the normal articular congruity is still present.

COMPLICATIONS OF OSTEOTOMY

Like any other surgical procedure, osteotomy is, of course not free of complications, and these can be either early or late. The most common early complication is hemorrhage, which in the occasional case requires reoperation with evacuation of the hematoma. However, if suction drainage is used the risk of postoperative hematoma is considerably reduced. Another early complication is infection, which may arise within a week after the operation. In a report on 123 osteotomized hips, Hirsch and Goldie[7] noted 12 infections, which was quite a considerable number. However, with an improved surgical technique and in particular with more rigorous hygienic principles in the operating theater, the rate of infection was reduced to 0.9 percent. Finally, an early complication is thrombosis, which in the aforementioned report was registered in 33 of 123 operated hips. It was thought that the postoperative immobilization was the cause of this thrombosis, since, when early mobilization was instituted, the frequency of thrombosis fell to less than 1 percent.

A late complication is nonunion of the osteotomy gap, which will be commented upon later in this chapter, since a difference was noted between half of a patient group which was operated upon without compression instruments, and other half, for which such instruments were used. Another late postoperative complication is loosening of the device owing to the screws coming loose. This, as a rule, appears when nonunion is established. Finally, some patients experience postoperative pain around the trochanter and down the lateral aspect of the thigh. This has most often been due to the device interferring with the movements of the muscles, and the pain has disappeared once the device has been removed. Occasionally, there is a fracture of the fixation device and, in the report by Hirsch and Goldie, this occurred in 7 of 123 operated hips. In 4 of these, reoperation was necessary, whereas in the remaining 3, no operation was carried out, since union of the osteotomy gap occurred. In the report mentioned, pseudoarthrosis occurred in 16 of 123 hips, and reoperation was performed in 10. Extraction of the fixation device because of pain was done in 32 of 123 hips.

Of specific interest from a biomechanical viewpoint are failures of the device employed to stabilize the osteotomy. The forces on such implants are high, and failures due to fracture of the device have been reported.[28] Undoubtedly, nonunion aggravates the situation, since internal fixation devices are not designed to withstand the full load of the body.

Naturally, complications can ensue if there are technical errors such as incorrect calculation of the displacement of the proximal and distal femur. This is emphasized below. However, in certain cases, long-term complications develop, sometimes after many years, which necessitate reoperation. Chief among the indications for further surgery is pain. Often the reason for the reoccurrence of pain and disability cannot be traced to any simple cause. This is illustrated in Case 1. It may be borne in mind that osteotomy is a conservative procedure, since there is later the option of a conversion to a total hip replacement in case of complication.

TECHNICAL HINTS IN INTERTROCHANTERIC OSTEOTOMY OF THE FEMUR

Preoperative planning of intertrochanteric osteotomy is essential. The angles of correction as well as the amount of medial displacement, should be calculated. Moreover, the decision must be made upon whether a valgus or varus osteotomy should be carried out. The most important correction is in the coronal plane—a varus or valgus alignment.

Preoperative drawings are important, since doing a medial capsulotomy and cutting through the femur in the intertrochanteric region results in a varus which can hardly ever be predicted preoperatively. A varus osteotomy usually amounts to some 20 degrees. The degree of correction in the sagittal plane depends on the amount of flexion contracture.

The medialization of the femoral shaft must be calculated preoperatively so that the physiologic angle between the mechanical and anatomic axes will be retained. If this is not done, troubles will develop in the knee joint, which can either go into the varus or valgus position. For details of operative techniques, reference is made to current orthopedic manuals.

THE RESULTS OF INTERTROCHANTERIC OSTEOTOMY OF THE FEMUR

The 1968 report of Hirsch and Goldie[7] included a 5-year followup of 110 hips in 102 patients operated upon with intertrochanteric osteotomy for osteoarthrosis of the hip. An overall good result was obtained in 83 percent. Weight-bearing pain was relieved in 73 percent, and an improved or unchanged range of motion was seen in 65 percent. A second followup was carried out in 1972,[8] comprising 81 hips in 72 patients of the earlier group. The average time of observation was 8.7 years (the longest time was 11 years, 2 months). The followup study was carried out by independent observers who had not been involved with the patients prior to the investigation. Relief of pain when at rest remained at 84 percent. Weight-bearing pain was relieved in 77 percent, and range of motion was improved or unchanged in 61 percent.

A comparison made with conservatively treated patients suggested a slightly more favorable outcome in the osteotomy series. In view of the new techniques being introduced, the results of the investigation indicated that intertrochanteric osteotomy still commands a good position in the surgical treatment of osteoarthrosis of the hip joint. The results reported in the literature much resemble those of this investigation.

A problem that has caused some concern in intertrochanteric osteotomy is the fixation of the osteotomy gap. In 1945, Blount[4] introduced internal fixation with plate and screws. Since then numerous types of fixation devices have been utilized, making immobilization in plaster unnecessary. With the advent of compression devices[15] an improvement of stable fixation was obtained, but it was uncertain how postoperative mobilization and healing would be influenced. For this reason studies have been carried out by Olsson involving radiographic evaluation of noncompression and compression methods. Two techniques were examined: the Wainwright noncompression method and the AO-compression osteosynthesis. The time of healing was studied and it was found to be equal for both groups. However, the rate of nonunion was higher in the Wainwright group than in the AO-group. The medial displacement of the femoral shaft was considerably less in the AO-group as compared to the Wainwright group, resulting in a larger area of contact in the line of osteotomy. The stability depends partly upon this, and the smaller contact area in the Wainwright procedure might consequently have been responsible for the higher rate of nonunion. The AO-compression osteosynthesis gave a stable fixation which enabled early mobilization of the patient.

Another difference was also noted in the two patient groups: the changes in radiographic appearance differed. In a 1972 study, Hirsch et al.[9] found that at 5 years following osteotomy the joint space, with the exception of the central part, did not show significant widening. Cysts in both the femoral head and the acetabulum were significantly decreased. Sclerosis in either the femoral head or the acetabulum was significantly less. Such favorable, progressive changes were not seen in patients operated on with the AO-compression ostteosynthesis; in these, cysts and sclerosis remained almost the same as preoperatively.

The possible explantation for this is in the hemodynamic events that occur in osteoarthritic bone. In osteoarthrosis of the hip there is an arterial hyperemia, as shown by Trueta[29] and Meriel et al.[14] Hulth[10] and Phillips[24] showed that there is an insufficient venous return, which contributes to an expansion of the sinusoids. Dilated veins can be demonstrated by phlebography and interpreted as a sign of increased intraosseous pressure;[23] the increased intraosseous pressure is supposed to produce pain. After osteotomy, Phillips[23] could show by phlebography that a normal venous flow pattern was restored. Arnoldi[1] has demonstrated a decrease in intraosseous pressure following osteotomy. It might be surmised that this decrease in intraosseous pressure would have been maintained for a longer time in the Wainwright group, where no compression was used and where the medial displacement was so large that considerable area of the intramedullary space was left open. With the compression induced in the AO-group—and also with the lesser medial displacement of the femoral shaft—the intraosseous pressure might return to its elevated value at a faster rate, thus inhibiting the reparative processes which diminish the cysts and sclerosis. Some support for this theory was obtained in a further study of the AO-compression group, which was divided into two subgroups with regard to the degree of translation at the osteotomy site. Radiographic signs of regression of osteoarthrosis were more prominent in the subgroup, in which translation was greater.

The results of intertrochanteric osteotomy in the treatment of osteoarthrosis of the hip thus indicate that the procedure is well justified. A factor, however, which seems to have been overlooked is the patient's walking ability. For this reason, an analysis was carried out of the walking habits of patients who had undergone intertrochanteric osteotomy.[8] Of 65 patients who had used a cane before operation, only 10 could manage without it postoperatively. Of 42 patients who did not use a cane before operation, 25 began to use one following the immediate postoperative period. These observations were interpreted to mean a decreased employment by the patient of the treated leg. Tests carried out on an electronic gait analysis walkway revealed that all patients limited the use of their operated hip to a level well below normal. A hypothesis was that such a decrease was secondary to muscular insufficiency of the hip abductors. The subjective satisfaction of the individual patient, however, outweighed the slight discomfort caused by the use of a walking aid.

CASE PRESENTATIONS

Case 1. B.G. is a male first seen in 1971 when aged 50 years. There was osteoarthrosis of the left hip, accompanied by severe pain. Preoperatively there was continuous pain for 2 years, upon both weight-bearing movement and when at rest. There was a flexion contracture of 20 degrees and in adduction contracture of 10 degrees. The degenerative changes are shown in Figure 4–1. An intertrochanteric osteotomy according to Wainwright was performed (Fig 4–2). After the osteotomy,

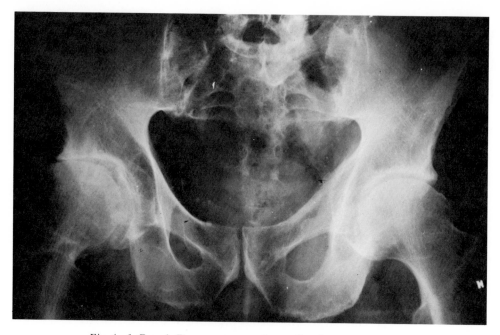

Fig. 4 – 1. Case 1. Roentgenogram immediately preoperatively.

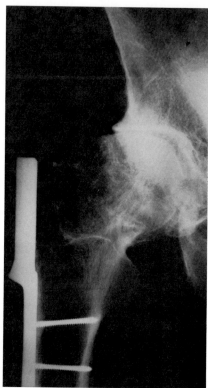

Fig. 4 – 2. Case 1. Roentgenogram postoperatively following Wainwright osteotomy.

the patient was pain-free. There was a near normal range of motion with a limp following release of the flexion and adduction contractures. There was a positive Trendelenburg sign on the left side. The patient did not use a cane. The patient was happy for 8 years postoperatively, at which point the pain began again. In 1979, reoperation was performed and a Christiansen (trunion bearing) total hip prosthesis was inserted (Fig. 4–3). Recovery was uneventful.

Case 2. S.V. is a female aged 57 years who presented in 1975 with moderate weight-bearing pain and pain at rest (Fig. 4–4). The patient continued to have difficulty, and in 1977 it was decided to perform an osteotomy. Figure 4–5 shows the condition some 2 months prior to surgery. A varus osteotomy was performed using the Müller compression technique (Fig. 4–6). The postoperative course was uneventful and the osteotomy healed. The patient has no pain and is very satisfied with the outcome. She does not use a walking aid and was able to resume full work.

Case 3. F.T., a woman born in 1900, had a diagnosis of osteoarthrosis of the right hip. In 1947, she had onset of pain in the right hip on loading the hip and also on moving it. She had no pain when at rest. She was treated with short wave diathermy, but without any results. Ten years later (in 1957) she received radiation treatment without any results. Later she received continuous physiotherapy without any improvement. An intertrochanteric femoral osteotomy was carried out in March, 1962, when the patient was 62 years of age.

The postoperative course was complicated by continuous pain in the right hip, and the patient used crutches. The reason for the continuous pain was initially difficult to

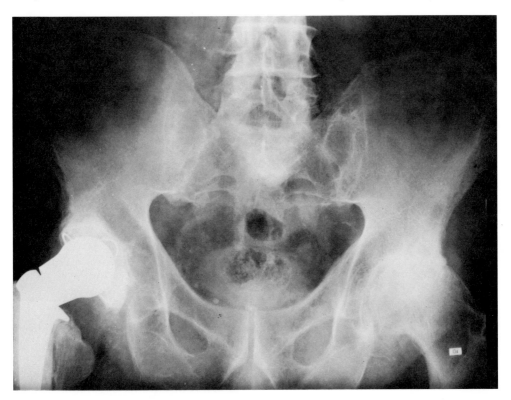

Fig. 4–3. Case 1. Roentgenogram following reoperation with a Christiansen total hip replacement (8½ years after Figure 4–2).

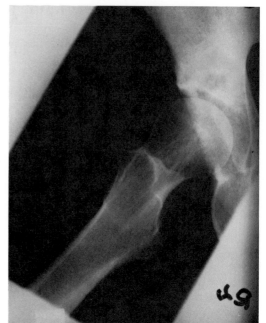

Fig. 4–4. Case 2. Roentgenogram at first presentation.

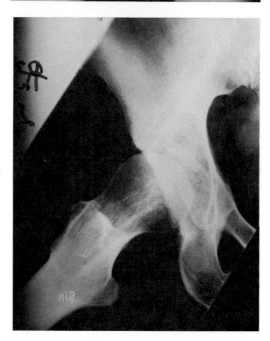

Fig. 4–5. Case 2. Roentgenogram of condition some 2 years later.

analyze, but there was a sign of nonunion and, in January 1963, x-ray disclosed that the fixation device had fractured. In the same month the device was removed and, at operation, despite the visible osteotomy gap, the hip appeared stable. The patient then continued with active mobilization, but had continuous pain in her left hip and had to use two crutches. Meanwhile the leg began rotating externally, and after a few years she had a 90 degree external rotation of the lower limb, owing to torque in the osteotomy gap. Since the patient did not improve, further surgical intervention

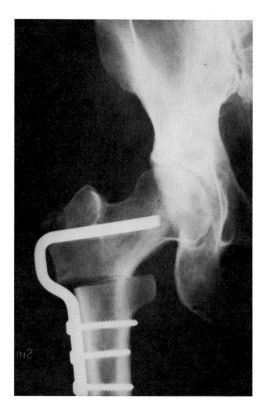

Fig. 4–6. Case 2. Roentgenogram following Müller compression osteotomy (10 weeks after Figure 4–5).

was suggested to her, but she declined. In 1971 a (McKee-Farrar) arthroplasty was carried out, and following this the patient has been completely free of pain and is now walking without any aids; she has no night pain, can go shopping without difficulty, and is 79 years of age.

Case 4. M.F., a male patient, was born in 1910. At the age of 35, pain developed in his left hip, and it was at that time (in the mid 1940s) suggested that the patient have a cup arthroplasty. The patient, however, declined the operation and therefore underwent a long series of conservative treatments including aspirin, antiinflammatory drugs, local steroid injections, and physiotherapy. With time, a very severe osteoarthrosis developed in the left hip, and in 1962 the patient was admitted for intertrochanteric osteotomy of the left femur. The procedure was performed and the osteotomy was stabilized with a device.

Initially, the results were good, but after a few months there developed in the left hip pain very much like the pain which previously caused the patient discomfort. By 1969, there was a considerable flexion contracture of the left hip with consequent shortening of the left leg, and the patient was given walking aids and adjustments to his shoe in order to aid walking. The patient refused all suggestion of surgical procedures and was quite satisfied with high doses of analgesics. After 2 years (1971), pain developed in the right hip joint and the patient requested a procedure that could guarantee him a better result than that obtained after the operation in the left hip. By this time, total hip arthroplasty had been introduced at our hospital, and the patient was therefore operated on in 1971, receiving a McKee Farrar arthroplasty. Unfortunately, technical difficulties arose in the perforation of the femoral stem through the femoral cortex, and the patient required reoperation—but not until

1974—owing to reluctance to be reoperated upon. The prosthesis was removed at that time; however, a new prosthesis loosened and the patient had once more to be reoperated upon, in 1977. This last hip is now performing well. However, the left hip, which was never improved, is the object for surgical consideration, and the patient is scheduled for surgery in the near future.

The failure of the left intertrochanteric osteotomy came from nonunion and later, fracture of the plate.

Case 5. F.F., a woman born in 1897, suffered from left osteoarthrosis of the left hip, which began in the early 1950s. In 1964 an osteotomy, fixed with a Wainright device, was done. Postoperatively there was never any relief of pain, and 1 year after the operation there was still nonunion and a fracture had developed in the Wainright plate. In 1965 the patient was reoperated upon, a new Wainright plate was inserted, and bone grafts were placed in the osteotomy gap. Postoperatively the patient improved slowly, and by 1969 she had no trouble whatever with her hip. Since then she has been on annual followup, and in 1979, at the age of 82, the patient was walking around without any discomfort whatever, moving as well as one can expect for a woman in her 80s.

Case 6. S.M. was born in 1894. In 1954 he began to develop increasing pain in his right hip, and had difficulty with mobility. In 1962 an intertrochanteric osteotomy of the right femur was performed, and following this there was continuous improvement, so that at followup in 1978, when the patient was 84 years of age, he was able to walk without any walking aids and claimed that he went for a 5 to 6 kilometer walk in the forest every day, and felt very well at that! On examination, he had slightly decreased movement in all directions of the right hip, but what more can be expected of a man of this age?

Case 7. M.S., born in 1902, developed pain in both hips in the early 1960s. He was first operated upon in the left hip joint in 1963 with a so called Voss or general release operation. However, this did not succeed, and in 1964 the patient therefore had an intertrochanteric osteotomy of the left femur, fixed by a Wainright plate and screws. Postoperatively, there developed thrombosis in both legs, as well as a pulmonary embolism which, however, improved with intensive care.

The patient never recovered completely, having continuous pain in his left hip and requiring two crutches There was no union of his osteotomy gap, and for this reason he was readmitted in 1964 for evaluation for a possible bone-grafting procedure. However, this was thought to be a bit premature, and the patient was scheduled for later surgery. One year following the osteotomy, he was reoperated upon for removal of the Wainright plate. At this operation it was noticed that the osteotomy gap was open, but for some inexplicable reason, nothing was done for this, leaving the patient with an open osteotomy gap without any fixation. Therefore, when the patient came to another surgeon, in 1965, he was reoperated upon with a bone-grafting procedure and had fixation with a MacLaughlin nail. Initially, the patient did very well, but with time he developed new pain, located in the lateral part of his left thigh. He claimed that there was no resemblance of this to the pain he had previously felt. It was therefore thought that the pain was not truly articular in origin, despite the fact that at x-ray examination there was still a gap in the osteotomy.

In 1970, the patient's MacLaughlin plate and screws were removed and at the same time the osteotomy gap was inspected, but the surgeon noted no true area of nonunion. Postoperatively, the patient's course was uneventful, and 1 year following the latest procedure the patient had no pain and was using only one crutch. The patient was reexamined in 1978 and claimed to be completely free of pain, was very happy with his leg, was able to walk without any walking aids, and enjoyed his old age as a pensioner sitting in the harbor looking at the ships passing by.

RATIONALE FOR INTERTROCHANTERIC OSTEOTOMY OF THE FEMUR IN THE TREATMENT OF OSTEOARTHROSIS

In treating osteoarthrosis of the hip joint by intertrochanteric osteotomy, it has been suggested that two principles prevail. The biomechanics of the hip become changed and biologic reactions alter in favor of a regression of the disease process. Müller[15, 16] has stated that an osteotomy is the only method of treatment that permits a fundamental change in the mechanics of the hip joint and thus gets at one of the basic causes of the osteoarthrosis process. Pauwels[22] regarded osteoarthrosis of the hip joint as a biomechanical problem, since there was a disturbance of the equilibrium between the load on the joint surface and the biologic resistance of the tissues in such a way that the load exceeded the biologic resistance. The main object in the Pauwels osteotomy is thus to diminish the stress on the joint surfaces. His aim was to distribute the load over a larger area, which could be achieved by an intertrochanteric angulation osteotomy. When the angle of inclination between the femoral neck and shaft is decreased, the areas of the articular surfaces that are subject to loading are increased. The lever arm of the abductors will become increased and the muscles acting on the hip joint will become relaxed. Thus, both the joint load and the local stresses become considerably diminished.

The planning of the angulation osteotomy is based on radiographs made with the hip in different positions of abduction and adduction. The position that will yield the best joint congruity is determined, and at the osteotomy a wedge is removed either medially or laterally in accordance with the radiographic determination. By increasing the length of the abductor lever arms, the force becomes directed more horizontally and is transferred to a more central part of the joint.

When a valgus osteotomy is necessary, a decrease of muscular force is obtained by separating the great trochanter and displacing it proximally and, by making the osteotomy oblique, the femoral shaft becomes displaced both medially and proximally. The change of direction of the active compressive and tensile stresses leads to a remodeling of the internal architecture of the femoral head and neck.

Osteoarthrosis may be regarded as a "wear and tear" phenomenon. Although chemical enzymatic and metabolic factors can lower the strength of articular cartilage, the removal of cartilage material requires a mechanical component. Cracks, tears, and ultimately cartilage removal depend upon the existence of stretching or tensile forces. If cartilage is compressed over its entire surface, as in a uniformly loaded joint, there are no tensile forces. However, joints are not uniformly loaded, and thus if one area of cartilage is compressed and another is not, the tissue between them is put in shear. Uneven compressive stresses may also arise from irregularities in the subchondral bone, permitting different degrees of deformation of the overlying articular cartilage. This is shown schematically in Figure 4–7. The repeated deformation and relaxation may cause the formation of a fatigue crack, which could enlarge to a tear and lead to detachment of a cartilage fragment.

Reduction of the overall load on a joint will reduce these forces, halting the progression of the degeneration and possibly providing the change needed for a healing

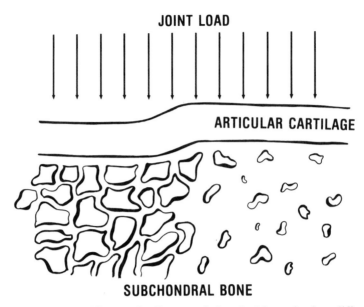

Fig. 4–7. Changes in the stiffness of adjacent subchondral bone lead to differences in deformation and the creation of tensile and shear stress in the cartilage.

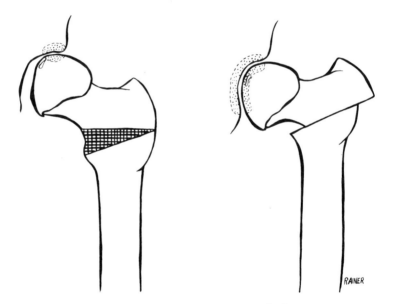

Fig. 4–8. The attainment of increased joint congruity by a varus osteotomy.

response. However, not only is the overall load important, but the distribution of the load also plays a role in determining the response of the articular cartilage. In fact, it is the surface stress rather than the load that is important. Roughly speaking, the stress may be regarded as the ratio of the joint load to the area of contact between the opposing joint surfaces. Thus, any procedure that increases the congruity of the joint will decrease the stress by increasing the area of contact. Maximum relief will be obtained when the load is reduced and the area of contact increased. These are, indeed, the points made by Pauwels.[13]

In order to obtain greater joint congruity, either a varus or a valgus osteotomy is performed. Figure 4–8 shows the increase of load-bearing surface at the hip with a varus osteotomy, while Figure 4–9 shows an increase in congruity for a valgus osteotomy. Whether a varus or a valgus osteotomy is performed depends on the relationship of the femoral head to the acetabulum, and is specific to the patient.

There is, as had been pointed out above, a change in the loading at the hip joint following osteotomy. The change in forces around the hip after osteotomy has been well treated by Phoenix,[25] using a two-dimensional analysis of one-leg stance that approximates the stance phase of slow walking. Figure 4–10 shows the forces acting, where R is the resultant on the femoral head, T the tension in the abductor muscle group, W_T the body weight less the weight of one leg, x the distance from the center of the femoral head to the line of action of T, and L the distance from the center of the femoral head to the line of action of W_T. The abductor action is taken as acting at an angle ϕ to the vertical, while R acts at an angle θ.

The free body diagram, in this approximation, consists only of the three forces R, T, and W_T. For equilibrium, these forces must all be parallel or must meet at a point. This is shown in Figure 4–10. The magnitude and direction of W_T is known, as is the direction of T. Thus the triangle of forces may be completed. The magnitude and direction of R, and the magnitude of T, may be obtained for the triangle mathematically or by graphical means. The magnitude of the femoral head reaction is given by:

$$R = W_T\sqrt{1 + 2\left(\frac{k}{x}\right)\cos\phi + \left(\frac{k}{x}\right)^2} \qquad (1)$$

and the angle of inclination is given by:

$$\tan\theta = \frac{(k/x)\sin\phi}{1 + (k/x)\cos\phi} \qquad (2)$$

Figure 4–11 shows a convenient method of recording the values of R and ϕ for

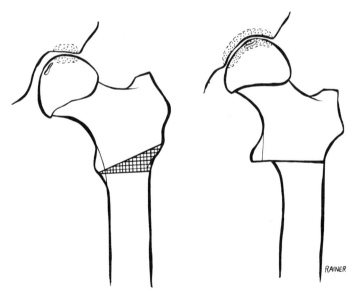

Fig. 4–9. The attainment of increased joint congruity by a valgus osteotomy.

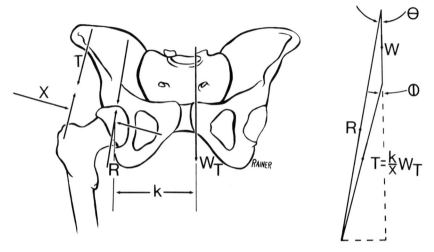

Fig. 4–10. One-legged-stance-forces acting in the coronal plane for pelvic equilibrium.

different values of (k/x) and ϕ. The force triangle may be drawn directly, and R and ϕ quickly estimated.

For a normal subject it is reasonable to take (k/x) equal to 2 and ϕ equal to 30 degrees. Triangle OBA_1 shows that $R = 3W_T$ and $\theta = 18$ degrees. For varus, (k/x) is smaller and ϕ larger, so that R tends to be smaller and less vertical. Triangle OBA_2 shows that $R = 2.4W_T$ and $\theta = 25$ degrees. For valgus, (k/x) is greater and ϕ smaller, so that R tends to be larger and more vertical. Triangle OBA_3 shows that $R = 3.5W_T$ and $\theta = 7$ degrees. It should be noted that the angle of pelvic tilt must be taken into account in estimating (k/x) and ϕ from roentgenograms.

Fig. 4–11. One-legged-stance-force triangle for pelvic equilibrium in the coronal plane. Information may be plotted from AP roentgenograms.

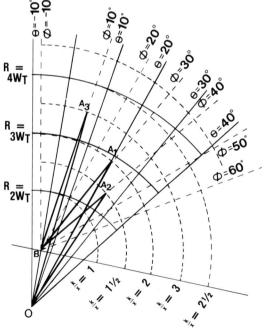
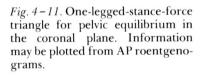

The varus osteotomy has the advantage of not only decreasing the load per unit area by increasing congruity, but of decreasing the overall load on the joint as well. Valgus osteotomy, on the other hand, leads to an increase in the joint load. It is often suggested that valgus osteotomy be accompanied by a surgical lengthening of the muscles around the hip, or that the muscle pulls be reduced. The force of contracture may be reduced by 20 percent. However, the decrease in load on the hip by this procedure is not straight-forward. To simplify matters, it will be assumed that T acts in a vertical direction, and equilibrium vertically is thus satisfied if:

$$T + W_T = R \qquad (3)$$

Decreasing T will decrease R, but rotational equilibrium must also be satisfied. Taking moments about the center of the femoral head gives, for equilibrium:

$$Tx = W_T k \qquad (4)$$

Thus T cannot be changed and equilibrium maintained unless x or k alters; this can be accomplished by a change in the pelvic tilt.

In the above discussion it will be noted that the change in load after an osteotomy is relatively small, amounting to about 20 percent. Since it is the load per unit area (stress) which is the probable determining factor in survival of the articular cartilage, the overriding criterion must be the degree of joint congruence, since changes in the area of contact of greater than 20 percent are easily accomplished by surgery. Thus the procedure must aim for congruity, with the consideration of whether the osteotomy should be varus or valgus as a secondary factor. In passing it may be noted that the joint load may be decreased through use of a walking aid. Calculations show that a cane reduces the load on the opposite leg by 60 percent (Fig. 4 – 12).

The stability of an osteotomy depends upon the type of osteotomy employed and, in particular, on the degree of shift of the proximal femur to the distal femur. Medialization of the femur below the osteotomy cut will give a reduced area of contact between the bone on either side of the cut, and an increase in the load per unit area. Too great a load can delay healing.

Figure 4 – 13 shows the forces acting about the osteotomy site. Not only is there compression and shear, but moments must be withstood. The force F may be resolved into a horizontal component F_H and a vertical component F_V. At the osteotomy site, F_H tends to slide the proximal fragment laterally with respect to the distal fragment. The vertical component F_V produces a force tending to compress the proximal fragment on the distal fragment. Friction between the fragments – that is, between the osteotomy cut surfaces – is the force opposing the force F_H. If μ is the friction coefficient for bone on bone, the value of the fixation force F_μ is:

$$F_\mu = \mu F_V \qquad (5)$$

For equilibirum (approximately, since F_V is not perpendicular to the cut and since F_H is not exactly along the line of the cut but acts horizontally):

$$\mu F_V = F_H \qquad (6)$$

Since F acts at about 20 degrees to the vertical, $F_V > F_H$, and would typically be

Fig. 4–12. Use of a cane reduces the load on the opposite leg. In the case shown the reduction is 60 percent.

about 3 times as large. Hence μ should have a value about 0.3. Where the osteotomy cut acts at an angle α to the horizontal it may be shown that the friction coefficient must exceed the value:

$$\mu = \frac{F_H + F_V \tan \alpha}{F_V - F_H \tan \alpha} \tag{7}$$

and, assuming that the joint force F acts at an angle θ to the vertical, gives the limiting friction coefficient as:

$$\mu = \frac{\tan \theta + \tan \alpha}{1 - \tan \theta \tan \alpha} \tag{8}$$

It should be noted that equilibrium cannot be maintained if the osteotomy cut is too steeply inclined to the horizontal.

However, it is the presence of moments which causes the biggest threat to stability. Both F_V and F_H have moments about the osteotomy which oppose each other, but the anticlockwise moment of F_V is the larger and can lead to the proximal fragment

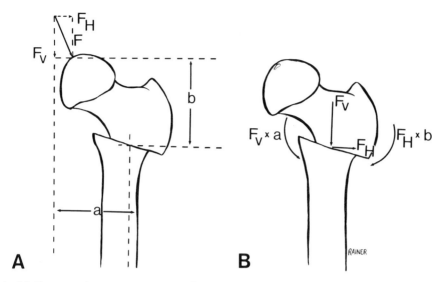

Fig. 4–13. Forces acting at an osteotomy site: (A) overall force system; (B) forces and moments at the osteotomy site.

being rotated into varus. This destabilizing moment is little opposed at the osteotomy cut, but may be opposed by muscular action (on the other hand, the action of muscles, in addition to the action of the joint force, may lead to destablization). It is clear that only by internal fixation can the osteotomy be adequately stabilized.

It is standard practice to employ a fixation device to stabilize the osteotomy, thus preserving alignment and allowing earlier patient mobilization. Whatever the osteotomy, the fixation device must withstand the forces and moments imposed on the hip. However, there will be differences, because the particular type of osteotomy will determine how much load is taken directly by the bone (the overall load may change for the reasons given earlier), and because there are different types of osteotomy fixation devices.

Figure 4–14 shows a simplified view of a nail and plate similar to a Nişşen/McKee device. It is assumed that the force W is concentrated at a point, and that the nail part of the device sustains the load—that is, the plate is sufficiently well fixed to the femoral shaft by screws that the nail/plate junction is completely stable. These are the same assumptions made to get a measure of the deflection of a nail/plate (solid) device in the case of femoral neck fractures. It may be shown that the deflection of the nail, B, is:

$$Y_B = \frac{F \cos \beta}{2EI} \ell^2 \left[L - \frac{1\ell}{3} \right] \tag{9}$$

and that the deflection at 0 is:

$$Y_O = \frac{F \cos \beta}{3EI} \ell^3 \tag{10}$$

where E is the modulus of elasticity and I is the moment of area. For minimum deflection and maximum stability, $\cos \beta$ should be small (β should be large), ℓ should be small, E should be large, and I should be large. Many of the foregoing factors are not capable of change over a wide range. The modulus of elasticity, for practical purpos-

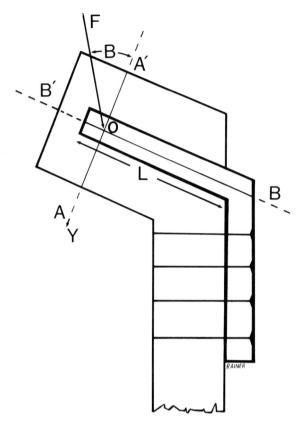

Fig. 4–14. Parameters influencing the deflection of a hip nail similar to a type used for internal fixation of an osteotomy.

es, is only variable by a factor of 2. As has been shown, the force F can only change by a factor of 2 unless the patient does not ambulate. However, rather large changes can be made by changing the shape of the cross-section, which will change I.

It may be remarked that the different designs of osteotomy fixation devices available, compared to a relative uniformity in configuration for rigid hip nails used for fracture fixation, can give quite different rigidities of device unless some allowance is made in the design. Figure 4–15 A shows a design of osteotomy plate typical of the Müller school. For this type of plate the angle β will be very small, so that the cos β approaches unity and the deflection is a maximum. The effect can be compensated for by increasing the section to increase I. The deflection of a Wainwright plate is obviously different owing to the different configuration (Fig. 4–15 B) of the plate. However, the same general considerations apply for minimizing deflection.

The calculations give an indication of those factors likely to influence the deformation of an implant device used for osteotomy fixation, and hence the stability of the osteotomy. One way to verify the calculations is by carrying out an experimental study of the deflection of the device under load. This has been done by Phoenix[24] for a Wainwright plate with the addition of a horizontal trochanteric screw (Fig. 4–15 C). Calculations indicated that the more rigid configuration was one in which there was little clearance between the screw and the plate, and this was verified experimentally, as shown in Figure 4–16.

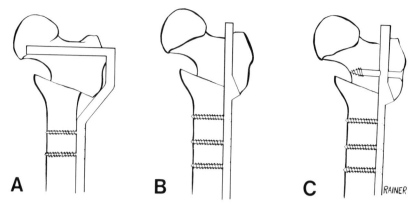

Fig. 4–15. (A) Fixation of an osteotomy with a Müller-type device. (B) Fixation of an osteotomy with a Wainwright-type device. (C) A Wainwright-type device with a transverse trochanteric screw.

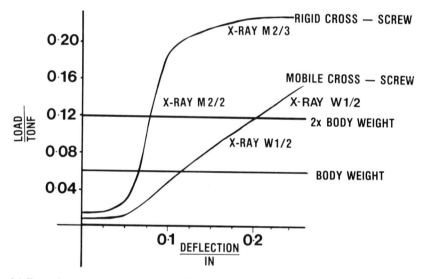

Fig. 4–16. Experimental determination of the deflection-versus-load characteristics of a modified Wainwright plate.

Some of the biomechanical principles have now been considered, but, as was stated earlier, the success of osteotomy is not entirely a biomechanical process. The biomechanical reasons for relief of arthritis after osteotomy include decreased load on the joint and increased joint congruity. Undoubtedly, a further factor is the redistribution of forces in the region around the osteotomy. However, the remodeling seen is believed to have other, more biologic, causes.

According to Olsson,[21] the significance of vascular and hemodynamic disturbances in osteoarthrosis of the hip is worthy of the greatest attention. By angiography on human postmortem and operative specimens, Harrison et al.,[5] in 1953, showed that osteoarthrosis of the hip joint was accompanied by arterial hypervascularity of the femoral head. In early stages of the disease, only the juxtachondral blood vessels were involved. With further progress, the whole vascular tree became affected in and around the bone. New arterial pathways also appeared, entering the femoral head at its junction with the neck. The findings of Müssbichler[17] confirmed the existence of arterial hypervascularity in osteoarthrosis of the hip. Phillips,[23] in 1966 (cited by Olsson[21]), demonstrated by phlebography that the venous drainage from the head and neck of the femur in osteoarthrosis attained a pathologic pattern. The vessels were dilated and varicose and carried the blood distally into the femoral shaft. After osteotomy a normal pattern returned. The disappearance of normal venous channels and delayed emptying of blood through tortuous veins into the femoral shafts in the hip osteoarthrosis was also demonstrated by Arnoldi et al.,[2] who found the intramedullary pressure of the head and neck of femur to be higher in osteoarthrosic hips than in normal hips. Intertrochanteric osteotomy causes an immediate fall of the intramedullary pressure, and it has been suggested that this may partially explain the relief of pain following this procedure.[1] Olsson[21] also stated that the marked retardation of emptying of blood from the femoral neck osteoarthrosis shown phlebographically by Arnoldi et al.[2] confirmed the observations of Hernborg,[6] who found an elimination rate of one-fifth in osteoarthrotic hips, as compared with controls, using $Na^{131}I$ as a transfer substance. The demonstration of increased vascularity and the indications of delayed blood flow may seem contradictory, but in pathologic states such conditions are not incompatible.

A reason for pain has been suggested to be lowered pH. This may develop under such circulatory circumstances as exist in the osteoarthritic hip and, as a matter of fact, increased acidity has been demonstrated in the intraosseous blood of the femoral head in osteoarthrosis.[26]

In osteoarthritis, relief of pain owing to osteotomy may thus be explained by biomechanical principles involving:

1. A medial shifting of the line of weight-bearing.
2. Reduced tension from correction of deformity and the alignment of the limb.
3. Improved congruity of the head and acetabulum with wider surfaces of contact.
4. Relief of pressure by release of muscle contracture.

Biologic changes due to the change of the vascular and hemodynamic events in the femoral head also play a role in relief of pain. The relative importance of biomechanical versus biologic factors is still in doubt, but a synergism between these factors cannot be ignored.

REFERENCES

1. Arnoldi, C. C., Lemperg, R. K., and Linderholm, H.: Immediate effect of osteotomy on the intramedullary pressure of the femoral head and neck in patients with degenerative osteoarthritis. Acta Orthop. Scand. *42*:357–372, 1971.

2. Arnoldi, C. C., Linderholm, H., Müssbichler, H.: Venous engagement and intraosseous hypertension in osteoarthritis of the hip. J. Bone Joint Surg., *54-B*:409–417, 1972.

3. Barton, J. R.: On the treatment of ankylosis by the formation of artificial joints, North Am. Med. Surg. J., *3*:279–300, 1827.

4. Blount, W.: Osteotomy in treatment of osteoarthritis of the hip. J. Bone Joint Surg. *46A*: 1297–1325, 1964.

5. Harrison, M. H. M., Schajowicz, F., and Trueta, J.: Osteoarthritis of the hip: A study of the nature and evolution of the disease. J. Bone Joint Surg., *35B*:598–613, 1953.

6. Hernborg, J.: Elimination of $Na^{131}I$ from the head and neck of the femur in unaffected and osteoarthritic hip joints. Arthritis Rheum., *12*:30–42, 1969.

7. Hirsch, C., and Goldie, I.: Osteotomy and osteoarthritis of the hip joint. A follow-up study. Acta Orthop. Scand., *39*:182–202, 1968.

8. Hirsch, C., and Goldie, I.: Walk-way studies after intertrochanteric osteotomy for osteoarthritis of the hip. Acta Orthop. Scand., *40*:334–345, 1969.

9. Hirsch, C., Goldie, I, and Ryba, W.: Intertrochanteric osteotomy for osteoarthritis of the hip. A radiological evaluation. Clin. Orthop., *86*:63–72, 1972.

10. Hulth, A.: Circulatory disturbances in osteoarthritis of the hip. A venographic study. Acta Orthop. Scand., *28:*81–89, 1959.

11. Kirmisson, E.: De l'osteotomie sous trochanterienne appliqué a certains cas de luxation congenitale de la hanche. Rev. Orthop., *5*:137–146, 1894.

12. Lorenz, A.: Uber die Behandlung der ireponiblen angebornen Hüftluxationen und der Schenkelhalspseudarthrosen mittels Gabelung (Bifurkation des oberen Femurendes). Wien. Klin. Wochenscher., *32*:997, 1919.

13. McMurray T. P.: Osteoarthritis of the hip joint. Br. J. Surg., *22*:716–722, 1935.

14. Meriel, F., Ruffie, R., and Fournie, O.: La phlebographie de la hanche dans les coxarthroses (á propos de 100 injections). Rev. Rheum., *22*:238–249, 1955.

15. Müller, M. E.: Die hüftnahen Femurosteotomien. George Thieme Verlag, Stuttgart, 1971.

16. Müller, M. E.: Intertrochanteric osteotomy in the treatment of the arthritic hip joint. p. 627. In Tronzo, R. Ed., Surgery of the Hip Joint. Lea and Febiger, Philadelphia, 1974.

17. Müssbichler, H.: Arteriographic findings in patients with degenerative osteoarthritis of the hip. Radiology *107*:21–29, 1973.

18. Nissen, K. I.: The arrest of primary osteoarthritis of the hip. J. Bone Joint Surg., *42B*: 423–428, 1960.

19. Nissen, K. I.: The arrest of early primary osteoarthritis of the hip by osteotomy. Proc. R. Soc. Med., *56*:1051–1063, 1963.

20. Nissen, K. I.: Un cas d'osteoarthrite primitive debutant de la hanche traite par osteotomie avec deplassement minime. Acta Orthop. Belg., *30*:651–662, 1964.

21. Olsson, S. S.: Intertrochanteric osteotomy of the femur for osteoarthritis of the hip joint. Tryckeri AB Litotyp, Göteborg, 1974.

22. Pauwels, F.: Atlas zur Biomechanik der gesunden und kranken Hute. Prinzipen Technik und Resultate einer kausalen Therapie. Springer Verlag, Berlin, Heidelberg, New York, 1973.

23. Phillips, R. S.: Phlebography in osteoarthritis of the hip. J. Bone Joint Surg., *48B*: 280–286, 1966.

24. Phillips, R. S., Bulmer, J., Hoyle, G., and Davis, W.: Venous drainage in osteoarthritis of the hip. A study after osteotomy. J. Bone Joint Surg., *49B*.:301–312, 1967.

25. Phoenix, O. F. Internal fixation in intertrochanteric osteotomy. pp. 10–16. In The Fixation of Fractures Using Plates. The Institution of Mechanical Engineers, 1974.

26. Richards, D. J., and Brookes, M.: Osteogenesis and the pH of the osseous circulation. Calcif. Tiss. Res. 2(Suppl. 93):1968.

27. Schanz, A.: Praktische Orthopädie. Springer, Berlin, 1928.

28. The Fixation of Fractures Using Plates, The Institution of Mechanical Engineers, 1974, pp. 1–8.

29. Trueta, J., and Harrison, M. H. M.: The normal vascular anatomy of the femoral head in adult men. J. Bone Joint Surg., *35B*:442–461, 1953.

5
Multiple Pin Repair of the Slipped Capital Femoral Epiphysis

S. M. K. Chung, M.D.; T. T. Hirata, Ph.D.

INTRODUCTION

A slipped capital femoral epiphysis (SCFE) results when the anatomic relationship of the femoral head with its neck and shaft changes as the result of an epiphyseal plate disruption. (Fig. 5–1). By definition, this condition can only occur before epiphyseal plate closure. In SCFE, proximal femur failure occurs at the epiphyseal plate rather than in the neck, as in developmental coxa vara, or below the trochanter, as in proximal femoral focal deficiency.

In almost all cases of SCFE the slip direction is inferior and posterior in relationship to the femoral neck. The femoral head remains in the acetabulum while the entire lower limb is externally rotated. In early stages of SCFE slipping may occur purely in a posterior or medial direction, but the posterior-medial position is the most common as the disease progresses. The femoral head is displaced superior and lateral over the femoral neck in 1 percent of cases; anterior slips rarely occur.

SCFE can occur in two ways. A large dynamic load can cause the femoral head to separate, with a crack through the epiphyseal cartilage-metaphyseal bone junction. Alternatively, a smaller excessive shear force applied intermittently and slowly over a longer period can cause a gradual slip. A normal epiphyseal plate can separate if the forces to which it is subjected exceed normal shear strength; normal forces can also separate a weakened epiphyseal plate. Chronic slips, which constitute 78 to 86 percent of the total, are more common than acute ones.

Acute slips may be further subdivided into two categories. In the first category, severe trauma, such as an automobile accident or a fall from a great height, can produce a slip without prodromal symptoms in young children from birth to age 9 years. Chronic slipping is rarely reported in this age group. Traumatic upper femoral epiphyseal separation occurring in newborns is usually the result of abnormal presentation.

In the second category of acute slips the patients are in the same age group (9 to 17 years) as those who develop chronic slips, and almost all have prodromal symptoms for a short time before the slip occurs.

UNRECOGNIZED SCFE

SCFE in mild forms may be more common than suspected. Murray and Duncan[17] suggest that trauma may initiate slipping, and note that SCFE in a mild form may be

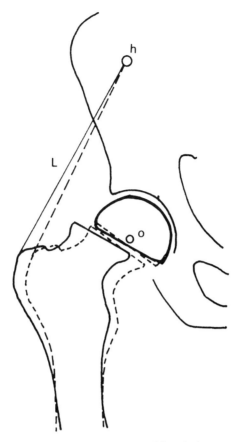

Fig. 5–1. Diagram of the upper femur showing positional changes of the femoral neck and shaft relative to the femoral head and acetabulum prior to (----) and after (——) slip.

seen on radiographs of normal adolescents. They feel that the "tilt sign" is an indication of minimal slipping, and conclude that excessive athletic activity causes this phenomenon and may be the cause of later degenerative hip disease. In our orthopedic practice, several adults examined for unrelated conditions were found to have decreased hip internal rotation. Radiographs showed a positive tilt sign. Further questioning revealed some of these patients to have had mild hip pain in adolescence, usually related to athletic activities.

ETIOLOGY REVIEW[12]

Theories proposed to explain SCFE include the abnormal hormonal,[11] genetic abnormal vascular, muscular imbalance, and abnormal chondroitin sulfate hypotheses. The most popular at present is the abnormal hormonal hypothesis, because experiments have shown that administering hormones to rats produced significant reduction in epiphyseal plate strength. Furthermore, SCFE does occur with several endocrinopathies. In the majority of SCFE cases however, the obesity observed is probably the typical exogenous type rather than a result of an endocrinopathy. Figure 5–2 shows that our SCFE patients (a finding confirmed by others[15]) are frequently overweight, but that their heights are within normal limits.

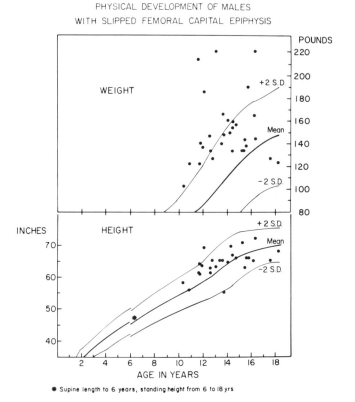

PHYSICAL DEVELOPMENT OF MALES
WITH SLIPPED FEMORAL CAPITAL EPIPHYSIS

* Supine length to 6 years, standing height from 6 to 18 yrs

Fig. 5–2. Height and weight measurements of 30 of our patients with SCFE reveals that many are overweight but almost all have normal heights. The increased body mass may load the capital epiphyseal plate beyond its safety factor during ordinary activities.

Our studies suggest that SCFE results from a combination of socioeconomic, dietary, anatomic, and mechanical factors. The increased body mass of these overweight children subjects the capital epiphyseal plate to a load that may exceed their structural safety factor during ordinary activites.

SCFE ASSOCIATED WITH FAMILIAL OBESITY

Since our endocrinologists have concluded that the obesity in our usual patient with SCFE is exogenous, we initiated a study with a dietician to determine family dietary patterns, our patients' understanding of nutrition, their physical activity, and the extent of obesity in their families. Most investigators have found exogenous obesity, a separate, well-known entity, to be a socioeconomic and familial phenomenon, frequently a result of excessive carbohydrate intake and of a poor understanding of optimum dietary requirements.

In our study we found obesity common among the parents and siblings of our SCFE patients—a pattern very similar to that noted in exogenous obesity (Fig. 5–3). Our patients with SCFE ate constantly, and their preferred foods were primarily carbohydrates. Neither the patients nor their mothers understood concepts of a balanced diet, and beyond required physical education courses in schools, most of our

FAMILIAL OCCURRENCE OF OBESITY (percent)

	One or Both Parents	Mothers Obese	Fathers Obese	Siblings Obese
Ordinary Exogenous Obesity	69-80	39-58	12-43	—
Obesity in SCFE (42 families in our study)	80	83	80	37-60

Fig. 5–3. A study of 42 families of children with SCFE shows the pattern of obesity closely resembling ordinary exogenous obesity.

patients did not participate in any physical activities, but preferred sedentary activities, such as watching television, in their leisure time. Further work remains to be done to clarify this relationship between exogenous obesity, activity, socioeconomics, SCFE, and adult osteoarthritis.

NORMAL CAPITAL FEMORAL EPIPHYSEAL SHEAR STRENGTH

The epiphyseal plate, a fibrocartilagenous structure, provides longitudinal growth but must also withstand load-bearing stresses. To better understand SCFE, we measured, in our laboratories,[4] normal human capital femoral epiphyseal shear strength. The shear strength and failure modes of this growth plate in hips obtained postmortem from children 5 days to 15 years of age were age dependent, primarily owing to a change in the perichondrial fibrocartilagenous complex. This structure is a caplike, thick fibrous layer adjacent to the epiphyseal plate. Resistance to shear was further provided by reciprocal bone and cartilage pegs (the mamillary processes) in adolescents (Fig. 5–4). When the perichondrial complex was excised, the epiphyseal plate strength was greatly diminished, especially in specimens from younger children.

We found a correlation between the growth plate shear strength and gross anatomic and histologic findings. As the femoral head and neck developed with age, gradual changes in gross and microscopic anatomy and mode of failure were evident (Fig. 5–5).

Birth to 13 Months (Fig. 5–6A and B). The capital growth plate is mainly extracapsular at birth, but with growth gradually advances superiorly. The head and neck are cartilaginous until 5 to 6 months of age. The perichondrial complex is extracapsular until the secondary center of ossification forms. During the first months of life, the growth plate with the surrounding perichondrial complex has two parts, one directed toward the femoral head and the other toward the greater trochanter. When tested with varying amounts of greenstick bending, these specimens failed, with some epiphyseal slipping. In all specimens tested, excision of the perichondrial complex was accompanied by a Salter-I type of growth plate failure.

Fourteen Months to 5 Years (Fig. 5–6C and D). The metaphysis advances to an intraarticular position, and the capital secondary ossification center gradually enlarges. The perichondrial complex is still prominent and plays an important role in capital

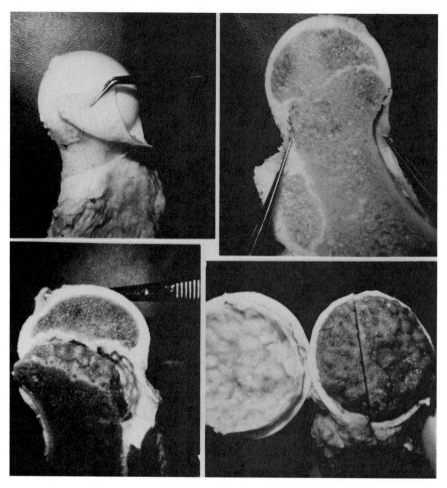

Fig. 5–4. The perichondrial fibrocartilaginous complex in a specimen from a child 10 years, 9 months old. Note: (1) the thick fibrous layer adjacent to the epiphyseal plate (top, right and left); (2) the cap formed by the complex which tightly surrounds the metaphysis (bottom right); and (3) the reciprocal pegs of bone and cartilage (mamillary processes) (bottom, right and left). (Chung, S. M. K. et al.: J. Bone Joint Surg., *58A*:94–103, 1976.)

growth plate support despite its decreasing size. The epiphyseal plate is now wavy and corrugated, and more resistant to shear. The perichondrial complex is composed of thick hyaline cartilage and includes both articular cartilage at the epiphyseal plate margins and fibrous tissue. On the lateral neck it extends as a continuous cartilage layer from the greater trochanter to the head. Specimens failed by neck fractures, Salter I type failure, or greenstick bending. In all specimens in which the perichondrial complex was excised, failure was of the Salter-I type. Histology of the bone-cartilage junction showed a weaving fracture which followed an irregular course, from the metaphysis to the zone of small-sized cells to the hypertrophic cell zone to the cell column layer and back to the metaphysis.

Six to 13 Years (Fig. 5–6E). During this period, the perichondrial complex decreases in volume to a mere rim at the periphery of the capital growth plate. The complex is composed of articular cartilage firmly attached to the periphery of the metaphysis at the border of the epiphyseal plate, and fibrous tissue which surrounds

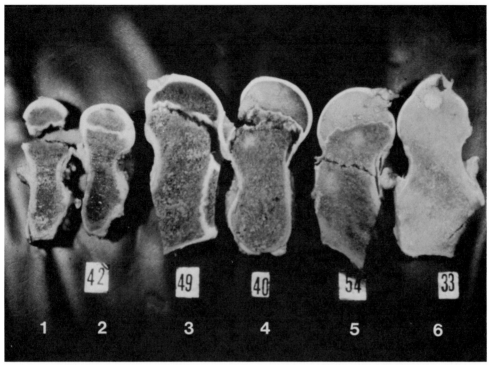

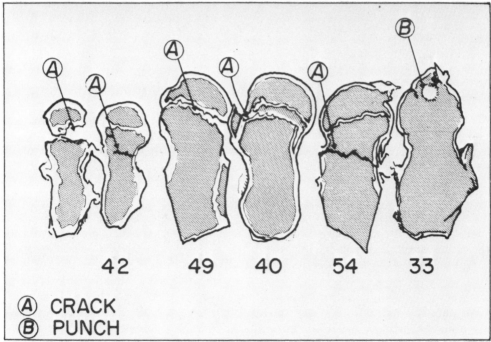

Fig. 5–5. Typical failure sites in six specimens, left to right (1 to 6). (1) At age 4 years failure occurred at one bone cartilage junction on the side with the perichondrial fibrocartilaginous complex excised (2) in the neck on the opposite control side; in (3) ages 10 years, 9 months and (4) age 12 years, 4 months, fracture occurred at the bone-cartilage junction as it does in SCFE; in (5) age 12 years when the plate was virtually closed there was a fracture of the neck; while in (6) age 15 years, 10 months when the plate was fused, the ram punching into the head. (Chung, S. M. K. et al.: J. Bone Joint Surg., *58A:*94–103, 1976.)

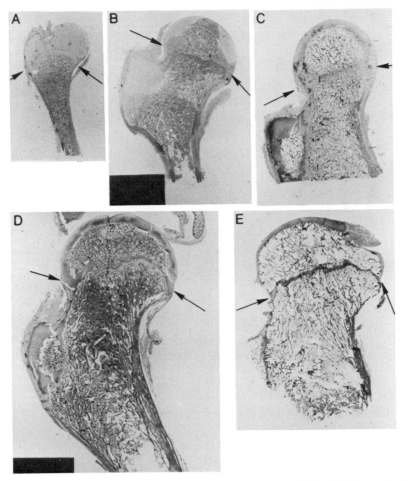

Fig. 5–6. Coronal histologic sections showing changes in the perichondrial complex (arrows) with age (see text). A=newborn, B=1 year, C=3½ years, D=10 years, and E=13 years (hematoxylin and eosin). (Chung, S. M. K. et al.: J. Bone Joint Surg. *58A*:94–103, 1976.)

the upper metaphysis. Within this age period the growth plate is thick, while the perichondrial complex is rapidly diminishing in size. When mechanically tested these specimens developed Salter-II or Salter-I failure or neck fractures. Perichondrial complex removal in these specimens had relatively little effect on the shear strength. Histologic studies usually showed a weaving failure. Although many authors have shown failure through the hypertrophic layer in epiphyseal injuries, their examples were taken from young rat growth plates. In human adolescent growth plates, failure would be virtually impossible through this thin layer because the mamillary processes would be sheared off, producing a weaving failure crossing many levels.

The support provided by the soft tissue "periosteum" surrounding the epiphyseal plate has been noted by others, but they failed to realize that the perichondrial fibrocartilagenous complex in infants and younger children — contrast to its role adolescents — has the major supporting function. As the child grows the complex becomes smaller and less important, while the reciprocal interdigitating mamillary processes provide more resistance to shear. In specimens from younger children failure occurred either by greenstick bending or by a neck fracture. This observation in the laboratory correlates well with the clinical findings; typical SCFE is only rarely observed in younger children.

The widened growth plate in patients with SCFE may be explained by an abnormally high compression load which stops calcification and inhibits normal degeneration of hypertrophic cells. Observations made in experimental animals[7] have shown that abnormally large compression loads cause the growth cartilage to increase to four times its normal height; the columnar arrangement becomes disorganized; the hypertrophic cells disintegrated; a degenerative fibrous tissue band extends across the plate from side to side; and, finally, if compression is maintained, growth plate fusion occurs. Similar pathology has been found in human SCFE.

LONG-TERM SEQUELAE OF MISMATCHED CONTACT SURFACES

Although the human hip joint is basically a ball and socket joint, neither the femoral head nor the acetabular articular surface is spherical.[5] The femoral head is somewhat egg-shaped, with its long axis along the femoral neck. Diameters in several planes are different. The acetabulum is less spherical than the femoral head, but in individuals over 18 years of age the acetabulum becomes more spherical while the femoral head does not.

The femoral head and acetabulum contact surfaces change with the magnitude and direction of an applied load.[3, 10] When, in young individuals, the load applied is from 20 to 50 percent of the body weight (as in the swing-gait phase), only the anterior and posterior femoral head surfaces contact. Acetabular and femoral head asphericity may account in part for this finding. The surfaces appear to fit better in slight abduction and internal rotation.[13] The joint incongruity may allow synovial fluid to flow over, nourish, and lubricate articular cartilage at light loads. No contact occurs in the inferior and perifoveal femoral head regions. When the entire articular acetabular surface is in contact, the contact area on the femoral head averages about 27 cm and covers about 70 percent of the femoral head articular surface in adults.

When SCFE occurs, acetabular-femoral head contact surface relationships change. This is suggested by changes in passive hip motion on physical examination, radiographic findings, and direct inspection of articular surfaces at the bone pegging operation. When the femoral head slips, its postero-medial surface makes less contact with the matching, posterior acetabular surface. In hip flexion, the anterior-superior exposed femoral neck metaphysis may touch the anterior acetabular articular cartilage, causing abnormal wear. Increased availability of contact surfaces during the activities of daily living may protect against the development of osteoarthritis and excessive wear — for example, the habit of squatting in Asian countries seems to distribute body load to a greater contact area, leading to less wear and subsequent osteoarthritis. Even slight degrees of SCFE may lead to a narrow articular cartilage many years later. There is some evidence that the malposition of the femoral head following SCFE may spontaneously correct with time, but whether the extent of such correction is enough to prevent excessive wear is yet to be determined.

CHONDROLYSIS AND AVASCULAR NECROSIS—UNEXPLAINED BIOLOGIC COMPLICATIONS

Chondrolysis is an acute articular cartilage necrosis of the femoral head and acetabulum, characterized by articular cartilage thinning while the subchondral bone remains unaffected. In this condition, the distance between the ossified acetabulum and the femoral head is less than the normal 5 to 7 mm on a standard AP radiogram. Chondrolysis, a common complication of SCFE, has been reported in from 1.1 percent to 40.4 percent of patients. Chondrolysis following SCFE appears more frequent in blacks, and occasionally occurs after valgus osteotomy. In contrast to osteoarthritis, which develops several years after SCFE, chondrolysis usually occurs within the first year. Patients with chondrolysis secondary to SCFE develop a progressive restriction rather than improvement of motion after treatment such as pinning or osteotomy.

The pathology of chondrolysis includes capsule and synovial thickening, mild chronic inflammation, and fibrosis. Adhesions occur between the femoral neck and synovial membrane. Necrotic cartilage patches are present on the femoral head and acetabulum. Chondrolysis is more frequent after severe slipping, but may also occur after a mild slip, prolonged immobilization, traction, and osteotomies.

Femoral head avascular necrosis can produce a poor result after SCFE treatment (Figs. 5–9 and 5–10). It is three times less common than chondrolysis after SCFE treatment. This complication is often encountered after a cuneiform femoral neck osteotomy performed to return the femoral head to an anatomic position. During this procedure the blood supply to the femoral head is interrupted. For this reason, such osteotomies have been abandoned by many surgeons. Avascular necrosis may sometimes occur after simple pinning, possibly by injury to the lateral ascending cervical vessels by the pins or by some other,unknown factor.

SCFE TREATMENT METHODS

SCFE treatment usually depends on: (a) The severity of the slip (mild, moderate, severe); (b) the duration of the slip (acute or chronic); and (c) the physician's experience and training. The most frequent treatment methods used at present include: (a) Threaded pin or wire fixation of the capital growth plate; (b) capital femoral epiphyseodesis with bone pegs; (c) osteoplasty (excision of a prominent anterior-superior neck hump) combined at times with (a) or (b); and (d) proximal femur osteotomy.

Knowles or other thin-threaded pins drilled across the capital growth plate prevent further femoral head displacement and are most useful for stopping mild or moderate chronic slips from progressing, or for maintaining reduction in acute slips. Narrow-diameter pins do not have the potential complication of further displacing the femoral head sometimes caused by larger Smith-Peterson nails. These extraarticular pins, introduced at the lateral femoral trochanteric flare, are advanced across the capital growth plate. The pins are removed after the capital growth plate fuses.

The use of Knowles pins has the following advantages: (a) They reliably prevent further slipping; (b) they may be easily and quickly inserted through a small incision; (c) neither a hip joint arthrotomy nor a deep dissection is necessary for their insertion. These pins have the following disadvantages: (a) Their sharp tips occasionally protrude undetected from the femoral head articular surface and damage the acetabulum; (b) the pins must be accurately placed using two-plane radiographs during the operation; a lateral view radiograph, however, is sometimes difficult to obtain in these obese children; (c) the pins gradually overgrow with new bone and are sometimes very difficult to remove; (d) removal of pins requires a second hospital admission; (e) the lateral protruding ends of the pins may cause a painful bursa before the pins are ready for removal; and (f) femoral head avascular necrosis or chondrolysis may occasionally occur with their use.

The Howorth bone peg epiphyseodesis is used for chronic mild and moderate slips. The advantages of the procedure are as follows: (a) A direct removal of abnormal growth plate and replacement with a bone graft bridge; (b) one definative procedure is done with none of the pin complications mentioned above, nor the necessity of another hospital admission and surgery for pin removal; (c) rapid (3 to 4 month) fusion of capital growth plate; and (d) osteoplasty may be combined during the same operation.

The disadvantages of the procedure include: (a) A more extensive dissection of the anterior hip and arthrotomy; (b) whereas postoperative infection which, while a potential complication of any operation, may involve the hip joint in these patients; (c) the theoretical possibility of further displacement of the femoral head during the early healing phases, although this has not occurred in our experience.

Proximal femoral osteotomies are used to correct severely displaced femoral heads by means of excision of an appropriate bone wedge at either the femoral neck base or the intertrochanteric or subtrochanteric level, followed by repositioning of the upper fragment. The osteotomy site is usually fixed internally with metal plates and screws. Although these osteotomies correct the femoral head position, one deformity is created to correct another. Anatomic repositioning of the femoral head by a femoral neck subcapital cuneiform osteotomy is unfortunately frequently associated with avascular necrosis, and most orthopedists have abandoned this procedure. An extensive review of proximal femoral osteotomies for SCFE is outside the scope of this paper.

CASE PRESENTATIONS

Case 1. An 11-year-old part-Hawaiian boy fell from a bicycle at age 6 years, injured his left hip, and developed persistent intermittent hip pain and a mild limp. His radiograph at the time of the original injury was "normal." About 1 month before admission his right hip became spontaneously painful without trauma. Examination showed an obese, 155-pound patient. For his age, his weight was over the 100th percentile and his height at the 94th percentile. He had a bilateral Trendelburg gait. His leg lengths were equal. His anterior hip was tender to palpation. Hip motion showed right flexion of 110 degrees and left flexion of 110 degrees. Internal rotation on both sides was 0 degrees. Right abduction was 40 and left was 30 degrees. Both hips rotated externally 30 degrees with flexion.

Radiographs (Fig. 5–7) showed a moderate SCFE on the right and a severe slip on the left. Both hips were treated with *in situ* pinning. Postoperative hip motion showed slight improvement. We anticipate continued femoral neck remodeling (this report presented by courtesy of the Shrine Hospital Honolulu).

Case 2. An 11-year-old obese Hawaiian girl fell and was mildly injured at school 1 month before admission. Subsequently she developed left knee pain and a limp.

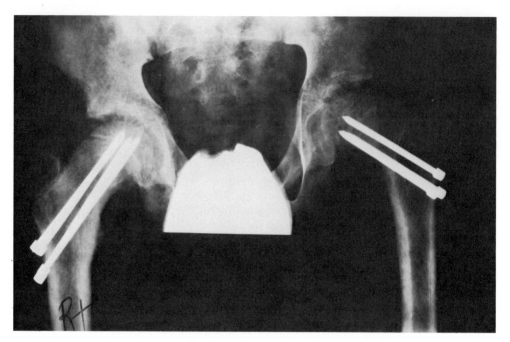

Fig. 5–7. Case 1. 11-year-old part Hawaiian boy with moderate slip on right and severe slip on left both treated with *in situ* pinning (case courtesy the Shrine Hospital of Honolulu).

Physical examination showed a girl above the 90th percentile in weight, with a left Tredelenburg limp and a positive Tredelenburg sign. Hip motion showed:

	Flexion	Abduction	Internal Rotation	External Rotation
Right	120	60	30	80
Left	120	60	10	70

Radiograms (Fig. 5–8) showed a mild left hip SCFE The slip was treated by

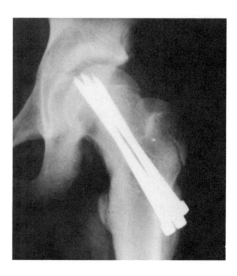

Fig. 5–8. Case 2. 11-year-old obese Hawaiian girl with a mild slip in the left hip treated by *in situ* pinning. One pin tip, within the femoral head articular cartilage, was recognized only on this film after surgery. See discussion in text.

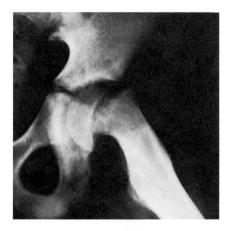

Fig. 5-9. Case 3. 11-year-old sustained on SCFE from a fall from a bicycle. This latent view shows the severe slip.

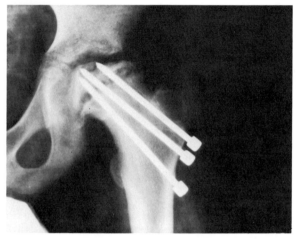

Fig. 5-10. Case 3. Six months after reduction and pinning, the femoral head is collapsed and deformed and the upper pin may be protruding into the joint.

Knowles *in situ* pinning. The patient did well after surgery, with no further pain, and the other hip has been asymptomatic.

Case 3. An 11-year-old obese Hawaiian boy fell from his bicycle and injured his left hip. He had severe pain. The left lower limb was shortened and externally rotated (Fig. 5-9). The hip was treated with bed rest and traction. The slip reduced after manipulation under anesthesia, and was pinned. However, 6 months later (Fig. 5-10), the femoral head became avascular and collapsed, with the upper pin protruding from the superior femoral head. Much further treatment will be needed.

BIOMECHANICS OF SLIPPED CAPITAL FEMORAL EPIPHYSIS

We believe that SCFE is a mechanically induced injury to the epiphyseal plate owing to overload. This section discusses the forces acting on the hip joint and their relationships to the stresses at the epiphyseal plate before and after placement of Knowles pins, as well as the failure mechanisms involved during slip and after pin placement. Possible hormonal, genetic, and nutritional factors leading to anatomic and material abnormalities are not specifically considered.

HIP JOINT AND EPIPHYSEAL PLATE FUNCTIONS

The hip joint, an eccentric ball and socket articulation composed of various biological materials with rate- and direction-dependent properties, transfers loads between the upper torso and the lower limbs, while maintaining structural integrity and accommodating a wide range of movement. It interacts with other parts of the body through several ligaments and muscle groups in both static and dynamic situations.

The epiphyseal plate, a fibrocartilagenous disc which transmits and distributes loads placed on the adjacent bony units, is the source of longitudinal bone growth during adolescence. The epiphyseal plate material is hyaline cartilage, which is porous, compressible, poorly resilient and viscoelastic.[6, 9] Hyaline cartilage exhibits creep under shear, which is usually small unless the epiphyseal plate thickens or its mechanical properties are altered by disease.[9] Hyaline cartilage has a low modulus of elasticity and a high mechanical compliance as compared to bone. The mechanical properties of hyaline cartilage are quantitatively not well known except for its tensile strength of approximately 1000 psi.[9]

Of the four epiphyseal plate zones described by Harris—the resting zone, the proliferating zone, the hypertrophied zone, and the zone of provisional calcification—the hypertrophied layer is the weakest, and once the supporting periosteal ring is removed, firm pressure will displace the plate at this level.[4, 11]

THE BODY'S RELATIONSHIP TO THE EPIPHYSEAL PLATE

The loading on the epiphyseal plate which leads to SCFE originates from the interaction of the body as a whole with the environment in the form of static, dynamic, or impact loads. Some of the load is absorbed by muscle action, some by bone and soft tissues, some is converted to forms of kinetic energy, while some of the load finds its way to the hip joint and to the epiphyseal plate. The immediate rate, magnitude, and direction of the load, as well as the prior loading circumstances, all determine the extent of injury to the epiphyseal plate.

Acute slip is associated with large dynamic or impact loading, resulting in a traumatic failure, while chronic slip is associated with large static or lower-magnitude dynamic or impact loading, resulting in the onset of injury and subsequent weakening and increased susceptibility to further displacement. Mild, moderate, and severe SCFE are functions of increasing permanent displacement relative to the plate diameter.

To show the interaction of loading on the body to loading at the epiphyseal plate, a static, one-legged stance is examined. Figure 5–11 is a body outline in the frontal plane showing the hip joint orientation and the two external forces acting on the body, the total body weight (W_t) and the ground reaction force (F). Equilibrium conditions require that these two forces be equal and opposite (act along the same line of action and have an opposite sense). Shown as a dashed arrow is the partial body weight (W_p, the body weight less the weight of the right leg), acting at its center of gravity, which is to the left and higher than the center of gravity of the total body weight.

Figure 5–12 is a free body diagram of the acetabulum, representing the body less the supporting leg. The purpose of this diagram is to determine the force (R') on the femoral head, which is equal and opposite to its reacting contact force (R) which acts

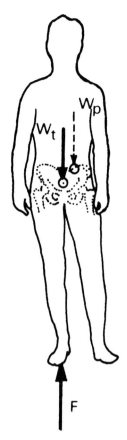

Fig. 5–11. Free-body diagram of a person in one-legged stance showing static loads.

on the acetabulum. The three forces acting on this free-body are W_p, acting at its center of gravity (g); the abductor muscle force (M), acting at its center of attachment (h) and having a known direction; and the contact force (R), acting at the acetabular surface and in line with the center of curvature of the femoral head (o). For static equilibrium, the three forces must be concurrent, thus establishing the direction of R, and all three must form a closed force polygon (shown in dashed arrows). The magnitudes of all of the forces can be found from the force polygon once any one of the forces (W_p, for instance) is known.

There is some disagreement in the literature[16, 18] as to which body weight to use and where is acts in determining loading on the femoral head. This is clarified once a suitable free-body diagram is selected, and all interactions with its surroundings are represented by equivalent forces and moments, including the effects of gravity. The free body is in static equilibrium when all of these *external* forces and moments balance. The forces and moments that are *external* in one free-body diagram may be *internal* in another diagram. In Figure 5–11, where the entire body is the free body, the total (not partial) weight must be used as one of the external forces acting on the free body. The abductor muscle force (M) and the loading (R) on the right femoral head cannot be found from this diagram since they constitute *internal* forces. Similarly, in Figure 5–12, the free body consists of the body less the supporting leg, and as such the partial body weight (W_p) (shown as a dashed arrow in Figure 5–11), acting at its center of gravity (g), is used. The interaction with the supporting leg

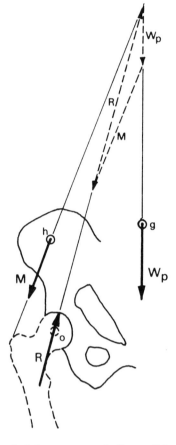

Fig. 5–12. Free-body diagram of the right acetabulum, representing the body less the supporting leg.

is now represented by the contact force (R) and abductor muscle force (M), which are *external* forces in this diagram and thus can be evaluated. A free body of the supporting leg could also have been used to evaluate R′ and M′ (identical to R and M, respectively, except for their sense).

The analysis for one-legged stance has to this point been used to relate the femoral head load (R), or its reaction (R′) on the acetabulum, to the total body weight (W_t).

The epiphyseal plate loading can now be related to the femoral head load(R′) by drawing a free-body diagram of the femoral head with a section at the epiphyseal plate (see Figure 5–13). The loading on the femoral head, although shown as a single resultant force acting at the center of curvature, is actually a distributed compressive load over a variable contact area. Stress and strain values in the proximity of the boundary have been shown to be affected by the loading conditions there.[2] However, for this study, a single force is used for illustrative purposes and not for obtaining magnitudes of stress and strain.

The most general state of stress at the epiphyseal plate consists of tensile or compressive stresses, bending stresses, shear stresses, and torsional stresses. Torsional stresses are negligible owing to the low-friction bearing surfaces at the femoral head and acetabulum. Figure 5–13 illustrates a situation of uniform compressive stresses and a linear distribution of bending stresses (shown combined), together with a shear component which represents a shear stress distribution with a maximum value at the plate center, zero at the edges, and acting parallel to the plate section.[19]

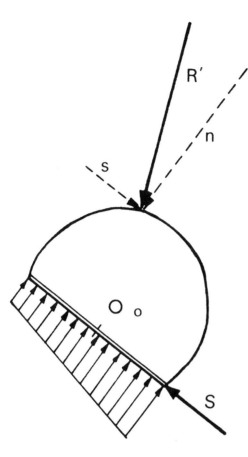

Fig. 5–13. Free-body diagram of the femoral head showing the stresses acting on the epiphyseal plate.

For a given loading condition, the magnitudes of tensile or compressive and shear stresses depend on the orientation of the selected plane upon which they act. At any location in the epiphyseal plate, there is an orientation that produces a maximum shear stress and another orientation that produces a maximum tensile or compressive stress. Local failure takes place in that orientation that produces failure stresses for the material, and which may not necessarily be in the direction of the epiphyseal plate. The global failure in the form of permanent displacement, however, takes place along the plate in SCFE.

DETAILS OF SLIP PROCESS

Bright, Burstein and Elmore[1] conducted variable strain-rate tests on the proximal tibial epiphysis of rats, and found that higher loads were required to fail the material at higher strain rates, indicating viscoelastic properties. During examination of the tibia sections after application of 50 percent failure energy (50 percent of the energy as indicated by stress-strain recordings of complete failure of the corresponding tibia in the same rat), short linear cracks were found whose orientations and locations suggested that they were caused by material shear failure. Observation of the complete failure of the corresponding tibial epiphysis indicated that a secondary and propagating crack began in the fibers that were furthest away from the center of the plate, and thus were in maximum tensile stress (bending). The propagating or failure cracks were found uniformly to extend through the physeal-metaphyseal junction in 15 percent of the cases, and at least partially through the upper proliferating or resting zone layer in 85 percent of the cases. The authors did not attribute any significant contribution of the periosteal ring to epiphyseal plate strength until after cartilage failure.

SHEAR STRENGTH OF THE EPIPHYSEAL PLATE

Experimental and analytical studies on human hip joint loading in static and dynamic situations have been two-dimensional and more recently three-dimensional.[21] The results range from 2.4 to 6.0 times body weight for the resultant load on the supporting hip joint in the one-legged stance, and from 0.8 to 6.5 times body weight for walking. Activities which involve impact loading, higher accelerations, or heavier equivalent body weights create even larger loads and greater susceptibility to plate slippage.

From Chung, Batterman and Brighton,[4] the two dimensional shear component of the resultant force on the femoral head (see Fig. 5–14) is:

$$F_s = F \cdot \sin (\psi - \alpha + \beta) \tag{1}$$

where α = neck-shaft angle
   β = plate-shaft angle
   ψ = resultant force-neck angle

Figure 5–15 is a plot of the resultant force (multiple of body weight) required to produce a shear component equal to the failure load (P_t), as a function of its angle, ψ, for two typical specimens 33 and 37:

Specimen 33: Weight = 12 kg; age = 5 years, 5 months

$\alpha = 42°, \beta = 13°$

$P_p = 41.5$ kg, $P_t = 70.0$ kg

Specimen 37: Weight = 20 kg; Age = 8 years, 8 months

$\alpha = 45°, \beta = 26°$

$P_p = 111.4$ kg, $P_t = 143.2$ kg

where P_p = failure load without perichondrial ring
   P_t = failure load with perichondrial ring.

Chung, Batterman and Brighton[4] state that the resultant force on the femoral head can vary from being parallel to more than 45 degrees with the neck axis. When ψ = 19 degrees and 29 degrees, respectively ($\psi = \alpha - \beta$), the resultant forces are perpendicular to the epiphyseal plate, so that there is no shear component. As ψ increas-

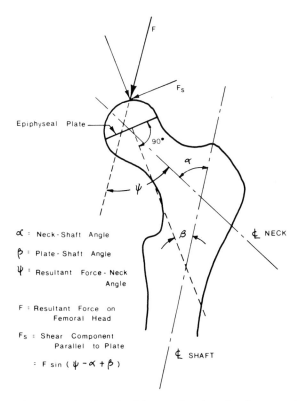

α : Neck-Shaft Angle

β : Plate-Shaft Angle

ψ : Resultant Force - Neck
Angle

F : Resultant Force on
Femoral Head

F$_s$: Shear Component
Parallel to Plate

: F sin (ψ − α + β)

Fig. 5 – 14. Diagram of the proximal end of femur showing the forces acting on the femoral head.

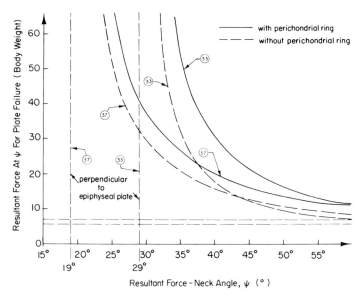

Fig. 5 – 15. Resultant force acting on the femoral head required to fail the epiphyseal plate as a function of resultant force-neck angle for two cases (Specimens 33 and 37).

es, the shear component increases, and the total resultant force required to fail the plate decreases, until $\psi = 119$ degrees, at which angle the resultant forces are parallel to the epiphyseal plate. Shown also in Figure 5–15 are the curves for epiphyseal plate failure without the perichondrial ring. It should be noted that Figure 5–15 is for a two-dimensional representation of a three-dimensional situation, and that failure due to combined compressive, bending, and shear stresses is not considered. At $\psi = 45$ degrees, which is 16 degrees from the perpendicular to the epiphyseal plate, Specimen 37 would require a resultant force of 16.3 and 12.7 times body weight, with and without the perichondrial ring, respectively. At the same value of ψ, which is 26 degrees from the epiphyseal plate perpendicular, Specimen 33 would fail at 21.2 and 12.6 times body weight, with and without the perichondrial ring, respectively.

Figure 5–15 shows that very large equivalent static loads must be generated—and at large angles from epiphyseal plate perpendiculars—for the plates to fail. Owing to the viscoelastic properties of biologic materials, loads are better resisted without damage at higher dynamic rates than at the slow loading rates under which this study was conducted. On the other hand, Bright, Burstein and Elmore[1] showed that for loads amounting to 50 percent failure energy, internal cracks were generated and some permanent damage was done which is not readily apparent on radiographic examination. Rat tibial epiphysis however, does not have mamillary processes like the human femoral capital epiphysis.

It is interesting to note Inman's[14] finding that for the two-dimensional (frontal plane) study of the one-legged stance in adults, the epiphyseal plate line is normally perpendicular to the medial trabeculae, which are in line with the reacting force on the femur. Consequently there is no shear, but only compressive stresses acting on the plate. Pathologic conditions or various physical activities can alter this situation to produce shear as well as compressive stresses at the epiphyseal plate.

BIOMECHANICS OF THE PINNED FEMUR

One form of treatment for SCFE is the insertion of several pins (typically Knowles pins) across the epiphyseal plate to provide shear support and prevent further displacement (Fig. 5–16). Knowles pins are stainless steel, with a typical tensile strength of 140,000 psi and a typical 0.2 percent offset yield strength of 115,000 psi.[22] They come in lengths (tip to collar) of 2½ inches to 6 inches, with a 2⅞ inch break-off length above the collar for insertion purposes. The front 1¼ inch is threaded and trocar-pointed. The shaft diameters are ⅛ and 5/32 inch, with thread diameters of 5/32 inch and 3/16 inch, respectively.

The usual procedure is to insert the pins, under radiographic control, across the slipped epiphyseal plate and 1 to 2 cm into the femoral head. For the usual 3½ inch long pins used, most of the remaining length lies within the femoral neck. Three to four pins are typically used to secure the plate.

Damage to epiphyseal plate cartilage and reduced attachment area between the femoral head and neck results in greater stress levels and reduced resistance to shear. Placement of the pins form a plate/pin/bone combination which has proven to be an effective means of preventing further displacement.

In analyzing the strength of this combination, the pins and cancellous bone in which they are imbedded are much more rigid than cartilage, and shear stresses

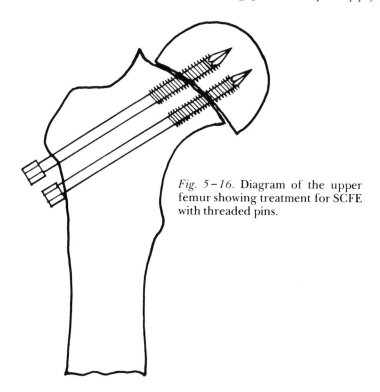

Fig. 5–16. Diagram of the upper femur showing treatment for SCFE with threaded pins.

placed on the femoral head are therefore resisted mainly by the pin/bone combination. The basis of the following analysis is that the pin/bone combination forms a bearing-type shear joint, in which the two modes of failure are the shearing of pins and the crushing of the cancellous bone, with the latter the more likely occurrence. An estimate of the magnitudes of failure loads involved in each form of failure can be made by making a conservative estimate of the shearing strength of the pin material and multiplying it by the nominal cross-sectional area of the pin:

$$\text{Shearing strength} \approx \tfrac{1}{2}\ \text{yield strength}$$
$$\approx \tfrac{1}{2}\ (115{,}000\ \text{psi}) = 57{,}500\ \text{psi}$$

Failure load for $\tfrac{1}{8}$ inch diameter pin (shear) $= 57{,}500\ \dfrac{\pi\ (\tfrac{1}{8}\ \text{inch})^2}{4} = 700\ \text{psi}$

Frost[9] gives a figure for the ultimate compressive strength of femoral-head bone of 575 psi. Multiplying this by the bearing area yields an estimate of the load required to crush the cancellous bone:

Failure load (cancellous bone) $= 575\ \text{psi}\ (\tfrac{1}{8}\ \text{inch})(1.5\ \text{cm})(0.3937\ \text{inch/cm})$
$$= 40\ \text{lbs}$$

So, by a factor of more than 15, the bearing material in the femoral head is by far the weaker. This is not to say that 40 lbs of shear applied to the femoral head will cause the plate to fail. There is additional resistance offered by the cartilage, periosteal ring, and mamillary processes. The calculations above assume that bending, tensile, or compressive and torsional stresses are small compared to shear stress, and do not affect the failure mode. Fatigue creep, material imperfections, and stress concentration effects, which would enhance failure rates, are also not included.

Frost[9] mentions that the pins or screws securing the epiphyseal plate are firmly imbedded in the adjacent osseous material and that, owing to the configuration and low friction of the hip joint, there is very little shearing or bending load exerted on the plate during ordinary activities. He therefore feels that the pins do not ordinarily provide very much shearing strength to the plate. They do, however, provide whatever strength is necessary to prevent further plate displacement until plate closure has occurred.

Future SCFE studies may include (a) three-dimensional, dynamic analyses, (b) obtaining of material property data such as load-deformation curves and the viscoelastic behavior of epiphyseal plate material, and (c) the performing of internal stress-strain studies using the finite-element method.

REFERENCES

1. Bright, R. W., Burstein, A. H., and Elmore, S. M.: Epiphyseal-plate cartilage. J. Bone Joint Surg., *56A*:688–703, 1974.
2. Brown, T. D., and Ferguson, A. B., Jr.: The development of a computational Stress Analysis of the Femoral Head. J. Bone Joint Surg., *60A*: 619–629, 1978.
3. Bullough, P., Goodfellow, J., Greenwald, A. S., and O'Connor, J.: Incongruent surfaces in the human hip joint. Nature *217:* 1290, 1968.
4. Chung, S. M. K., Batterman, S. C., and Brighton, C. T.: Shear strength of the human femoral capital epiphyseal plate. J. Bone Joint Surg., *58A*: 94–103, 1976.
5. Clarke, I. C. and Amstutz, H. C.: Human hip joint geometry and hemiarthroplasty selection. p. 63–89. In: The Hip. Proceedings of the Third Open Meeting of the Hip Society, 1975. St Louis, C. V. Mosby and Co., 1975.
6. Dumbleton, J. H., and Black, J.: An Introduction to Orthopaedic Materials, Charles C. Thomas, Springfield, Illinois, 1975.
7. Ehrlich, M. G., Mankin, H. J. and Treadwell, B. V.: Biochemical and physiological events during closure of the stapled distal femoral epiphyseal plate in rats, J. Bone Joint Surg., *54A:* 309–322, 1972.
8. Frankel, V. H., and Burstein, A. H.: Orthopaedic Biomechanics. Lea and Febiger, Philadelphia, 1970.
9. Frost, H. M.: Orthopaedic Biomechanics. Charles C Thomas, Springfield, Illinois, 1973.
10. Greenwald, A. S. and Haynes, D. W.: Weight-bearing Areas in the Human Hip Joint, J. Bone Joint Surg. (Brit), *54(1):* 157–163, 1972.
11. Harris, W. R.: The endocrine basis for slipping of the upper femoral Epiphysis An experimental study J. Bone Joint Surg., *32B:* 5–11, 1950.
12. Howorth, Beckett: Etiology, slipping of the capital femoral epiphysis Clin. Orthop., *48:* 49–52, 1966.
13. Johnston, R. C.: detailed analysis of the hip joint during gait. In: Proceedings of the Second Open Scientific Meeting of the Hip Society, C. V. Mosby Co., St. Louis, 1974.
14. Inman, V. T.: Functional aspects of the abductor muscles of the hip. J. Bone Joint Surg., *29:*607–619, 1947.
15. Kelsey, J. L., Acheson, R. M., and Keggi, K. J.: The body build of patients with slipped capital femoral epiphysis. Am. J. Dis. Child., *124:*276–281, 1972.
16. Merchant, A. C.: Hip abductor muscle force. J. Bone Joint Surg., *47A:*462–476, 1965.
17. Murray, R. O., and Duncan, C.: Athletic activity in adolescence as an etiological factor in degenerative hip disease. J. Bone Joint Surg. *53B:* 406–419, 1971.

18. Pauwels, F.: Biomechanics of the Normal and Diseased Hip. Springer-Verlag, Berlin, 1976.

19. Popov, E.: Introduction to Mechanics of Solids. Prentice-Hall Inc., Englewood Cliffs, NJ, 1968.

20. Williams, M., and Lissner, H. R.: Biomechanics of Human Motion. W. B. Saunders Co. Philadelphia, 1962.

21. Williams, J. F., and Svensson, N. L.: A force analysis of the hip joint, Biomed. Eng., *3:* 365–370, 1968.

22. Zimmer Product Encyclopedia. Zimmer USA, Inc., Warsaw, Indiana, 1978.

6
Repair of Intertrochanteric Fractures with a Sliding Nail

L. S. Matthews, M.D.; D. A. Sonstegard, Ph.D.;
J. H. Dumbleton, Ph.D.

INTRODUCTION

Reduction and internal fixation is the generally accepted treatment technique for intertrochanteric hip fractures in the elderly. The usual patient is in his or her seventies and in general ill health because of serious concurrent chronic disease. Frequently living alone, or with an aged mate, a patient with increasingly poor balance and coordination may have stumbled and fallen to the floor.

Frankel and Burstein[2] have shown that a 110 lb. woman with a center of gravity 0.86 m above the floor will generate 373 N.m of energy in a simple fall not involving strenuous activity. The velocity of impact is approximately 4 m/second. The energy generated should be compared to the energy of about 6 N.m required to cause fracture of the femoral neck. Thus, essentially, all of the energy generated in the fall must be absorbed or otherwise directed in order to avoid a fracture; it is especially difficult to avoid fracture of the hip when the impact is over the hip area. Energy may be absorbed by the muscles of the upper and lower limbs, by elastic strain in the soft tissue, and by elastic and plastic strain in the skeletal system. The nature of the floor also plays a role in the energy absorption process. Lack of coordination, while being responsible for the fall in the first place, may also result in the energy absorption and redirection processes being delayed or largely absent. Thus, while the energy of the fall may not appear to have been great, the patient will very frequently have sustained an intertrochanteric fracture of the hip.

The resultant inability to walk and the loss of both social and, in fact, physical independence are likely to be fatal to this patient, who may die from any of the recognized complications of bed rest in the elderly – pneumonia, pulmonary embolus, or progressive skin ulceration and infection. The surgical treatment goal for this patient is to relieve pain and thereby to maximize the patient's activity while the fracture heals. Relief from pain, the ability to walk, and even survival itself appear directly related to postoperative fracture stability. Interfragmentary motion is extremely painful. Because of this relationship among pain, activity level, survival, and fracture fragment stability, stable reduction and secure internal fixation are the keys to successful treatment of intertrochanteric fractures in the elderly.

Intertrochanteric fractures vary in their extent, displacement, fragmentation, and associated soft tissue dissociation. Figure 6–1 presents many features of such frac-

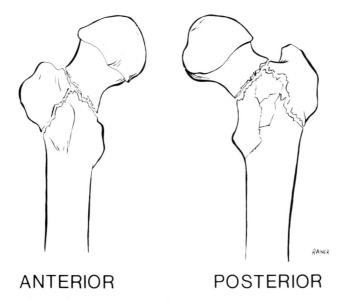

ANTERIOR   POSTERIOR

Fig. 6-1. A typical unstable intertrochanteric fracture. The greater and lesser trochanters have broken away from the intertrochanteric mass. There is considerable posterior wall comminution.

tures which must be considered in a mechanical description of their treatment. The major fracture line is nearly always angled from the most lateral extent of the superior neck down toward the superior extent of the lesser trochanter. While the anterior fracture line is usually simple, the posterior main fracture component is rarely straight or simple. Usually, there is considerable comminution and fragmentation of the posterior intertrochanteric region. The greater trochanter is usually avulsed as a large, free fragment. The somewhat posterior lesser trochanter comes off as a large or small piece, but always in association with the posterior comminution. This fragmentation of the posterior wall causes the greatest difficulties with reduction and stabilization. Healing is not generally a problem, since the major fragments have an excellent blood supply and since the fracture region is surrounded by vascular soft tissue.

Three general types of reduction are used to maximize the bone's contribution to internally fixed fracture stability. In the ideal situation, the fracture has only two fragments, there is little displacement, and some of the surrounding attached soft tissue has remained intact. In this case, and even at times with some four fragment fractures, it is possible to achieve a very nearly perfect anatomic reduction, as seen in Figure 6-2. The fracture surface areas directly oppose each other, the normal hip neck shaft is restored with satisfactory internal fixation. A small amount of resorption with rapid healing, in a nearly normal position, would be expected. Normal hip anatomy and function should follow.

In many cases the fracture has been more comminuted, with separation of the lesser and greater trochanters from the intertrochanteric mass. The posterior wall may have been fragmented. Under these circumstances a truly anatomic reduction is rarely accomplished and, without perfection, the reduction of the fracture fragments is always compromised and stability is rarely achieved. Recognizing the impossibility of achieving a stable anatomic reduction, two types of special rearrangement

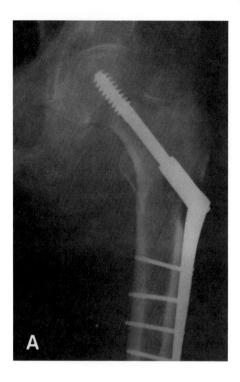

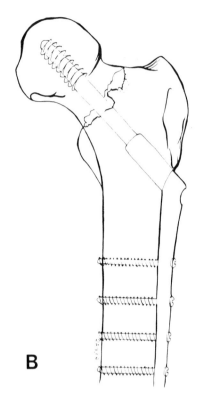

Fig. 6–2. (A) A simple, relatively uncomminuted intertrochanteric fracture reduced to an anatomic position. (B) Note the accurate fit of the fragments, which allows for an even distribution of interfragmentary compression forces across the fracture site.

of the fracture fragments have been advocated. The Wayne County General Hospital (Fig. 6–3A and B) reduction involves a medial displacement of the proximal head and neck fragment, relative to the femoral shaft. This head and neck fragment is also slightly tipped into valgus. The fragments are sculpted and the limb rotated to achieve maximum bony contact. The objective is to provide initial valgus against the inevitable varus deforming forces, and to provide a solid, medial calcar to cortical shaft resistance against collapse of the head-neck fragment.

The final reduction technique is based on the experience that unstable fractures always collapse, and that stability and not anatomic shape is the most important goal. First Rowe[7] and then Dimon and Hughston[1] recommended that for very comminuted fractures a surgeon recognize the inevitability of collapse. They suggest that the surgeon remodel both of the major fragments until the lateral fracture face of the proximal fragment can be inserted into the remaining intertrochanteric region of the distal fragment (Fig. 6–4A and B). The proximal fracture face should, in this reduction, rest securely against the medial side of the lateral cortex of the femoral shaft. Thereby, there can be no further collapse. The neck of the proximal fragment is supported by a natural or sculpted yoke which acts as a supporting fulcrum medially. This construct provides the minimal possible bending moment arms acting to deform a device and disrupt fracture stability. While a study of fracture stability[3] did not indicate a strength benefit for this reduction type, as compared to perfect anatomic reduction, clinical experience has indicated that it appears to realize the theo-

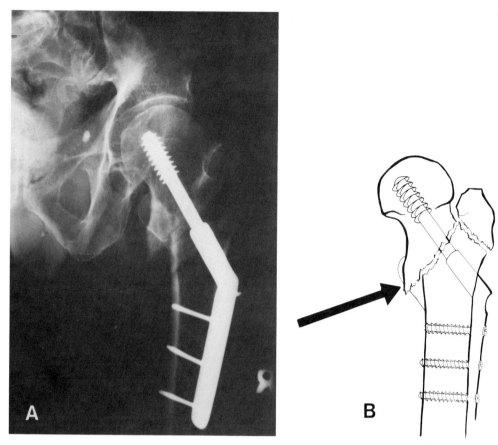

Fig. 6–3. (A) A Wayne County General Hospital reduction with relative valgus proximal fragment position. (B) The inferior neck cortex (calcar) of the proximal fragment is displaced medially to act as a buttress against collapse and a fulcrum support to the fracture and hip nail.

retical benefits in practice. It must be recognized that, just as the deforming moment arms are shortened, so too is the abduction lever arm for pelvic stabilization.

Thus, while several difficult methods of rearrangement or reduction of the separate fracture fragments are used, all seek to oppose the fracture faces so as to obtain maximum interfragmentary stability in a position compatible with normal hip function. Once the fragments have been repositioned to this optimally stable state, an internal fixation device is used to augment the stability of the construct. Various types of devices are used for fracture fixation at the hip. These include screws, nails, pins, sliding devices, and side-plated nails. Whatever the type of device, a stable opposition of fracture faces, with maximal fragment interdigitation, is desirable.

Figure 6–5 gives load-deflection curves for several intertrochanteric fixation devices averaged for several different reductions, tested under laboratory conditions.[3] It is obvious that devices of a wide range of stiffness are available. The fact that all of these devices will produce satisfactory results under appropriate conditions emphasizes that it is essential for the fracture faces to bear a large part of the applied load. The result of having the load carried largely or entirely by the fixation device is often permanent deformation of the device, or its outright failure.

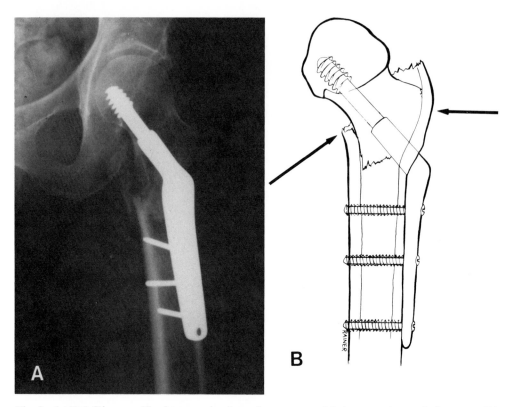

Fig. 6–4. (A) A Dimon – Hughston reduction of a very unstable intertrochanteric fracture. (B) Note the cortical fulcrum, yoke support of the screw shaft, and proximal fragment. The proximal fracture face is buttressed against the lateral femoral cortex to prevent further collapse of the proximal into the distal fragment.

If the reduction is not optimal, or during the course of healing, settling of the bone may occur. Thus, a device is required which can adjust to these movements so that the load may still be borne to a substantial extent by the fracture surfaces. Such a device is the telescoping or sliding nail. For example, in the Wayne County General Hospital reduction, the calcar against the medial cortex buttress is expected to provide a fulcrum to maximize the strength contribution of the fixation device against varus deformation of the fracture site. *In vitro* studies,[3] however, have indicated that the calcar spike is deceptively weak in this construct and breaks off at low loads. The fracture then settles. The total amount of collapse that will occur depends on the new geometries and the strength of the remaining bone. However, a telescoping device can adjust to such settling, provided the fracture is not inherently misaligned and, after settling, that stability ensues.

The sliding hip nail to be described in this chapter is a metallic telescoping screw implant having two major component parts. The screw portion itself is a shaft with coarse protruding threads at the tip (Fig. 6–6). The second major component is a combination sleeve-side plate (Fig. 6–6), designed in such a way that the screw shaft fits loosely within the sleeve. The sleeve-side plate component is screw-attached to the femoral shaft. At operation, the screw is twisted into the proximal femoral head-neck fragment and an attempt is made to have the nail lie along the longitudinal axis of the neck-head fragment. The tip of the screw should, optimally, be positioned to lie just under the subchondral bone of the femoral head.

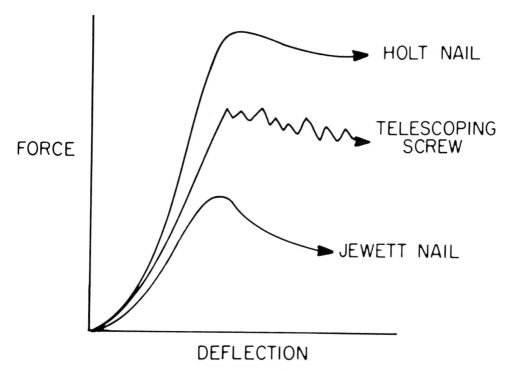

Fig. 6–5. Force-deflection curves for three fracture fixation devices in the hip (averaged for three reductions: anatomic, Dimon-Hughston, Wayne County). (Adapted from Kauffer, H., Matthews, L., and Sonstegard, D.: Stable fixation of intertrochanteric fractures: A biomechanical evaluation. J. Bone Joint Surg., *56A*, 899–907, 1974.)

The sleeve is installed through a hole made in the lateral femoral cortex, in such a way that the screw shaft freely enters the sleeve. After final positioning of the sleeve-side plate, the neck-shaft angle should be nearly normal, at 135–140 degrees. The proximal and distal fracture fragments should oppose firmly, and in fact be impacted together if possible. The side plate should easily, without pre-stress,,fit the shaft and be secured there with screws. For sliding to be easily possible, the screw shaft must reach to 1 cm or less from the outer sleeve opening. Thus, the device will resist varus angulation at the fracture site while allowing the fragments to settle toward each other as needed to maximize stability and opposition between the fracture surfaces.

A major consideration for the understanding of intertrochanteric fracture biomechanics is the physical characteristics of the bone fragments that must be stabilized by the device. Laros and Moore,[4] in an important clinical study, concluded that the quality of the trabecular bone, as assessed by the technique of Singh,[9] is the most important single factor bearing upon internal fixation stability in the clinical setting. Usually there is substantial trabecular bone in the subchondral region of the head itself. The intertrochanteric region is much less substantial even in normal femurs. However, the cortical femoral shaft nearly always provides an adequate source for sideplate stability.

In summary, for a successful end result, the construct of a reduced and internally stabilized hip, including the new, selected geometry, a specific stabilizing device, and maximum use of bony support, must have enough composite strength to allow the patient's weight to be partially supported without pain while the fracture heals.

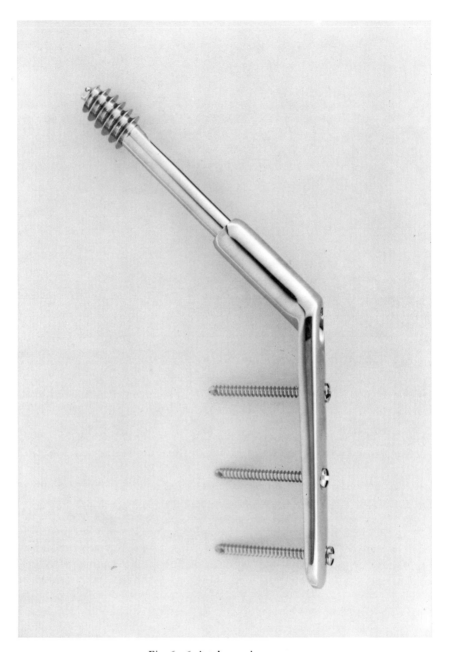

Fig. 6–6. A telescoping screw.

Failures can usually be traced to non-optimal reduction of the fracture, very poor bone quality, or overloading by the patient. Nonoptimal reduction can lead to failure of the device due to overloading, or to excessive settling of bone that cannot be accommodated by the device. Poor bone quality impairs the transfer of load between the implant and the bone, and can result in collapse of the bone overlying the screw section of the nail. Overloading by the patient can result in failure of the device even when the bone shares a good part of the load. The cases that follow have been chosen to illustrate some of the above points.

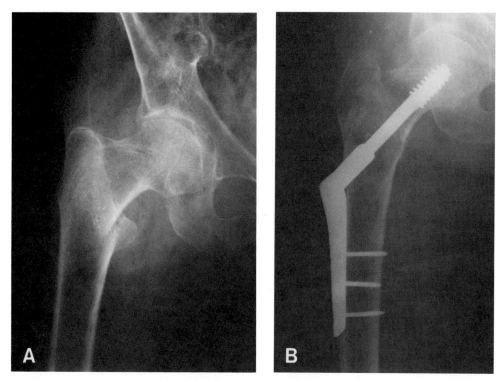

Fig. 6−7. Case 1. (A) The minimally displaced, minimally comminuted intertrochanteric fracture. (B) Anatomic reduction and sliding nail internal fixation.

CASE PRESENTATIONS

Case 1. The radiograms of Figure 6−7 present an 81-year-old, generally healthy woman who had been living alone at her home. She tripped on a rug, fell to the floor, and immediately had severe pain in the right groin and greater trochanteric region. Admission radiograms demonstrated her minimally displaced, minimally comminuted intertrochanteric fracture. After a routine preoperative evaluation she was treated with a closed anatomic reduction using a fracture table and internal fixation with a telescoping screw. She was discharged 2 weeks after her surgery to her daughter's home. After 2 months, she is again living at her home with a painless, nearly healed fracture. She has been supporting most of her weight with the uninjured limb for the past 10 weeks.

Case 2. This (Fig. 6−8) 91-year-old retired plumber was admitted from his nursing home, disoriented and with a history of "being found on the floor by his bed." When his cardiac arrythmias and failure had been controlled with digitalis, his comminuted, displaced intertrochanteric fracture, (Figs. 6−8A and B) was successfully reduced to a near anatomic position, and was stabilized with a sliding hip nail. A completely anatomic reduction was not possible (Fig. 6−8C) because of posterior intertrochanteric region comminution. The surgeon recognized the lack of complete stability, but because of the patient's increasing arrythmias and blood loss decided that a reasonable compromise had been achieved. After the operation the patient's wound healed, but was never pain free when the patient used the limb. He pro-

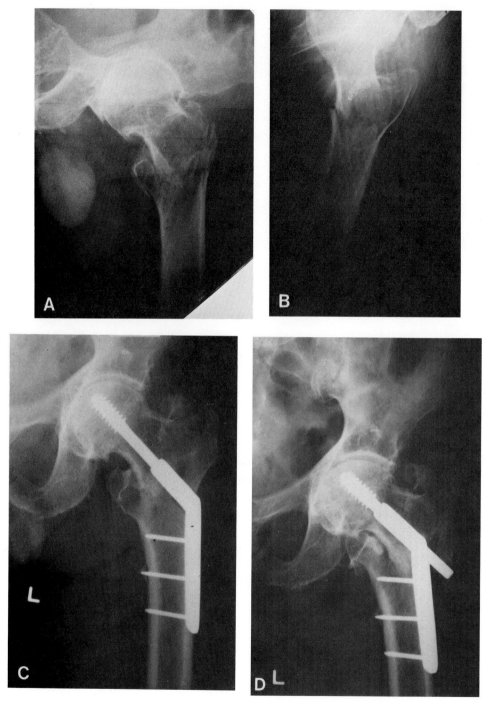

Fig. 6-8. Case 2. (A and B) The comminuted, displaced intertrochanteric fracture. Note the considerable intertrochanteric mass fragmentation. (C) Because of comminution the fracture faces could not be precisely opposed. (D) Collapse of an "anatomic" reduction, with telescoping of the device to a new, stable position. Note the similarity to a Dimon-Hughston reduction.

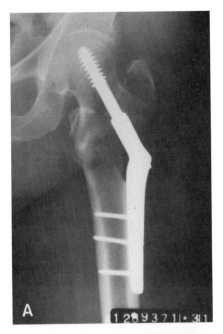

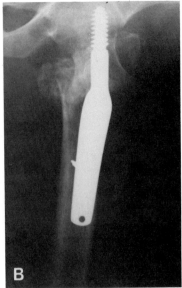

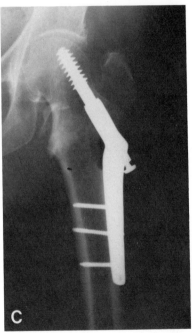

Fig. 6–9. Case 3. (A and B) Reduction and fixation of a comminuted intertrochanteric fracture in an elderly, osteoporotic woman. (C) Anticipated settling of the fracture.

gressed slowly in physical therapy and was discharged with a walker to the nursing home. Outpatient radiograms a few months after surgery demonstrated that the "anatomic" reduction had been lost and that the end of the screw shaft was protruding from the sideplate. The patient's chronic heart disease precluded any operative therapy.

Two months after discharge the patient was returned to the hospital after a large "lump" was noted on his lateral proximal thigh. This mass was covered by red and

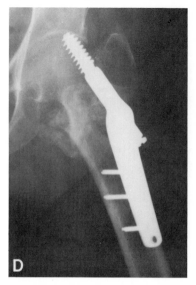

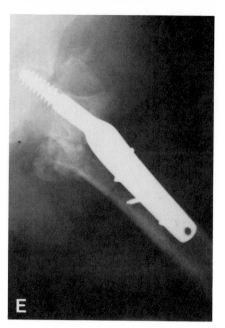

Fig. 6-9. (cont.) Case 3. (D and E) Continued collapse to the end of possible travel of the telescoping screw, followed by migration of the screw through the femoral head into the hip joint.

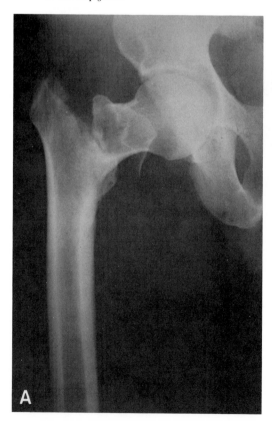

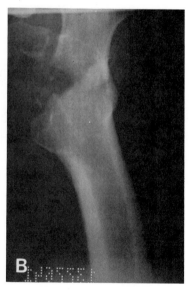

Fig. 6-10. Case 4. (A and B) Very displaced, fragmented intertrochanteric fracture in a healthy man. (C) Failure to accomplish an anatomic reduction. (D) Resorption at the superior neck contact point. (E) Sideplate stabilization failure. (F) The healed fracture.

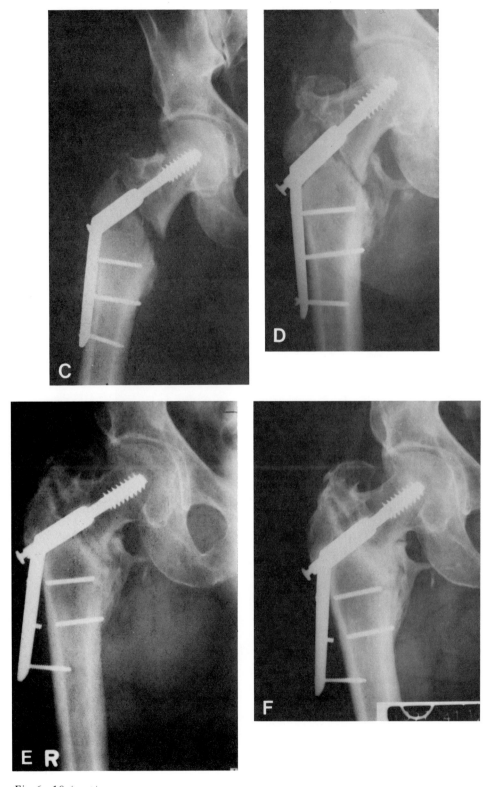

Fig. 6–10. (cont.)

crusting skin that appeared as though it would ulcerate. Radiograms (Fig. 6–8D) demonstrated continued settling of the fracture fragments with gross protrusion of the shaft of the sliding screw from the side plate. The patient's heart condition was felt to be much worse and he was returned to the nursing home disoriented and unable to walk.

Case 3. This elderly, retired school teacher, living at home with her family, had had a diagnosis of postmenopausal osteoporosis for several years. Previous radiograms had demonstrated marked radiolucency of her bones and several dorsal and lumbar compression fractures. She had been treated with fluorides and calcium supplementation. She fell at home and was unable to stand because of hip and groin pain. At a local hospital, radiograms demonstrated a comminuted intertrochanteric fracture of the left proximal femur. At surgery, her intertrochanteric bone was very fragile. The reduction demonstrated in Figures 6–9A and B was achieved with good medial calcar support. A moderately anterior placement of the threaded portion of the screw in the femoral head was accepted. Her fracture settled somewhat during the weeks after surgery (Fig. 6–9C), and she was able to walk with crutches

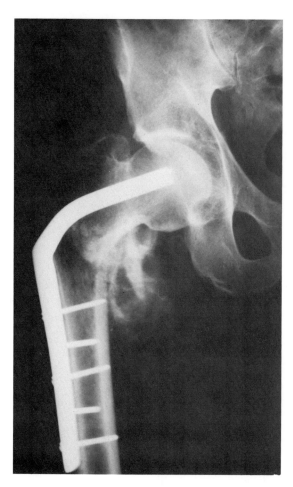

Fig. 6–11. Case 5. Bending failure of a Jewett nail in an intertrochanteric fracture site in an uncontrollable patient.

with minimal pain. Unfortunately, a few months later, her groin pain increased considerably. Radiograms (Figs. 6–9D and E) showed progressive migration of the screw through the femoral head. The sliding nail had telescoped maximally before the subchondral bone was penetrated. Her general health prevented any major reconstructive surgery. The nail will be removed when her fracture has healed.

Case 4. Figure 6–10 presents the radiograms of a 54-year-old man who was involved in an automobile accident in which he sustained multiple injuries. After two units of whole blood were administered, a splenectomy was performed. His right intertrochanteric hip fracture (Fig. 6–10A and B) was treated in traction. At 2 weeks after the injury he had a reduction on the fracture table and internal fixation with a telescoping screw. An anatomic reduction was not accomplished (Fig. 6–10C). Instead of an accurate fit, the proximal fragment was fixed in valgus with contact at the superior fracture face. This reduction would allow the bone to contribute little to the construct stability. With time and resorption at the contact point (Fig. 6–10D), telescoping of the screw occurred. A year later, the patient, now quite active and walking with moderate pain without crutches, had radiograms which demonstrated that his sideplate had separated from the femoral shaft. This construct failure allowed the medial fracture faces to firmly oppose each other. Healing progressed rapidly and the patient is presently active, nearly pain free, and again employed as a laborer.

Case 5. A 52-year-old alcoholic epileptic fell during a seizure and sustained a severe intertrochanteric fracture. After neurologic consultation, seizure control, and pre-operative preparation, his fracture was reduced anatomically and stabilized with a Jewett nail. He rapidly recovered from the surgery and left the hospital against medical advice. He returned 7 months later with severe groin and hip pain. His radiograms (Fig. 6–11) demonstrate that the nail is bent and must have been subjected by usage, accident, or seizure to a force maximum near 2,800 N, the force at which similarly stabilized specimens failed in *in vitro* studies.[3] The patient refused reconstructive surgery and left the hospital without further treatment.

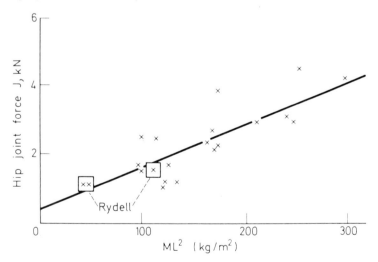

Fig. 6–12. Relationship between hip joint force and ML^2, where M is the subject body mass and L is the double stride length. (Paul, J. P.: pp. 53–70 In Schaldach, M., and Hohman, D., Eds.: Advances in Artificial Hip and Knee Joint Technology. Springer-Verlag, New York, 1976.)

BIOMECHANICS

The intertrochanteric region of the proximal femur has a complex shape and internal trabecular structure which must resist the unique and constantly changing combinations of forces that are imposed. While these imposed loadings are unique to patient, limb position, muscle activity, and many other momentary influences, information obtained both analytically and experimentally is available to guide understanding of intertrochanteric fracture fixation with a sliding hip nail.

Rydell[8] experimentally determined the directions and magnitudes of resultant forces acting *across the hip joint* for two post-hip-arthroplasty patients during "normal" activities. The force curves showed maxima at heel-strike and toe-off, with force maxima of 2.3 times body weight. These maxima are similar in magnitude to the value obtained for a simple two-dimensional calculation of forces acting at the hip during one-leg stance, assuming that only the abductors are active.[10] The form of the curves determined by Rydell is very similar to that of curves obtained by Paul[5] for the force at the hip versus position in the walking cycle for normal persons. However, Paul obtained force maxima between 3.39 and 4.46 times body weight. Later work by Paul[6] has indicated force maxima of between 4.9 and 7.6 times body weight, depending upon the speed of walking. Paul has also shown that the hip joint force is related to the body mass and double stride-length (Fig. 6–12). In this way, the values of Rydell and the considerably higher values of Paul may be reconciled.

The angular inclination of the resultant hip force may be obtained from the work of Paul and also from the simple calculations of one-leg stance. For practical purposes, the resultant force may be taken to lie in the frontal plane at an angle of approximately 15 degrees to the vertical, acting in a medial-to-lateral direction. It should be pointed out that no determinations were made in the above force calculations for dyscoordinate activities or for falls. The forces generated during such activities are

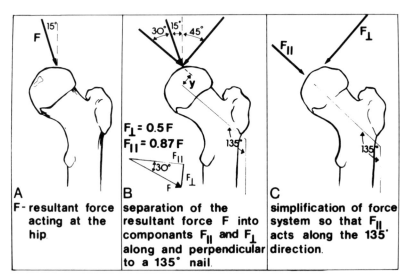

A	B	C
F - resultant force acting at the hip.	separation of the resultant force F into componants $F_{||}$ and $F_\perp$ along and perpendicular to a 135° nail.	simplification of force system so that $F_{||}$ acts along the 135° direction.

Fig. 6–13. (A) The resultant hip joint force acting in the frontal plane at an angle of 15 degrees to the vertical. (B) The resolution of the resultant F into a component $F_{||}$, parallel to the femoral neck, and a component $F_\perp$, perpendicular to the femoral neck. (C) Simplification of the force system where $F_{||}$ is taken as acting *along* the femoral neck.

likely to be considerably higher than those generated during normal ambulation.

Another general source of appropriate biomechanical information is from *in vitro* loading experiments.[3] On preserved human femurs, average loads of over 7,200 N, or about 10 times average body weight, are required to fracture the proximal femur when the loading is applied in a physiologic direction. It is interesting to note that reported fractures are in the femoral neck and not the intertrochanteric region, which can sustain an even greater load.

Thus, while there is considerable information regarding the loading specifications of the joint proper, various anatomic and physiologic features of the intertrochanteric region prevent such an accurate characterization there. Thus, for instance, while tension generated by the gluteus medius muscle acts in a line nearly parallel to the long axis of the femoral neck, its action may cause loadings at the intertrochanteric region of the femur which are entirely different and much more complex. Undoubtedly, excessive and inappropriate contracture of this muscle is related to fracture. Similarly, the iliopsoas inserting at the lesser trochanteric develops nearly axial compressive forces along the femoral neck, but at its site of insertion the effect has not been defined. Likewise, the hip joint capsule, a very strong structure that posteriorly incorporates the iliofemoral ligament, inserts at the intertrochanteric region. The capsule and ligament are normally relaxed at the end range of motion during a fall, but they may act as a tether fulcrum and contribute to the initiation of fracture or disruption of an internally fixed fracture. Finally, whereas the line of normal weight bearing may be determined during walking for normal people, it bears little relationship to the direction of forces across the intertrochanteric region during a fall.

The intertrochanteric region has a unique anatomy which, in a sense, parallels its protean function. Its relatively great peripheral substance, its large dimensions, and its cortical surface provide considerable strength for resisting combinations of bending, axial, and torsional loadings imposed by normal or unusual activities. This region also resists the tension generated by the major muscle groups attached there, and at the same time its protrusions act as beams or lever arms for these same muscles. The intertrochanteric trabecular bone pattern, again in its complexity, reflects the function of the region. Infrequently, permanent deformation or fatigue failure of a known and well characterized implant will provide an idea of the direction of magnitude of loads which the device has sustained. And, since the device was partially substituting for the normal intertrochanteric region, its failure can give a clue regarding the loads that had been imposed on it.

Under normal circumstances, the gluteus medius muscle is a major contributor to axial compressive loads along the femoral neck. Likewise, after a simple, undisplaced two-part intertrochanteric fracture, the medius tension remains a compressing and stabilizing factor. Unfortunately, many intertrochanteric fractures include separation of the greater and lesser trochangers from the intertrochanteric mass. This muscle-release effect of the injury eliminates the stabilizing effect of the medius tension, disturbs the topology, disrupts the internal supporting trabeculae, and grossly changes the mechanics of normal proximal femoral function. Little information is available regarding the residual loads which must be resisted by the reduced and internally stabilized hip fracture fragments. Clinical experiences and observations can, however, contribute some useful mechanical insights. If intertrochanteric fractures are treated by traction, cast immobilization, and supported walking, the result is nearly always a varus deformity of the healed proximal femur. If hip nails are found

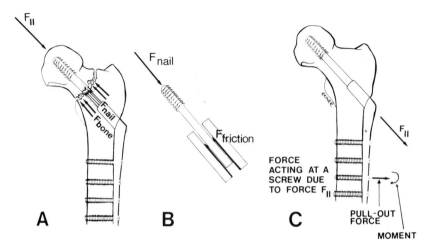

Fig. 6–14. (A) Force balance at the fracture site showing the load sharing between nail and bone. (B) Forces acting between the nail shaft and sleeve. For sliding, $F_{nail} > F_{friction}$ (C) Screw forces and moments due to the force $F_{||}$. An excessive value of $F_{||}$ will give screw pull out.

broken or bent, the result is always varus at the fracture site. In failures caused by inadequate strength of the proximal fragment, with migration of the fixation device through the head-neck fragment, the resultant final position is in varus. The internal fixation device is always resisting a varus-directed load. The magnitude of the load can be grossly estimated by laboratory experiments which produce a deformity similar in appearance to that which is at times seen clinically. Historically, Jewett nails were noted to bend in the clinical setting with some frequency. Bending of the telescoping screw shaft is rare, but has been seen. Therefore, extrapolating from the study of Kaufer, Matthews and Sonstegard,[3] the loads must at times reach maximums as high as 4,000 N.

Although the exact forces acting on the normal or on the fractured and stabilized intertrochanteric region have not been experimentally determined, the available information allows at least a qualitative estimate of the biomechanics of the sliding nail treatment of intertrochanteric fractures. Figure 6–14A shows the resultant force F acting at the hip. If the anteroposterior force component is neglected, the resultant F may be assumed to lie in the frontal plane. By drawing a triangle of forces, the components $F_{||}$ and $F_\perp$, parallel to and perpendicular to a direction of 135 degrees may be obtained (Fig. 6–13B). In this way it is found that $F_{||}$ equals 0.87 $F_\perp$, and that F equals 0.5 F. A further simplification results in Figure 6–14C, where it is assumed that force $F_{||}$ acts *along* the 135 degree direction (in other words, the moment of $F_{||}$ about the nail, $F_{||} \cdot y$, is neglected).

The force $F_{||}$ tends to settle or collapse the major fracture fragments together as it slides the screw shaft into the sleeve. Figure 6–14 illustrates the balance of forces at various locations. At the fracture site the bone and device share the load and are under compression (Fig. 6–14A). At the junction of the nail and sleeve, the force acting on the device is opposed by the friction force acting between the nail and the sleeve. Note that the force acting on the nail is less than $F_{||}$, owing to the force supported by the bone. In fact:

$$F_{||} = F_{nail} + F_{bone}$$

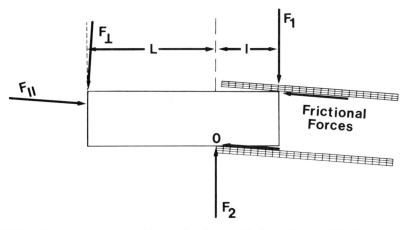

Fig. 6–15. Detailed arrangement of forces for the equilibrium of the nail in the sleeve.

and if

$$F_{nail} > F_{friction}$$

the nail will slide in the sleeve (Fig. 6–14B).

The force $F_{||}$ also has an effect on the side plate and screws, as shown in Figure 6–14C. Both a moment tending to bend the screws and a force tending to pull the screws out of the bone are generated. Excessive values of $F_{||}$ will lead to bending of the side plate, pulling out of the screws, or both.

The force $F_{\perp}$ has a bending effect on the construct. Figure 6–15A shows the force balance at the fracture site. Both a shear force and an internal moment must be generated to maintain equilibrium. Thus the bone must transmit shear, and this is aided by interdigitation of the fracture fragments. The force $F_{\perp}$ will develop a maximum bending moment at the sleeve–side-plate junction, as shown in Figure 6–15B. The application of force $F_{\perp}$ also affects the sliding action of the nail, as shown in Figure 6–16. The nail may be regarded as being pivoted at point 0. Taking moments, the force F_1 is given as follows:

$$F_1 \cdot \ell = F_{\perp} \cdot L$$

or

$$F_1 = \frac{L}{\ell} F_{\perp} = nF_{\perp}$$

where n is the ratio L/ℓ. For equilibrium the force acting at 0 is:

$$F_2 = F_{\perp} + nF_{\perp}$$

The friction force which must be overcome for sliding is:

$$F_{friction} = \mu(F_1 + F_2)$$
$$= \mu(2n + 1)F_{\perp}$$

where μ is the friction coefficient.

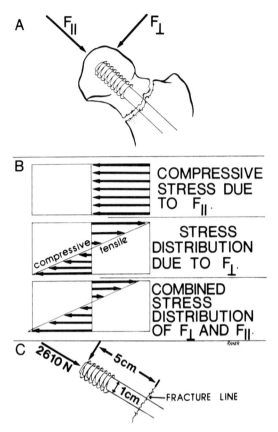

Fig. 6–16. (A) Forces acting on the nail in the limiting case of no bony support. (B) Stress distribution across the nail due to $F_{||}$, $F_\perp$ and the combined effect of $F_{||}$ and $F_\perp$. (C) Specific values used to calculate the stresses in the nail at the fracture site for the limiting case.

As a limiting case, the force acting on the nail may be taken as $F_{||}$. Since $F_{||}$ equals 0.87F and $F_\perp$ equals 0.5F, then:

$$1.74 \; F_\perp \geq \mu(2n + 1)F_\perp$$

for sliding, or:

$$\mu(2n + 1) \leq 1.74$$

The nail should, therefore, be inserted well into the shaft so that n is reduced.

Force $F_\perp$ tends to deform the stabilized hip fracture construct into varus, with bending of the internal fixation device and further fragmentation of the bone – a consequence of excessive force magnitudes. Since the force $F_\perp$ is transmitted to the screw thread region by the femoral head, it can be seen that the trabecular bone, even the cortical bone of the femoral head, can deform and fail as a result of this force.

It is of interest to determine the stresses on the fixation device itself for the limiting case of no bone support. The nail may be regarded as a cantilever under the combined loading of $F_{||}$ and $F_\perp$ (Fig. 6–16A). The Force $F_{||}$ produces a compressive

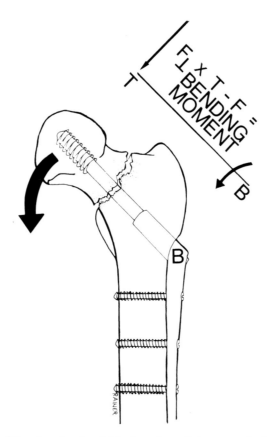

Fig. 6 – 17. Diagram of the situation in which there is no fulcrum at the inferior fracture interface and in which the internal device is assumed to be unsupported along its length.

stress $F_{||}$/(cross-sectional area of nail), while the force $F_\perp$ produces a stress distribution giving a maximum stress at the surface of:

$$\frac{(Moment) \cdot (Distance\ from\ center\ line\ to\ surface)}{(Moment\ of\ area)}$$

Figure 6 – 16B schematically shows the distribution of stresses. To consider an actual case, suppose that the nail is of a circular section, with a diameter, d, of 1 cm; that F = 3,000 N (about 4 times body weight); and that the distance from the fracture to the applied force is 5 cm (Fig. 6 – 16C). Then:

$$\text{The cross-sectional area} = 7.85 \times 10^{-5}\ m^2$$

and

$$\text{The compressive stress} = \frac{2610}{7.85 \times 10^{-5}}$$
$$= 3.32 \times 10^7\ N/m^2$$

Since $F_{||} = 0.87\ F = 2{,}610\ N$

The moment = $1,500 \times 0.05 = 75$ N.m, since $F_1 = 0.5F = 1,500$ N

The distance from center line to the surface = 0.005 m

and

$$\text{The moment of area} = \frac{\pi d^4}{64} = 4.91 \times 10^{-10} m^4$$

giving a maximum surface stress of 7.64×10^8 N/m^2.

There are two points to be noted. First, the tensile stress generated by $F_\perp$ is much larger than the compressive stress generated by $F_{\parallel}$. Second, the magnitude of the tensile stress is greater than the yield stress of a typical nail and would result in failure. Yet again this illustrates that bony support is essential for adequate performance of the fixation device.

Returning to the stability of different constructs, the moments generated should be considered. Figure 6–17 diagrams the case in which no fulcrum exists along the screw shaft or sleeve. The force component of the moment is F_1 and the moment arm length is the distance from the midscrew threaded region, T, to the sleeve side plate angle, B. The moment tending to disrupt the construct is F_1 times the

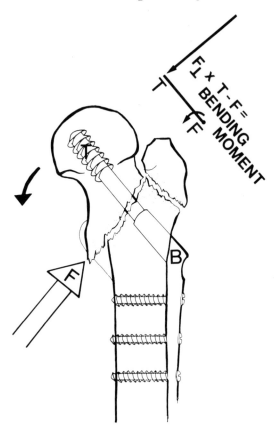

Fig. 6–18. Mechanics of the Wayne County General Hospital reduction with an unsubstantial fulcrum.

distance T–B. Given this situation, excessive forces would be expected either to bend the screw shaft at the shaft-sleeve intersection or to bend the device at the intersection angle of the sleeve and side plate. For any deformation or bending to take place, a combination of shear interaction or calcar-cortical compression—or both—would be required at the fracture site.

The Wayne County General Hospital reduction construct appears to alter the analysis somewhat. The overlap (Fig. 6–18) medially of the calcar on the medial femoral shaft would act as a fulcrum, F, which would change the length of the bending moment arm from T–B to T–F, a much shorter distance. This much-decreased moment should protect the device and therefore the construct from varus deformation failure under the load $F_\perp$. The fact is that in laboratory experiments,[3] the calcar was found to break free from the proximal-head neck fragment at low loads. This fracture then converted the situation to the one previously described. The slight valgus angulation that would be accomplished by this reduction technique would be expected to contribute little to resistance to deformation.

The Dimon-Hughston reduction (Fig. 6–19) substantially decreases the length of the screw shaft-sleeve combination. Distance T-B is therefore greatly reduced and the possible moment values are similarly affected. In fact the femoral neck is cradled in the yoke of the medial femur. This provides a fulcrum, and the moment arm T-F for bending of the device and disruption of the fracture reduction is therefore extremely short. The bending moments that can be generated are much reduced, and

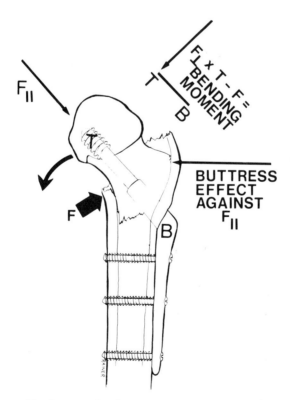

Fig. 6–19. The Dimon-Hughston reduction offers a decrease in the bending moment arm and, in addition, a fulcrum support of the femoral neck and the internal device. It prevents further collapse by buttressing the proximal fragment against the lateral femoral shaft cortex.

the clinical incidence of such failure is miniscule. While the referenced *in vitro* experiments did not document the value of this reduction technique, it has proved of great clinical value when accurate anatomic reduction could not be accomplished. Perhaps the fragility of the bone experimentally became the controlling factor for this mechanically attractive technique.

Case 1 is a clinical representation of a situation in which the strength of the bone, the device, and the reduction fixation construct was sufficient to allow for nearly normal activities. In Case 2 the fragility of the remaining bone of the intertrochanteric region was not recognized. The force component F could not be resisted by the imperfectly opposed fracture faces. The small areas of contracting bone fragmented and the frictional resistance of the shaft-sleeve was easily overcome, allowing the construct to collapse. In retrospect, the Dimon-Hughston reduction, with its resistance to further collapse and its short lever arms, might have been a more appropriate choice. Again in Case 3, the fragility of the bone was the major factor leading to mechanical failure. In Case 4 the lever or moment arm T-B was very great and in no way supported by a central fulcrum. Astoundingly, neither the shaft nor the sleeve-side plate angle bent as the fracture failed to heal, its surfaces held apart by the construct. The final failure of fixation of the side plate to the femoral diaphysis is unusual, and speaks only to the great magnitude of the moment that was generated and was resisted by the remainder of the device. Healing sometimes occurs against all odds. Case 5 is included to emphasize the relative lack of strength of the Jewett nail and to document the bending moment that can exist at the fracture site.

Clinical treatment of intertrochanteric fractures is, in fact, applied biomechanics. A consideration of the physical qualities of the patient's bone, of the mechanical function of the reduction-fixation construct, and of the biomechanical implications of the postoperative management of the patient is essential for an optimal result.

REFERENCES

1. Dimon, J. H. and Hughston, J. C.: Unstable intertrochanteric fractures of the hip. J. Bone Joint Surg., *49A*, 440–450, 1967.

2. Frankel, V. H. and Burstein, A. H.: Orthopaedic Biomechanics. pp. 77–89. Lea and Febiger, Philadelphia, 1970.

3. Kauffer, H., Matthews, L., and Sonstegard, D.: Stable fixation of intertrochanteric fractures: A biomechanical evaluation. J. Bone Joint Surg., *56A*, 899–907, 1974.

4. Laros, G. S., and Moore, J. F.: Complications of fixation in intertrochanteric fractures. Clin. Orthop. Rel. Res., *101*, 110–119, 1974.

5. Paul, J. P. Forces transmitted by joints in the human body. Proc. Inst. Mech. Engrg., *181*, pt. 3J: 8–15, 1966–67.

6. Paul, J. P.: Loading on normal hip and knee joints and on joint replacements. In: M. Schaldach and D. Hohmann, Eds. Advances in artificial hip and knee joint technology. Springer-Verlag, Berlin, 1976.

7. Rowe, C. R.: The management of fractures in elderly patients is different. J. Bone Joint Surg., *47A*, 1043–1059, 1965.

8. Rydell, N. W.: Forces acting on the femoral head prosthesis. A study on strain-gauge-supplied prostheses in living persons. Acta. Orthop. Scand., (Suppl.) *88*: 1966.

9. Singh, M., Nagrath, A. R., Maini, P. S., and Rohtak, M. S.: Changes in trabecular pattern of the upper end of the femur as an index of osteoporosis. J. Bone Joint Surg., *52A*, 457–467, 1970.

10. Tronzo, R. G.: Hip nails for all occasions. Orthop. Clin. North Am., 5:479–491, 1974.

11. Williams, M., and Lissner, H. R.: Biomechanics of Human Motion. pp. 109–110. W. B. Saunders Co., Philadelphia, 1962.

12. Williams, D. F., and Roaf, R.: Implants in Surgery. pp. 403–404. W. B. Saunders Company Ltd., London, 1976.

7
Femoral Head Arthroplasty
for Aseptic Necrosis

E. M. Lunceford, Jr., M.D.; A. M. Weinstein, Ph.D.;
J. B. Koeneman, Ph.D.

INTRODUCTION

Avascular necrosis (aseptic necrosis) is a condition which develops subsequent to compromise of the blood supply in the femoral head. This condition may be seen in childhood and is known by the eponym of Legg-Calve-Perthes disease; its severity in childhood varies with the amount of femoral head involvement. The avascular necrosis with which we are concerned in this discussion is seen in the adult patient and is most often associated with a fracture of the neck of the femur. The condition may occur subsequent to treatment of the fracture by open reduction and internal fixation. The incidence of avascular necrosis arising from this particular condition varies from 6 to 50 percent, depending upon the series reviewed.

The methods of fixation employed and the treatment of femoral neck fracture may have some bearing on the development of avascular necrosis. The size of the fixation device—that is, the triflange nail, compression screw, or similar device—that fills the central portion of the femoral head and a large portion of the medullary canal of the femoral neck can produce a mechanical interruption of the blood supply to a given segment of the femoral head. Coleman[3] has shown that a wedge-shaped segment of bone in the weight-bearing portion of the femoral head may become completely avascular following the insertion of a nail in the central portion of the neck and into the weight-bearing portion of the femoral head. Thus, a fixation device of sufficient size to produce compression of the medullary vessels and cause multiple fractures of the trabecular bone in the femoral head can interrupt the blood supply. Multiple pin fixation, as recommended by Moore,[9] Knowles,[7] and Deyerle,[5] has less of a chance to produce damage to as large a portion of the femoral head and neck. The peripheral fixation obtained with the multiple pin technique tends to produce smaller areas of ischemia when and if a blood vessel is damaged, and the resultant area of avascular involvement is of such size that it can be revascularized with less difficulty than the large weight-bearing segment produced by the larger internal fixation devices.

Nontraumatic avascular necrosis of the femoral head has been seen in gout, corticosteroid therapy, Gaucher's disease, caisson disease, SS hemoglobinopathy, and alcoholism. In these conditions, the cause for interruption of the blood supply is not known, but the theoretical hypotheses suggest that the effect on blood coagulation by

140

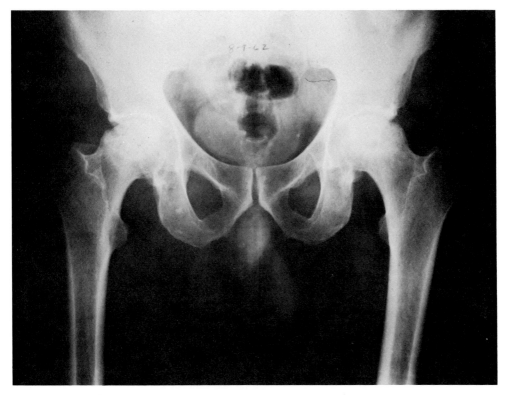

Fig. 7 – 1. Typical roentgenographic appearance of avascular necrosis.

the disease process, leading to sludging or some alteration in lipid metabolism, may cause obstruction of the vascular supply from thrombus embolization, by fat embolization, or both.

By far the largest number of nontraumatic avascular necrotic femoral heads are those associated with alcoholism, whereas the number of cases secondary to other causes such as gout, caisson disease, sickle-cell hemoglobinopathy, and Gaucher's disease are smaller in number and can readily be distinguished from the idiopathic variety by their different roentgenographic manifestations. Idiopathic avascular necrosis is usually seen in males, with a ratio of approximately 4 to 1 males to females, and is most often seen in patients aged 30 to 55. The condition is often bilateral, some authors citing an 80 percent incidence of bilaterality.[6]

Progressive changes with involvement of the femoral head have been divided into six categories. These are based on roentgenographic, clinical, and gross pathologic findings. The lesions can be further delineated with histologic, microradiographic, radioscanning, and intravital tetracycline labeling techniques to better define the limits of the femoral head necrosis.

With interruption of the femoral head blood supply in the interstices of the femoral head, resorption of bone will occur at the periphery of the infarcted zone. When this occurs, continued compression of the femoral head during normal gait results in transmission of forces through the articular cartilage into the bony matrix of the femoral head, and can cause actual infarction of the trabecular pattern, resulting in displacement of the avascular segment of bone and subsequent subchondral loosening of the bone-cartilage interface. The avascular segment of bone can then recede into the femoral head, where osteoclastic resorption at the viable bone–necrotic bone interface has occurred. This permits further compression and compromise of the blood supply at this juncture.

The zones of avascularity adjacent to zones of increased vascularity, where healing is occurring, produce a roentgenographic picture of patchy ossification and resorption, resulting in the all too familiar pattern of avascular necrosis (Fig. 7–1). Stage 1 and 2 lesions are usually clinically asymptomatic and do not show the collapse of the femoral head as described. They have a mottled appearance and are usually detected only in a routine x-ray examination for some other reason. The degree of viability of necrotic areas is very difficult to determine, and the viability of the femoral head in these stages is quite elusive to the usual methods of investigation.

Calandruccio[2] and others have not been successful in accurately predicting which femoral heads would remain viable following fracture of the femoral neck and which would succumb to avascular necrosis. More recent reports using radioscanning techniques are encouraging, but do not fulfill the tenet of accurately predicting the outcome of management of femoral neck fractures coupled with the viability of the femoral head. The use of computerized axial tomography (CAT) may prove beneficial in establishing the extent of avascular necrosis, but its value is unknown, since it is too early in the development of the CAT scan technique for any valid conclusions to be drawn from this particular procedure.

The patchy areas of increased bone density and radiolucency in the femoral head, along with the zone of avascular involvement, depend upon the degree of interruption of the blood supply. The zones of increased density, compared with the adjacent zones of radiolucency, are produced architecturally by decreased vascularity and adjacent areas of increased vascularity. Thus, osteoclastic and osteoblastic activity are occurring simultaneously. The new bone formation occurs concomitantly with bone destruction, and there is a constant turnover of bone in the remodeling process of the femoral head. The subchondral bone undergoes resorption, and flattening or collapse of the femoral head is then evident roentgenographically. Subsequent to these changes, there is narrowing of the joint space and roughening of the articular cartilage of the femoral head and the acetabulum, with the progressive development of coxarthrosis with exostosis.[6]

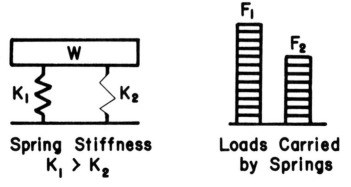

Fig. 7–2. Spring model for load transfer as a consequence of avascular necrosis in the femoral head. The smaller spring constant represents necrotic bone; load transfer produces greater stresses in normal bone.

The adjacent pockets of cancellous bone of different stiffness, which occur in the initial stages of avascular necrosis, cause load redistribution in the femoral head which can increase the severity of the defect. When two parallel springs share a load, as shown in Figure 7–2, the stiffer element carries a higher percentage of the load. Thus load sharing is a factor which can accentuate the difference in stiffnesses. A decrease in stiffness in a given region, as a result of avascular necrosis, causes more load to be carried by adjacent areas. The higher loads in these areas result in bone remodeling and increased bone density. This causes more load to be carried by the viable areas. Stiffer segments also cause increased dynamic loads on localized regions. The remodeling of the acetabular cancellous bone in Case 1 shows the orientation of the trabeculae to carry the principal stresses. Such a trabecular pattern occurs in the femoral head. Disruption of these arches causes a process that seeks a new supporting mechanism, which however, may not be possible within the available geometry of the femoral head.

One method that has been attempted to help in the fight against progressive deterioration is osteotomy. Mechanically, osteotomy does three things:

1. It changes the areas and region over which the joint reaction force is distributed.
2. It changes the stress distribution in the femoral head.
3. It changes the moment arm of the abductors.

CHANGE IN AREA

If the shaded area in Figure 7–3 is necrotic, a varus osteotomy will place the more viable, lateral portion of the head into the loaded beams area of the acetabulum. In this case there is a possibility for the bone to remodel to a structure that can carry the load. Such an osteotomy may also increase the load bearing area and thus decrease the average stress on the femoral cartilage.

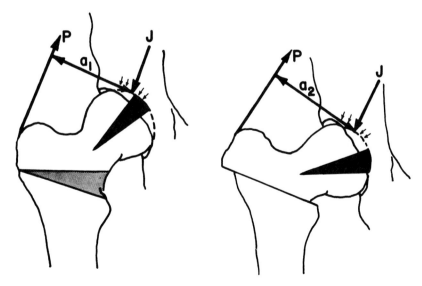

Fig. 7–3. Change in contact area by varus osteotomy of the femur.

STRESS IN THE HEAD

Figure 7–4 shows the head and neck modeled as a beam, with the joint reaction force as an equivalent point load and the connection to the femoral shaft as a fixation point.

The varus osteotomy causes the equivalent joint reaction force to be located at a greater distance from the fixation point. The result is that for the same load the stresses become more tensile on the superior surface of the neck after the osteotomy. More sophisticated finite-element models show the same qualitative increase in stresses as shown by this simple model. Several conventional modeling techniques were used in this model, as follows:

1. The stress distribution on the surface was replaced by an equivalent load. This procedure gives good stress distribution at distances away from the contact area. The magnitude and direction of the equivalent forces are determined by requiring the reactions at the fixed end of the beam to be the same for the distributed load and the equivalent load.

2. The principle of superposition was used by dividing the equivalent load into its components perpendicular to and parallel to the beam. The stresses due to these loads were calculated separately and then added for the final result. The magnitude and direction of the equivalent joint reaction forces and the abductor pull have been estimated in numerous papers, for example, Rydell.[12]

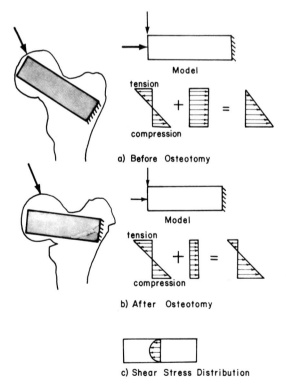

Fig. 7–4. Stress distributions in the femoral neck, modeled as a beam before and after osteotomy.

CHANGE IN ABDUCTOR MOMENT ARM

The increase in stress for any given load is counteracted by a decrease in stress caused by an increase in the moment arm of the abductors from a_1 to a_2 in Figure 7–3. Static equilibrium of the pelvis is shown in Figure 7–5. The weight of the body is balanced by the pull of the abductors. The further the abductors are away from the femoral head, the less their force has to be to counteract the moment of the body weight times its moment arm. Lowering the pull of the abductors reduces the magnitude of the joint reaction force.

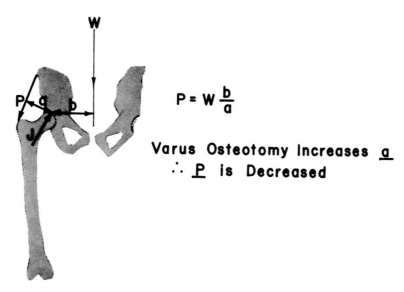

$$P = W \frac{b}{a}$$

Varus Osteotomy Increases $\underline{a}$
$$\therefore \underline{P} \text{ is Decreased}$$

Fig. 7–5. Static equilibrium of the pelvis.

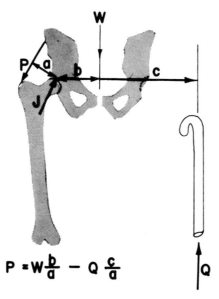

$$P = W\frac{b}{a} - Q\frac{c}{a}$$

Fig. 7–6. Effect of the use of cane on opposite side on static equilibrium of the pelvis.

The moment equilibrium diagram of the pelvis, in Figure 7−6, also shows the effectiveness of a cane on the opposite side in reducing the joint reaction force. The joint reaction force is a direct function of body weight. A necrotic femoral head or a loose prosthesis can thus be protected by using ambulatory aids or by reducing body weight or activity, as indicated in Case 2.[10]

If the disease process is recognized early in stages 1 and 2, nonoperative or conservative treatment may be instituted with limited weight-bearing range of motion exercise programs and with antiinflammatory drugs such as phenylbutazone, oxyphenbutazone, indomethacin, ibuprofen, or similar agents. Once the diagnosis has been established, operative procedures for correction of avascular necrosis consist of bone grafting of the femoral head; cancellous and cortical bone grafts have been utilized for correction of this problem. Bonfiglio[1] has recommended cortical bone grafts be placed accurately into the subchondral region to support the femoral head and prevent collapse of the normal articular surface. This form of treatment must be instituted prior to deformation of the femoral head to obtain the optimal results.

Once deformity of the femoral head has occurred, it is usually best managed with a procedure that replaces the femoral head or produces an arthrodesis of the hip. Arthodesis of the hip is a time-honored procedure that will produce lasting benefits and thus relieve pain in the individual so treated. Patients receiving replacement of the femoral head fall into two categories: (a) Those with a relatively intact acetabulum, being treated effectively by unipolar endoprosthetic replacements; and (b) those with a significant degree of acetabular involvement, being treated with a total hip arthroplasty type of reconstruction to achieve relief of pain. The endoprosthetic replacement will be considered in this chapter, specifically illustrated with the Moore prosthesis for management of avascular necrosis of the femoral head. The use of the endoprosthesis is indicated in patients who have avascular necrosis with minimal derangement of the acetabular cartilage. The best results obtained to date in patients with avascular necrosis have been those in which there has been no damage to the articular surface of the acetabulum or to the subchondral bone in this region. If the subchondral bone is damaged, there will usually be a depression with fragmentation of the articular cartilage.

CASE PRESENTATIONS

Case 1. This 39-year-old white female had a previous fracture of the neck of the femur treated by Smith-Petersen nailing, with subsequent avascular necrosis of the femoral head. She was treated for this by insertion of a Moore cobalt-chromium alloy prosthesis in 1953 (Fig. 7−7). Her postoperative convalescence has been quite satisfactory, and she has had no pain in her hip for 25 years, with an excellent result having been obtained. She had developed increased bone density about the stem of the prosthesis in the lateral aspect of the femur and also in the lesser trochanteric region (Fig. 7−8). Her x-rays reveal increased bone density about the stem of the prosthesis, with a radiolucency in the trochanteric (tensile area) portion of the femur. A joint space has been maintained. She has been able to walk without support, beginning approximately 1 year following surgery, and has continued to function effectively since that time. This case illustrates convincingly the ability of bone to form in response to applied stress, and to strengthen in the characteristic manner as defined by Wolff's law.

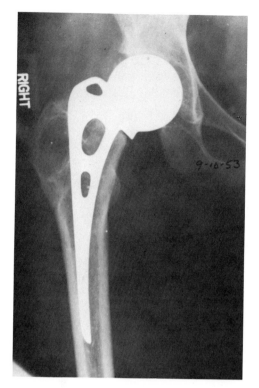

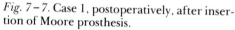

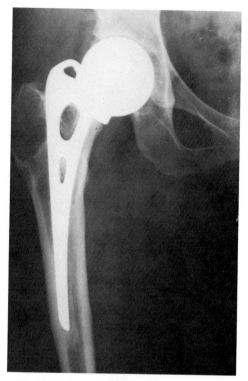

Fig. 7–7. Case 1, postoperatively, after insertion of Moore prosthesis.

Fig. 7–8. Case 1, 25 years postoperatively.

<u>Case 2.</u> This is a 78-year-old white female who was seen in 1960 because of avascular necrosis of the head of the femur following a fracture of the neck of the femur. She was treated 5 years previously by multiple pin fixations of an impacted fracture of the femoral head and neck. She gradually developed increasing pain with restrictions of motion, and x-ray revealed an increasing deformity with avascular necrosis of the femoral head. She was treated by insertion of a Moore prosthesis in 1960 (Fig. 7–9), and initially had a very satisfactory result. However, after the first 3 to 4 postoperative years, she began developing increasing pain in the hip and thigh with weightbearing and also on rotation of the hip. Her x-rays revealed gradual loosening of the stem of the prosthesis and proximal migration of the femoral head into the acetabulum, producing a protrusio acetabulae (Fig. 7–10). Eleven years after surgery, she exhibits very marked loosening of the femoral prosthesis, with migration of the prosthesis proximally in the acetabulum and distally in the femur (Fig. 7–11). The bone has resorbed about the femoral stem. It is necessary for her to ambulate with crutches or a walker to help relieve her pain.

A qualitative evaluation of the stress distribution in the stem of a Moore prosthesis can be obtained by modeling it as a beam. Figure 7–12A shows the simplified model. Figure 7–12B shows how the bending moment changes along the beam. Figure 7–12C shows the variation of the stresses. If the beam had not been tapered, the stress distribution would have the same shape as seen in Figure 7–12B.

The reason for the peak in the stress along the stem length, despite the continuous decrease in the bending moment resisted by the beam, is that the amount of material

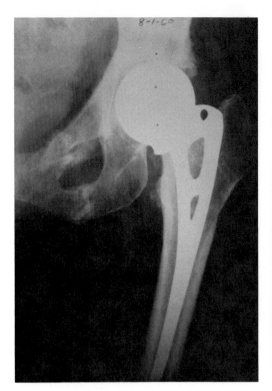

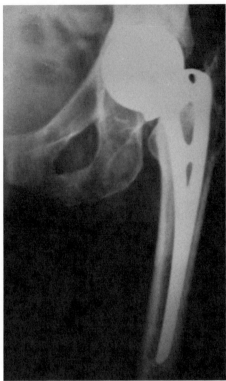

Fig. 7–9. Case 2. postoperatively, after insertion of Moore prosthesis.

Fig. 7–10. Case 2, 4 years postoperatively. Note protrusio acetabulum.

resisting the bending moment is continuously decreasing distally along the stem. The beam theory equation used to calculate the stress in the beam is:

$$\sigma = \frac{Mc}{I} \tag{1}$$

where σ = stress on lateral aspect of stem
     M = Bending moment at any particular cross section
     I = Area moment of inertia
     c = Distance from the neutral axis to the outside fibers

Radiographs for Case 1 (Figs. 7–7 and 7–8) show an increase in bone in the calcar region and in the lateral cortex in the region of the distal stem. These regions correspond to the support areas R_1 and R_2 of the model and demonstrate Wolff's Law. The prosthesis in Case 2 (Figs. 7–9, 7–10, and 7–11) was not initially wedged between the calcar and lateral cortex, and subsequently became loose.

The area moment of inertia, I, is a geometric factor, and relates the amount and distribution of material available to resist the applied loads. Equation (1) shows that the larger I is, the lower the stress will be. An equation for any geometry can be derived for the calculation of I. It does not depend on any material property but involves only dimensions. The magnitude of I depends not only on the magnitude of

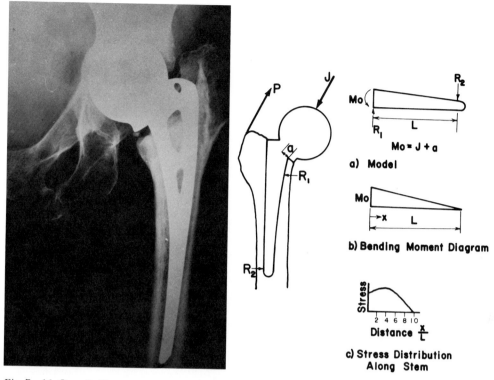

Fig. 7–11. Case 2, 11 years postoperatively.   *Fig. 7–12.* Stress analysis of the Moore prosthesis.

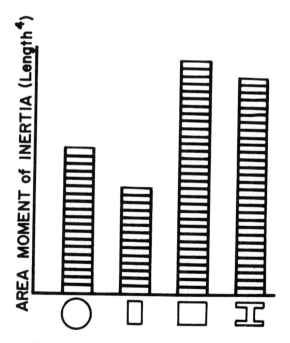

Fig. 7–13. Relative areal moments of inertia.

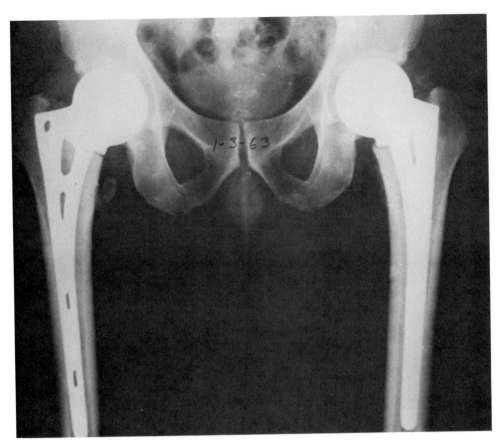

Fig. 7 – 14. Case 3, postoperatively, after bilateral insertion of Moore prostheses.

the beam cross sectional area, but on how it is distributed. A 1-square-millimeter area is much more effective (larger I) if it is located 6 mm from the neutral axis than if it is right on the axis. Figure 7 – 13 shows the respective values of I for various cross-sectional areas. If a beam is made up of two or more sections, such as a bone with a prosthesis in it, the I values of the two cross-sectional areas are added together.

Case 3. This is a 40-year-old white male with bilateral avascular necrosis of the femoral heads secondary to collagen disease with steroid administration. The patient was first seen in 1960 because of pain in both hips, and developed increasing discomfort with further deformity of the femoral heads. In November and December 1962, he had arthroplasty of the hips with replacement of the femoral heads using the I-beam Moore prosthesis, size 1¾ inches. His acetabulae were in good condition, with the articular surface intact (Fig. 7 – 14). His postoperative course was satisfactory. He was initially weight-bearing with a four-point gait using crutches, and continued to function effectively with this. He did not put excessive stress on his hips in this manner. After approximately 6 months, he ambulated without support and continued to do so until 1976. He functioned in a sedentary-type occupation until further loosening of the prostheses occurred in both the acetabulum and femur, with proximal and distal migration. A zone of radiolucency developed about the stem of the prosthesis, and this resulted in increasing pain (Fig. 7 – 15). In 1977, the patient had bilateral

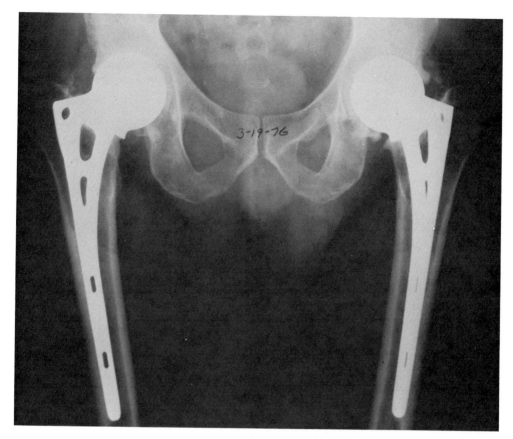

Fig. 7–15. Case 3, 13 years postoperatively.

total hip arthroplasty, with revision of the prosthetic replacements. Subsequent to this, he has resumed his occupation and is functioning well at the time of this writing.

The current Moore prosthesis has an I-beam shape. Note that the area moment of inertia is only slightly less than a full rectangle, while the weight of the prosthesis is significantly less with grooves present for possible stabilization by bone and revascularization of the medullary canal. Thus the stresses in a structure depend not only upon the loads applied, but on the geometry of the load-bearing elements. Beam theory is a useful method for calculating the stresses in bone and in the long stems of prosthesis. This method gives relatively good results away from the load application areas, and can be used to investigate the significance of variation in the values of important variables. However, it cannot be used to examine the interfaces of stress concentrations.

A structural member that is relatively long compared to its lateral dimensions, such as the femur, and whose principal function is to support transverse loads, is called a beam. The main mode of deformation of a beam is bending. In beam theory, a model of the structure to be analyzed is first made by representing all support areas, whether restrained or free. The distribution of tension and compression stresses (bending stresses) in a beam is shown in Figures 7–4 and 7–12. The distri-

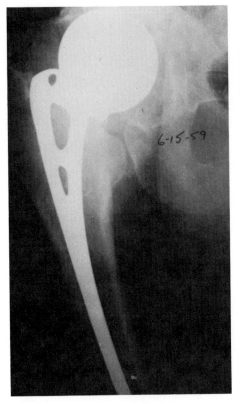

Fig. 7–16. Case 4, postoperatively, after insertion of Moore prosthesis.

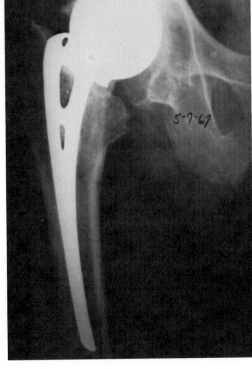

Fig. 7–17. Case 4, 8 years postoperatively.

bution of shear stresses is also shown. Note that any point in the beam has both a bending stress and a shear stress. Which stress causes failure first depends on the relationship of the bending stresses to the bending strength of the material at that point, and on the magnitude of the shear stress to the shear strength of the material at that point. In most situations the tensile stresses will cause failure first.

Deflection of a beam depends not only on I but also on the modulus of elasticity, E, and some function of the linear dimension of the beam. For example, the end deflection of an end-loaded cantilever beam is $\delta = PL^3/3EI$. The EI combination appears in all deflection calculations and is called the modulus of rigidity. This points up a major distinction between the calculation of stresses and the calculation of displacements. Stress distribution does not depend on the modulus of the material, while displacement does. However, in composite beams, as will be discussed later, the ratio of the elastic moduli of the two materials will affect the allocation of the loads between the materials.

Case 4. This 61-year-old white male presented in 1959 with avascular necrosis of the right hip following dislocation of the femoral head. The dislocation of his hip occurred in 1938, and was treated by a closed reduction. He subsequently developed increasing discomfort in his hip, and by 1959 the pain had reached a point that required the patient to have an arthroplasty of his hip. A Moore prosthesis was inserted in 1959 (Fig. 7–16), and 8 years later, in 1967, he was still ambulatory with inter-

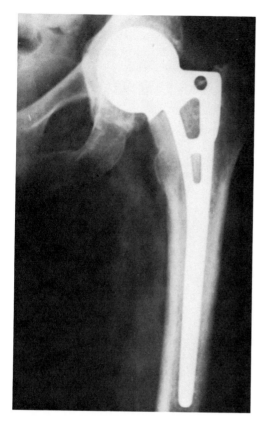

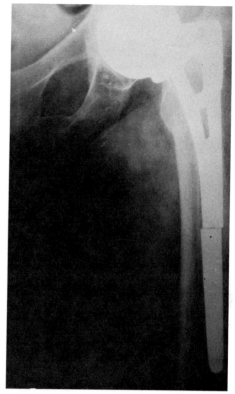

Fig. 7–18. Case 5, postoperatively, after insertion of Moore prosthesis.

Fig. 7–19. Case 5, 9 years postoperatively,
after sudden onset of pain in thigh.

mittent episodes of discomfort in his hip, but able to function effectively, even though
he weighed 220 pounds and put considerable stress on his hip. He was not using a
cane or crutch for assistance in ambulation when last seen in 1967 (Fig. 7–17).

Case 5. This 53-year-old white female was seen for pain in her hip, with increasing
discomfort. She was initially treated with a hanging hip procedure in an attempt to
alleviate some pain in her left hip, but subsequently continued to experience pain. In
1965, she was treated with insertion of a Moore prosthesis (Fig. 7–18), and subsequent to this progressed quite satisfactorily until 1974. She then developed sudden
onset of pain in her thigh. Subsequent films revealed a fracture of the stem of the
prosthesis (Fig. 7–19). The Moore prosthesis was removed approximately 1 year later, and a total hip arthroplasty was performed. Subsequent to this, she has gotten
along satisfactorily. The patient is 5 feet, 11 inches in height, weighs 160 pounds,
and puts considerable stress on her hip in ambulation.

For a proper approximation of the magnitude of the stresses in the femoral stem
and femoral shaft the sharing of the load by the femur and the prosthesis must be
included in the model.

The ratio of the modulus of elasticity of the stem to that of the bone is about 15 to
1. However, the bending load is distributed in the ratio of the factors EI prothesis/EI
bone. Since the bone is further from the neutral axis, its area of moment of inertia is
greater, and in most systems a larger percentage of the load is carried by the bone

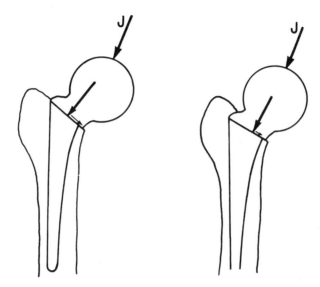

Fig. 7–20. Components of hip resultant force, J, at flange as a function of flange-stem angle.

along the shaft. However, if the prosthesis becomes loose, as in Cases 3 and 5, the load transfer between the bone and the prosthesis is reduced and a larger percentage of the load is carried by the prosthesis.

There is a difference in the flange angle of the Moore prosthesis used in Case 1 and the flange angles of those used in the other cases. A reason for this change in design is shown in Figure 7–20. Any force can be broken into component forces. Shown are the components of the joint reaction force normal to and parallel to the flange surface. Note that in the later design the component of force acting in a shearing mode between the prosthesis and femur shoulder is greatly reduced and even changed in direction.

There are several levels of biomechanical analysis—static free body diagrams, beam theory, and the more sophisticated techniques such as finite element analysis. All of these techniques involve assumptions of some type. Biomechanical analyses are good for indicating the effects to be sought in clinical experience and to point out possible trends. But the results should be accepted only after confirmation by experience or experiment. Beam theory models can be extremely powerful in generating hypotheses. A good model is one that includes the major effects and ignores the secondary influences. Two-dimensional and three-dimensional finite element analyses are essential for predicting stress concentration factors and detailed stress distributions of already-developed designs, but may not be as useful in the creative stages of hypothesis and device development as other techniques.

The success of the first case, followed for 25 years, undoubtly depended on the ability of the patient to gradually apply stress to the adjacent bone and prosthesis in such a manner that they were able to react favorably to the stresses and strains applied. As discussed, the modulus of elasticity of the metal and of the bone obviously differs greatly. Thus, loosening occurs unless appropriate shock-absorbing mechanisms are present to relieve these stresses. In some individuals this can be accomplished with alterations in gait pattern, utilization of ripple-type soles on their shoes, and a cane to assist in relieving some of the stress on the hip.

Quite the contrary, failures will occur with increased stress being applied to the hip and further difficulties being encountered at the prosthesis-bone interface because of the difference in elasticity of, and the varied pressures applied at, this interface. Obviously, when the prosthesis becomes loose in its environment it can no longer function as it was originally designed to do.

Movement that occurs at the prosthesis-bone interface can produce changes that exceed the body's potential for adjusting to this through increased tissue strength. Failure will then occur. Attempts to prevent such failures, based on the principles presented here, are being investigated.

REFERENCES

1. Bonfiglio, M., and Bordenstein, M. D.: Treatment of bone grafting of asceptic and closure of the femoral head and nonunion of the femoral neck. J. Bone Joint Surg. *40A*; 1, 329, 1948.

2. Calandruccio, R. A. and Boyd, H. B.: Further observations on the use of radioactive phosphorous (P^{32}) to determine the viability of the head of the femur: Correlation of clinical and experimental data in 130 patients with fractures of the femoral neck. J. Bone Joint Surg., *45A;*445–460, 1963.

3. Coleman, S. H.: Personal Communication, 1969.

4. Crenshaw, A. H.: Campbell's Operative Orthopaedics. C. V. Mosby Co., St. Louis, 1971.

5. Deyerle, W.: Internal fixation of femoral neck fractures with peripherally placed pins. and a side plate. In Crenshaw, A. (Ed.): Campbell's Operative Orthopaedics. 5th Ed. C. V. Mosby, St. Louis, 1971.

6. Enneking, W. F., et al.: The Hip, pp. 3–18. C. V. Mosby, St. Louis, 1975.

7. Knowles, F. L.: Fractures of neck of femur. Wisconsic Med. J., *35:*106–109, 1936.

8. Lunceford, E. M., Jr.: Use of the Moore self-locking vitallium prosthesis in acute fractures of the femoral neck, J. Bone Joint Surg., *47A:*832, 1965.

9. Moore, A. T.: Hip Joint Fracture (a mechanical problem). In: Edwards, J. W., Ed.; American Academy of Orthopaedics Surgeons Instructional Course Lectures, Vol. 10, C. V. Mosby, St. Louis, 1953.

10. Moore, A. T.: The self locking metal hip prosthesis. J. Bone Joint Surg., *39A:*811, 1957.

11. Moore, A. T.: The Moore self locking vitallium prosthesis in fresh femoral neck fractures. A new low posterior approach (the southern exposure). American Academy of Orthopaedic Surgeons Instructional Course Lectures, Vol. 16. C. V. Mosby, St. Louis, 1959.

12. Rydell, N. W.: Forces acting on the femoral head prosthesis. Acta Orthop. Scand., *37* (Suppl. 88): 1, 1966.

8
Total Hip Replacement with a Collarless Femoral Stem

A. J. C. Lee, Ph.D.; R. S. M. Ling, M.D., F.R.C.S.

INTRODUCTION

Between 1965 and 1969 some 334 metal-on-metal total hip replacement operations were performed at the Princess Elizabeth Orthopaedic Hospital in Exeter, England. In 1969 a significant number of these implants were becoming loose, and it was apparent that all was not well with the implant being used.

A survey was made of the two alternative products available at that time, but for one reason or another neither was considered ideal. It was therefore decided that a new implant would be designed and developed in Exeter, and a basic specification for it was prepared. The specification included the usual mechanical, surgical, and biological requirements for a total hip joint. The design that was originated to meet this specification in late 1969 has not changed in essence, although some detail changes have been made, and there have been many instrument, insertion technique, and cementing technique developments. The special feature of the design that had not been specifically emphasized in any other design was that it was conceived from the beginning as an integrated system involving the prosthetic components, the surgical instrumentation and technique, and a better understanding of the way in which bone cement should be handled in order to achieve the best possible fixation.

Specific features of the prosthetic components of the new implant will now be described. The instrumentation and surgical technique will be described, as appropriate, in other parts of this chapter.

THE EXETER FEMORAL COMPONENT

The most obvious feature of the femoral component of the Exeter system (Fig. 8–1) is the lack of a collar between the neck and the stem. When the femoral component was designed, one of the prime objectives was to transmit load, via bone cement, into the femur in the most nearly physiologic, or natural, manner possible. A cursory glance at the structure of the proximal end of a natural femur shows that load is transmitted through the femoral head, into the trabecular bone system, and into the cortex of the femur over a length of the femur; in general terms, also, load is transmitted from the inner trabecular system to the outer cortical bone. Load is first transmitted into the calcar femorale, and distally from this point.

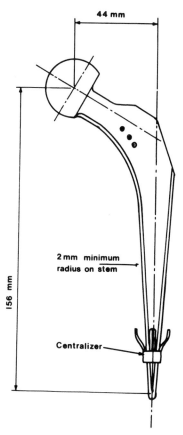

44 mm

2 mm minimum
radius on stem

156 mm

Centralizer

Fig. 8 – 1. Diagram of Exeter femoral component with centralizer.

Therefore, a femoral component should be designed to transmit load over an area of the femur, and not at any specific level, and should transmit that load via the internal trabecular bone system into the calcar femorale and distally from this point. The presence of a collar on an implant implies load transmission of some real magnitude at a specific section of the femur. This was felt to be wrong, and to be justified only in an implant that was inserted without bone cement, as were some of the early head replacement endoprostheses. Indeed, it was speculated that many collars appeared on femoral components because they had always been there, even though the basic reason for their need (no cement), no longer applied. (Specific reference will be made later in this chapter to the appearance of the calcar femorale after the Exeter implant has been in use for periods of over 5 years.)

The absence of a collar on the femoral component of an implant has other implications: (a) The position and angle of cut across the femoral neck is not critical. (b) The depth of insertion of the implant into the femur can be varied to achieve optimal muscle tightness at the hip or to correct leg length. (c) The cement around the stem can be controlled by the surgeon throughout the insertion procedure and, in particular, on the medial side as the implant is pushed into its final position. (d) if fixation is, for some reason, inadequate, the stem can sink into the femur, self-tightening on its taper with no chance of "hang-up" on the neck of the femur.

The second feature of the Exeter femoral stem is its straight taper on all surfaces.

The resulting wedge shape minimizes the chance of voids being created in the cement during insertion, and tends to pressurize the bone cement into the bony spaces, thus enhancing fixation.

Third, a small centralizing jig is placed on the lower end of the stem and helps to maintain the stem in the middle of the medullary canal during insertion.

Finally, the stem has no sharp edges, which can act as stress raising foci in the cement. There are five sizes of stem available: lightweight, narrow, standard, heavy duty and ultraheavy duty, all with a 29.75 mm diameter head. Special stems are available for congenital dysplasia of the hip (CDH) and for cases in which a long stem is needed. Stems are currently manufactured in LVM stainless steel; their shape is such that they can be machined all over and, therefore, are reproducible in quantity at the correct size.

THE EXETER ACETABULAR COMPONENT

The acetabular component (Fig. 8–2) of the Exeter implant is made from ultra-high-molecular-weight polyethylene. As can be seen from the diagram, the outer surface is just larger than a hemisphere and is grooved for location in the bone cement. The center of the outer surface is at point 0, and the surface is set on a center line that is at 40 degrees to the vertical when inserted into a patients acetabulum, with the patient standing. The inner surface is also hemispherical but is centered at point A on an axis that is at 20 degrees to the vertical. Distance OA is 5 mm. The center point variation and axis tilt variation give the cup its distinctive skirt when viewed from the outside. The variation also means that the cup thickness is maximum in the expected wear path, and that this thickness is large compared with that of a conventional, symmetrical cup of the same diameter.

The reason for the difference in axis angle is that the implant is designed to be inserted by the southern or posterior approach. To compensate for the lack of posterior soft tissue support for the hip in the immediate postoperative period, great inherent ball-in-socket stability is needed. This is achieved by presenting the femoral component with a cup whose face is angled at 20 degrees to the horizontal, while still presenting the acetabulum with a cup whose axis is at 40 degrees (Fig. 8–2).

An indicator wire is placed circumferentially around the cup to allow postopera-

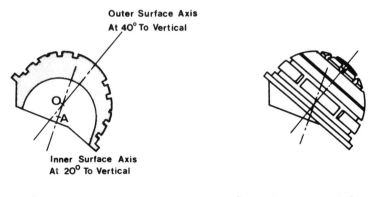

Cross Section of Cup   **Outside View of Cup**

Fig. 8–2. Diagram of Exeter acetabular component.

tive radiographic checking of the cup position and rate of wear. Four sizes of cup are available, ranging from 44 mm to 56 mm external diameter. Both the acetabular component and the femoral component are inserted into place with special instruments, and both are located in position with bone cement.

INDICATIONS FOR TOTAL REPLACEMENT OF THE HIP JOINT BY THE EXETER SYSTEM

Indications for total replacement of the hip joint using the Exeter System are the same as for any conventional total replacement of the hip joint. The most common cause is osteoarthritis, a degenerative joint disease that leads to progressive destruction of the joint, with consequent pain and loss of mobility to the patient. The system has been used to treat hips following trauma, to treat CDH in young adults, and to treat many of the results of other degenerative joint diseases. The shape of the Exeter implant, particularly that of the straight femoral stem, makes the implant good for treating hips that have had previous surgical procedures, such as failed osteotomies, in which the fairly large offset of the head of the implant from the center line of the stem is also an advantage. When salvage operations are attempted following, for example, stem fracture, the surgeon is very commonly presented with a femoral medullary canal that is almost devoid of cancellous bone and is, in effect, a cortical tube that tapers from top to bottom. The use of the straight tapered Exeter stem in such a case can provide very good fixation, the simple taper of the stem in the cement and the cement in the bone providing anchorage by means of the "engineering taper" effect, so common in engineering practice (very large direct and torsional loads are taken by tapers, such as the well-known Morse taper[10]). A collared implant cannot take full advantage of the taper for fixation, owing to the danger of "hang-up" on the cut surface of the femoral neck.

By July 1979, more than 5,000 Exeter hips had been implanted in the United Kingdom and elsewhere. More than 1,000 had been implanted in Exeter, and the longest had been in use for more than 8½ years. (The first patient to receive an Exeter hip died in 1978; the second, a very active patient, completed 9 years with the hip in November 1979 and substantial numbers have been in use for 5 years or more.) From this clinical experience four cases have been selected for discussion, one good result and three that have been involved in difficulties of one sort or another. It was felt that comparatively little could be gained from presenting further good or excellent results (and the very large majority of cases fall into this category, in common with results with other hip joints), while much could be learned from a study of failures. Comments on the cases presented are given in the sections following the descriptions of the four cases.

CASE PRESENTATIONS

Case 1. W. J. F. was an electrician. At age 58, he found it increasingly difficult to continue his work, because of considerable difficulty climbing steps and ladders. Osteoarthritis (OA) of the right hip was diagnosed, and a McMurray osteotomy was performed. The patient continued to complain of pain, and the osteotomy plate was removed 2 years postoperatively. Total hip replacement was recommended 2 years

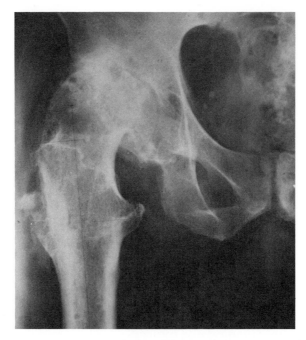

Fig. 8-3. Case 1, preoperatively.

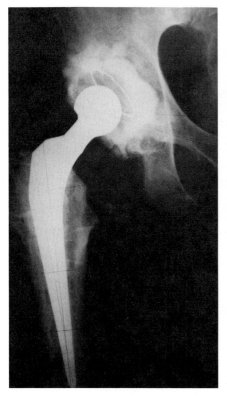

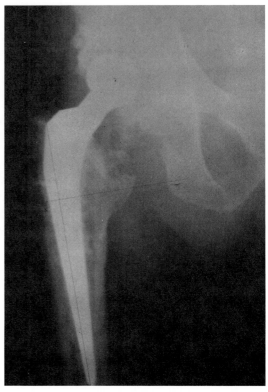

Fig. 8-4. Case 1, postoperatively.   *Fig. 8-5.* Case 1, 99 months postoperatively.

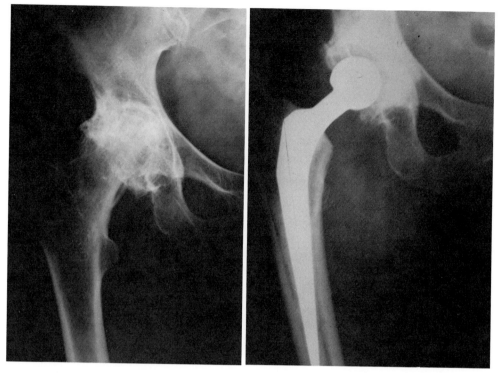

Fig. 8 – 6. Case 2, preoperatively.   *Fig. 8 – 7.* Case 2, postoperatively.

later (Fig. 8 – 3), and was carried out with the patient aged 62 (Fig. 8 – 4). The operation was uneventful and, postoperatively, the patient was totally pain free. The patient has now retired and has progressed well since his operation (Fig. 8 – 5). He plays golf regularly and walks without a stick. Clinically and radiologically the patient's condition is very satisfactory 8 years after total hip replacement.

Case 2. J. T. is a housewife. When she was aged 63, a diagnosis of OA of the right hip was made, and total hip replacement was recommended (Fig. 8 – 6). A narrow Exeter stem was inserted – one of the original stems with a machined-down neck (see the discussion for Case 4). The operation and recovery were uneventful (Fig. 8 – 7). On routine examination 14 months postoperatively, a sinkage of 2 mm of the femoral component was noted on x-ray. The patient felt no difference in leg length or range of motion, and expressed herself happy with the result. A crack in the cement, that appeared to encircle the femoral stem, was seen in the midstem area. The crack appeared to be about 3 mm wide.

The patient was brought back for annual review, at which time x-rays were taken and a clinical assessment made. Figure 8 – 8 shows a typical x-ray, taken 31 months postoperatively. It was apparent that sinkage of the implant was progressive until about 5 years postoperatively, when the implant stabilized in the femur. The crack in the bone cement has become more difficult to see, and the latest x-ray, at 97 months postoperatively (Fig. 8 – 9) almost suggests that the crack is being obliterated by bone. Although the implant has moved, and the cement has moved in the bone, the patient is clinically satisfactory 8 years after total hip replacement, with a good range of movement and no pain. She is quite active and walks without a stick. Radiologically, the cement mantle as well as the implant in the femur are less than optimal.

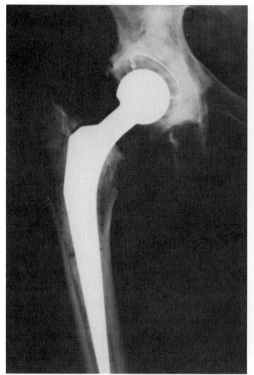 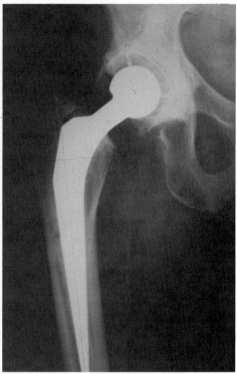

Fig. 8-8. Case 2, 31 months postoperatively.   *Fig. 8-9.* Case 2, 97 months postoperatively.

<u>Case 3.</u> A. S. is the owner of a guesthouse. She was examined at age 67, when she was found to have a fixed flexion deformity of her right hip at 50 degrees with only 30 degrees of movement, fixed in adduction and external rotation. Total hip replacement was carried out, and a narrow Exeter stem was inserted (Fig. 8-10). The operation was uneventful and recovery good. Postoperatively the patient had 90 degrees of flexion and walked comfortably with a stick. At 16 months postoperatively her grandson knocked into her and she fell over. The leg became swollen and very painful. The femoral component was found to be fractured (Fig. 8-11). It was replaced with another component and the patient has progressed to an uneventful recovery.

<u>Case 4.</u> P. B. is a self-employed male who was diagnosed to have OA of the left hip at age 61; a total replacement was advised. This was done at age 63, when an Exeter implant was inserted (Fig. 8-12). The operation and recovery were normal. At 10 months postoperatively, the patient was readmitted to the hospital, complaining of intense pain in his groin. On x-ray it was found that the neck of the femoral component had fractured (Fig. 8-13). An exchange procedure for the femoral component was advised. The new implant was inserted 1 month later (Fig. 8-14), and the patient was discharged home after a normal recovery. The patient was admitted to hospital for a third time 34 months after the exchange, again complaining of severe pain. It was found that the shaft of the femoral component had failed (Fig. 8-15). The fractured component was removed and a new component was inserted (Fig. 8-16). Recovery from the operation was normal, and at the last contact with the patient, 2 years after his third operation, he was asymptomatic and leading a normal life.

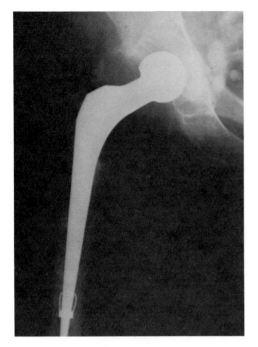

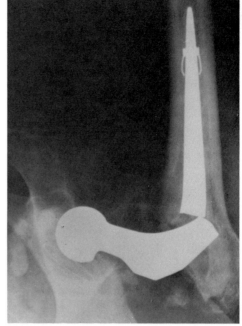

Fig. 8–10. Case 3, postoperatively.   *Fig. 8–11.* Case 3, 16 months postoperatively.

Discussion of the Clinical Cases

W. J. F. (Case 1) received one of the early Exeter implants that had a relatively sharp corner machined on the top surface of the neck (see the discussion of Case 4 below). Implantation of the femoral component was not ideal, it being in about 4 degrees varus from the ideal neutral position. Cement technique was adequate, a reasonable mantle around the stem being present with only minor deficiencies at the tip. After 8 years in the patient, the implant is still functioning very well with no sign of loosening, sinkage, or deformity. In particular, the calcar femorale is still present and functioning. Signs of ectopic bone formation are clearly present.

It is perhaps worthwhile pausing for a moment to consider the mechanisms of force transfer in the femoral component of a replacement joint, and to speculate on the effects of the presence or absence of a collar.

The forces acting on a hip joint are shown crudely in Figure 8–17. In this diagram only the abductor muscles are represented (by force F), none of the many other muscle groups are shown. The reaction R between the implant and pelvis is shown, as is the body weight W. The magnitude of these forces depends on the distances a and b, to a first order approximation, as moment equilibrium must be maintained. The distance b depends on the extent to which the surgeon deepens the acetabulum and medializes the cup, and to some extent on the design of the cup. The overall effect can change distance b by as much as 2 cm. The distance a will vary according to the offset of the center of the prosthetic head from the center line of the stem. In the varus Exeter implant this is fairly large, in a valgus implant, such as the Computer Aided Design (CAD), it is smaller. Therefore the magnitude of the forces R and F

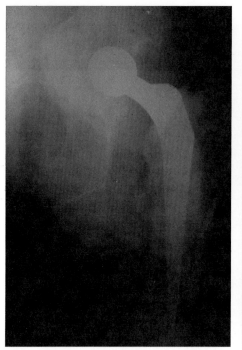
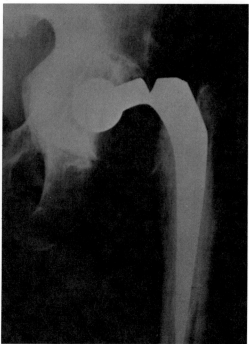

Fig. 8–13. Case 4, 10 months postoperatively.   *Fig. 8–12*. Case 4, postoperatively.

will vary between the two types of stem. For the varus stem, R will be relatively large and F relatively small; for the valgus stem, R will be relatively small and F relatively large. The force R is transmitted from the implant into the bone, via the bone cement, as a reaction force and a bending moment. The magnitude and direction of all forces varies throughout a loading cycle; consequently, the forces and bending moments put into the bone are variable and complex in nature.

It has been estimated that the maximum value of force R can be reduced by about 22½ percent by going from a varus to a valgus stem.[6] The way in which this larger or smaller force and bending moment is transmitted into the bone of the femur is still a matter of speculation, since there have been no *in vivo* studies of force transfer from implant to bone, and the *in vitro* studies that have been made with strain gauges[11, 15] reflect a very simplified version of the clinical situation. However, one conclusion from the *in vitro* studies is noteworthy: the stress in a prosthetic component can be raised to 100 times the normal stress if the cement mantle around the component is incomplete.[1] This means that surgical technique has two orders of magnitude more effect on the stress situation in an implant (and hence on the stress in the bone supporting the implant) than does the design of the implant itself. If, then, there is no actual measurement of force or bending moment available, the only way in which the investigator can determine how load is transferred in any particular joint is to observe the effect that the joint has on the bone surrounding it.

According to Wolff's law, bone reacts to reflect the load put upon it: if it is unloaded, bone will be resorbed, if it is loaded "normally," it will maintain its "normal" density; if it is overloaded, more bone will be laid down to take the load. If bone is grossly overloaded, it will be destroyed.

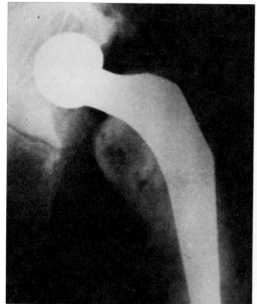 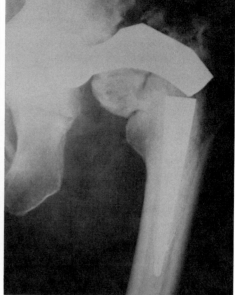

Fig. 8–14. Case 4, second implant, postoperatively.

Fig. 8–15. Case 4, second implant, 34 months postoperatively.

If load is being transmitted effectively from the implant into the bone, then the bone should, in the absence of destruction from extraneous causes, be maintained and not resorbed. The high loads imposed on an implant stem by the body must be transmitted to the bone by the implant, to protect the relatively small stem sections from overload and subsequent fatigue failure. Even the very largest stems are probably not large enough to survive without support—for example fractures in the St. George stem are not rare. The calcar femorale is the part of the femur that is loaded first by any implant, and has been studied by a number of authors because of its importance in maintaining support of the implant. The collar on an implant stem is widely considered as being the means by which load can be effectively transmitted to the calcar. As has been stated in the introduction to this chapter, the calcar is loaded by the medial trabecular bone system, which transmits load from the femoral head to the cortex. As can be seen from Figure 8–18, the calcar is the region in which load *starts* to be inserted into the cortex, and this is the region of the cut surface of the neck of the femur in a conventional total hip replacement. Therefore, a significant load should not be put on the calcar at this cut plane; rather, load should begin to be put on the bone from this plane, if physiological conditions are to be maintained in any way at all.

The implantation of an intramedullary stem always creates nonphysiological loading of the upper femur; however, the transmission of loads directly to the cut surface of the calcar through the collar can only magnify the nonphysiological situation. Thus, the best an intramedullary stem can do is to load the calcar from its endosteal aspect, starting at the cut surface of the femur and extending down the femur. If the bone is loaded in this way it should be maintained in a viable and healthy condition, able to support the implant over many years.

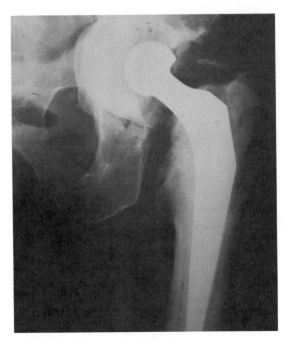

Fig. 8-16. Case 4, third implant postoperatively.

Looking at clinical evidence, many authors have reported disappearance of the calcar.[3, 4, 5, 13] Charnley and Cupic suggested a "normal" resorption of 3 to 4 mm of bone below the collar in implants in which cement is used, occurring in the first 2 to 3 years. An important finding of a study recently reported by ourselves, based on experience with the first 107 Exeter implants inserted in Exeter (all of which have exceeded 5 years since operation), is that no hip showed absorption (disappearance) of the calcar and that no hip showed cystic change, cavitation, or destruction with erosion of the calcar.[2] This must mean that in no case was the calcar unloaded, since if it were, it would have been absorbed. This must call into question the validity of the strain-gauge tests conducted by Harris and Oh,[14] who reported zero or near zero strain in the calcar when collar-to-calcar contact at the cut surface of the femur was not attained. Since the Exeter implant has no collar, such collar-to-calcar contact could clearly never have been present; yet, equally clearly, the calcar has been loaded.

The total absence of destructive changes in the region of the calcar in the Exeter series also calls for comment. It seems likely that the destructive changes reported[3, 4, 5, 13] are always associated with some degree of bone cement fragmentation. The circumstances by which cement fragmentation can be initiated must be examined. Local areas of very high stress in the cement are needed if the cement is to crack and form fragments. Such high stress areas can be formed in unconstrained cement (see Case 4 below), by high-stress-concentrating features in the cement, and by loading thin layers of cement (typically found between the collar and calcar) at high strain rates. Local stress-raising features can be found on many implants. Sharp edges to the collar, if buried in cement, can easily crack the cement; similarly, the cross-sections of many implant stems have sharp edges. Drill holes, or wires in the

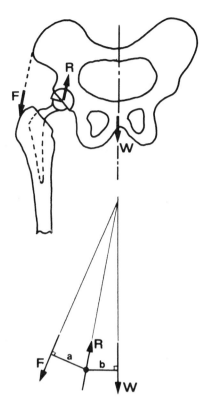

Fig. 8–17. Diagram of forces on hip joint.

cement can produce stress concentrations sufficient to cause cement failure. The formation of a thin layer of cement between a collar and the calcar is inevitable when a collared implant is pushed into its final position. The thinnest part of the layer will receive the greatest load, and at a higher strain rate than other parts of the layer. It is liable to crack and then be available for repetitious movement against bone, leading to bone destruction.

Considering the above, it seems likely that the lack of calcar destruction found with the Exeter stem is due to the greater control the surgeon has over the cement in the medial calcar region, permitting the surgical avoidance of stress-concentrating effects and thin layers of cement. Lack of absorption of the calcar would also appear to arise from the fact that cement is controlled in the medial region, allowing good contact between the implant, cement, and calcar to be ensured, coupled with the varus aspect of the Exeter stem that ensures a bending moment in the calcar region that is sufficient to ensure that the calcar is loaded.

Case 2 has now had her hip implant in place for over 8 years, and is very happy with the result. A circumferential crack appeared in the cement shortly after implantation, accompanied by sinking of the implant. Obviously, to allow this sinking to take place, longitudinal cracks must also have been present in the cement. The sinking progressed during a 5 year period until a stable situation was reached – a situation that has now been constant for 3 years. Two conclusions can be drawn from this case. The first is that if an implant is designed so that it can sink, which means that its

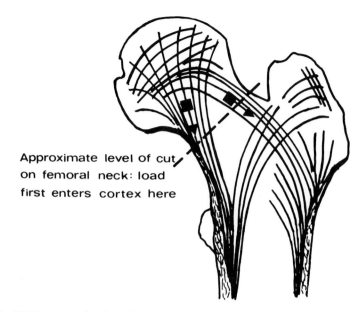

Approximate level of cut
on femoral neck: load
first enters cortex here

Fig. 8–18. Diagram of trabecular bone in femoral head showing load trajectories.

stem should have as near a straight and uniform taper as possible, then the taper can self-tighten upon sinking, with no clinically noticeable ill effects. The presence of all but the smallest collar can prevent such sinking by "hang-up" on the cut surface of the neck of the femur.[7] Obviously, cement technique must be much poorer than optimum to allow sinking at all, but if sinkage does occur it must do so in a safe way. Avoidance of repetitious movement of cement against bone is essential if bone destruction is to be avoided. Simple sinking, with self-tightening each time movement takes place, does not give the harmful repetitious movement of cement against bone; such repetitious movement can easily take place if a collar pivots the implant about the neck section.

The other important conclusion to be gained from Case 2 is that failure of bone cement does not necessarily mean failure of the replacement joint. The mechanism by which load is transferred through the bone cement, and the stress situation in the bone cement, obviously changes when the cement cracks, but this is not harmful if the cement is properly constrained by bone that forms an enclosure around the cement.

Case 3 was quite active after her hip had been replaced. It can be seen from the postoperative x-ray (Fig. 8–10) that the neck of the femur has been cut very low, and that the upper end of the femur in the region of the greater trochanter has not been opened properly. Consequently, a narrow stem was inserted in the belief that a larger stem would not fit. The cementing technique in the upper part of the stem is considerably less than optimum and provides very poor support for the stem in this region.

One of the dangers of the Exeter collarless stem, particularly if the top of the femur is not opened properly, is that the neck of the femur will be cut back to make insertion easier. With no collar, there is no large and obvious feature on the stem to

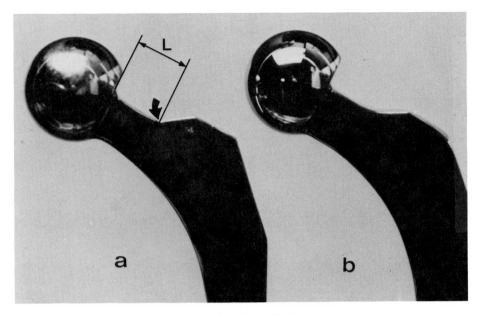

Fig. 8–19. Original and modified neck design.

which the cut on the neck of the femur must be matched. If, as in this case there is no proper support to the stem given by good quality bone, then the danger of fatigue failure is obvious. Such a failure took place in the implant in Case 3, the final fracture occurring when the patient was knocked over by her grandson. It is interesting to note that when the case was revised and the medullary canal opened properly, a heavy-duty stem could be inserted.

Case 4 has had a most unfortunate series of failures of his hip. The first implant (Fig. 8–12) was inserted in a technically defective manner, in that it was in marked varus owing to the fact that the canal had not been properly opened. The stem was one of the very early Exeter implants that had a somewhat sharp corner on the machined upper surface of the neck of the femoral component. The original neck design, which was modified after 52 stems had been used clinically, is shown in Figure 8–19A. Of the 52 stems manufactured to this design, all but the first 10 (that is, numbers 11 to 52) had excessive machining on the dorsal part of the neck (that is, length L was made greater than was designed and the neck depth was slightly reduced; see Fig. 8–20). The sharp angle on the dorsal surface was clearly an exceedingly bad feature of the original design and was made worse by the excessive machining. Such stress-concentrating features as this type of corner should never be present on an implant, and the current version of the stem (Fig. 8–19B) has a large radius curve at this point.

None of the first 10 stems have fractured despite a service life of up to 9 years. Of the 42 defective stems, 14 have now fractured, all at the sharp corner marked with an arrow in Figure 8–19A, and typified by the failure of the first implant in case 4 shown in Figure 8–13. Extensive fatigue testing of the neck section of the implant has led to the view that the failures were due, not only to a reduction in the neck section and increased bending moment, but also to individual machining marks on each

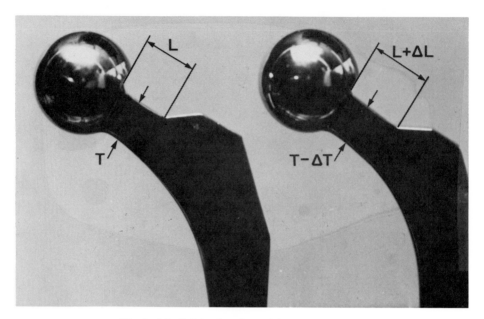

Fig. 8–20. Original and machined-down neck.

implant. There is no correlation, in the 14 failures, between weight and activity level of the patients and the time to failure of the implant neck. No original pattern stems were ever used outside Exeter, and there have been no neck fractures in any of the subsequent five-thousand-odd stems of the current design. The current design has been fatigue tested at eight times body weight in each load cycle, for up to ten million load cycles, many times, without ever failing.

Although these 14 failures are disasters of the first magnitude, the reexploration of the hips at revision provides an almost unique opportunity to observe the quality of bone and cement in a hip joint in which fixation of the femoral component has remained very good for a significant number of years. It has been found that it is not possible to insert even the point of a very fine scalpel blade between the implant and the cement, or between the cement and the bone, in the medial calcar region. This is a most encouraging finding, especially in the case of the last failure to be reoperated upon; a male who had had his implant for 7½ years before failure occurred.

This male patient, who weighs 97 kg, is exceptionally vigorous. It was interesting to see that his calcar really did not change at all during the life of the implant. In this case a thin section of bone and bone cement was removed from the cut surface of the femur when the failed implant was removed, and was examined to determine the quality and viability of the bone in the calcar region. The bone was found to be of excellent, healthy appearance, with cells in the lacunar spaces and no evidence of infarction. Thus, after 7½ years of pain-free active life, this patient displayed both radiographic and histologic evidence of the preservation of normal calcar.

The failed implant in Case 4 was removed, and a new implant inserted into the old track of cement already in the femur. It could not be inserted to the full depth required, owing to minor difference in shape between the old and new implant stems.

Consequently, the stem protruded excessively at the top end, and some attempt was made to support it by placing a buttress of cement on the top of the calcar and under the medial aspect of the stem (Fig. 8–14). This cement was not of good quality in itself but, more importantly, was entirely unconstrained by bone. When the patient walked after the operation, the upper end of the stem would flex under the load, putting a large stress on the cement at the junction of the cut surface of the femur and the cement buttress. Unconstrained cement cannot stand this type of stress, and the buttress cracked away from the cement in the medullary canal. The upper end of the stem was now entirely unsupported and, in due course, failed by fatigue. A third implant was then inserted and the patient recovered normally.

CONCLUSIONS

A number of conclusions of general interest can be made as a result of the experiences with the implant system described in this chapter:

1. A collarless hip implant has been in clinical use for periods of up to 8½ years. The clinical success of the implant appears to be at least as good as conventional collared implants.

2. No calcar resorption or destruction has been seen in the first 107 implants inserted at Exeter, all of which have been in service for over 5 years. Direct observation of the calcar in cases in which the neck of the implant has failed has shown the calcar and cement fixation to be good.

3. In the first 920 hips operated on in Exeter since 1972, there has only been one operation for nonseptic loosening of the acetabular cup. This is of great importance and significance. Since 1972,[8] the Exeter system has used an acetabular pressurizer that enables the surgeon to force cement into the bony spaces of the acetabulum before inserting the prosthetic acetabular cup. The resulting penetration of cement into bone shows a generally unchanged appearance over many years, with the radiolucent line, seen by many investigators at the boundary of cement and bone, not being present in a large number of the Exeter cases. It would, therefore, appear from the clinical results that the practice of forcing cement into open trabecular bone with the cement pressurizer is a definite advance in the technique of achieving permanent acetabular cup fixation. The principle of pressurization of cement is now being extended to include the femoral component and, by other investigators, the knee.[12]

4. Failure of cement by cracking does not necessarily mean failure of the implant. If sinkage and self-tightening can take place, bone erosion is not a problem and clinically good results are achieved.

5. Sharp edges on the implant, or on the bone, in contact with the cement should be avoided, in order to avoid cement fragmentation which can lead to bone destruction.

6. Sharp stress concentrations on the implant in highly stressed areas should be avoided, since they can lead to failure of the implant.

7. The implant must be properly supported by bone cement of good quality that is itself constrained by bone. The technique of using and handling bone cement and preparing the bone surface for cement is vital.[9]

REFERENCES

1. Amstutz, H. C.: Personal communication.

2. Black, J., Lee, A. J. C., Ling, R. S. M., Sew-Hoy, A. L., and Vangala, S. S.: A study of the radiological appearance of the calcar femorale after total hip replacement using a collarless stem. British Orthopaedic Association Spring Meeting, Exeter, April, 1979.

3. Blacker, G. and Charnley, J.: Long-term study of changes in the upper femur after low friction arthroplasty. Clin. Orthop. (In press)

4. Bocco, F. Langan, P. and Charnley, J.: Changes in the calcar femoris in relation to cement technology in total hip replacement. Clin. Orthop., *128:*287, 1977.

5. Charnley, J. and Cupic, Z.: The nine and ten-year results of low friction arthroplasty of the hip." Clin. Orthop., *95:* 9 – 25, 1973.

6. English, T. A. and Kilvington, M.: A direct telemetric method for measuring hip load. Personal communication, 1978.

7. Gruen, T. A., McNeice, G. M., and Amstutz, H. C.: Modes of failure of cemented stemtype femoral components – A radiographic analysis of loosening. Clin. Orthop., *141:* 17 – 27, 1979.

8. Lee, A. J. C., and Ling, R. S. M.: A device to improve the extrusion of bone cement into the bone of the acetabulum in the replacement of the hip joint. Biomed. Eng., November, 522 – 524, 1974.

9. Lee, A. J. C., Ling, R. S. M., and Vangala, S. S.: Some clinically relevant variables affecting the mechanical behaviour of bone cement. Arch. Orthop. Trau. Surg., *92:*1 – 18, 1978.

10. Machinery's Handbook, 17th Ed. pp. 1467 – 1478, The Industrial Press, Brighton, England.

11. Markolf, K. C., and Amstutz, H. C.: A comparative experimental study of stresses in femoral total hip replacement components: The effects of prosthesis orientation and fixation." J. Biomech., *9:* 73, 1976.

12. Miller, J., Burke, D. L., Krause, W., Keleboy, L., Tremblay, G. R., and Ahmed, A.: Improved fixation of knee arthroplasty components to prevent loosening 46th Annual AAOS Meeting, San Francisco, February, 1979.

13. Nicholson, R.: Total hip replacement: A radiological review, Proceedings of the 25th Annual Meeting, NZOA, New Plymouth, J. Bone Joint Surg., *57B:* 256, 1975.

14. Oh, I.: Effect of total hip replacement on the distribution of stress in the proximal femur: An in-vitro study comparing stress distribution in the intact femur with that after insertion of different femoral components. Proceedings of the 5th Open Scientific Meeting of the Hip Society. C. V. Mosby Co., St. Louis, 1977.

15. Weightman, B.: Stress analysis. In Swanson, S. A. V., and Freeman M. A. R., Eds. The Scientific Basis of Joint Replacement. Pitman Medical Publishers, London, 1978.

9
Gait of Patients with Hip Pain or Loss of Hip Joint Motion

M. P. Murray, Ph.D.; D.R. Gore, M.D.

INTRODUCTION

The ability to walk is compromised in thousands of individuals with hip joint disease. Thus, gait evaluation is included in the complete clinical examination of the patient with problems of the hip joint. This evaluation is not done to establish a specific etiologic diagnosis, but to provide the clinician with information on the extent to which the disease has affected the patient's ability to ambulate normally. Additionally, gait evaluation helps the clinician determine what factors, such as pain or loss of motion, play a major role in the gait abnormalities that are observed. This chapter is intended to: (a) help the clinician to interpret observations and learn how to analyze the gait of a patient with complaints related to the hip joint; (b) describe the pain-avoidance maneuvers used by patients with hip pain and the compensatory mechanisms used by patients with loss of hip motion; (c) describe, using mechanical principles, why these maneuvers are of value to the patient; and (d) to synthesize this information so as to help the clinician understand the gait of the individual with both pain and restricted motion.

That a patient has an abnormal gait is obvious even to the casual observer; however, the characteristics that distinguish the pathologic from the normal gait, or that distinguish one pathologic gait from another, are not always as obvious. Through studies of the kinematics of locomotion, ranges of normal variability have been established for movement patterns of multiple body segments during walking. These ranges of normal variability provide a means for identifying those specific aspects of gait that are disordered in different disease states, and also provide a means of assessing the degree of gait abnormality of disabled patients. Although there are a variety of methods for recording the kinematics of locomotion, the values presented in this chapter were obtained using interrupted-light photography (Fig. 9–1). This technique allows accurate measurement of the simultaneous displacement patterns of multiple body segments in three planes of space without using devices that encumber the patient.[11, 12, 14, 15]

DESCRIPTION OF NORMAL LOCOMOTION

In order to recognize gait abnormalities, it is important to understand normal patterns of movement. Patients with joint disabilities of the lower limbs have abnormal

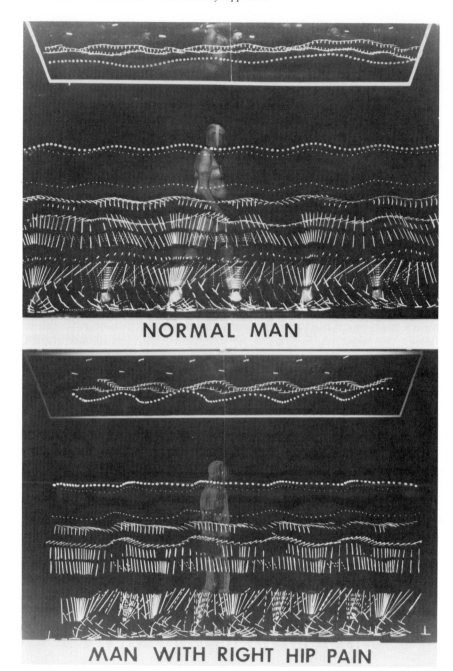

NORMAL MAN

MAN WITH RIGHT HIP PAIN

Fig. 9–1. Displacement patterns during free-speed walking for a normal man and a patient of similar age and height with right hip pain. A slotted disc rotates in front of the camera lens, allowing the recording rate to be varied, depending on the walking speed of the subject. The exposure rate was 1,200 per minute for the normal man and 900 per minute for the patient. The targets on the medial aspect of the thigh and leg present as broken lines, while those on the lateral aspect present as solid lines. Images projected in an overhead mirror allow measurement of simultaneous displacement patterns in three planes of space.

patterns of motion, not only of the involved joint but also of uninvolved joints and the trunk. These abnormal displacement patterns reflect attempts by the patient to offset limited joint motion, instability, or the inability to generate, tolerate, or coordinate the forces necessary for normal walking. This section contains a description of terms used in discussing locomotion and a brief description of relevant aspects of normal locomotion. Normal patterns of motion of the major body segments are described. Velocity and related factors are discussed, since the inability to achieve a functional walking speed can limit the rehabilitation potential of some individuals. Forward, lateral, and vertical trajectories of the head are described, since these parameters have important implications in terms of the mechanical energy expenditure of the patient.

Traditionally, all of the events of locomotion are time-based to a walking cycle. The cycle begins at heel-strike of one foot and ends with the next heel-strike of the same foot. For a given lower limb, there is one period of stance and one period of swing within each walking cycle. The stance phase, approximately 60 percent of the walking cycle, is that period when the foot is in contact with the floor; the swing phase is that period when the foot is swinging forward to make the next step. In normal walking, the right and left stance phases tend to be of equal duration, as do the right and left swing phases. Within each walking cycle there are two periods of single-limb support during which one limb provides support while the other limb swings forward to make the new step, and two brief periods of double-limb support when the right and left stance phases overlap. Step length is defined as the distance between the successive contact points of opposite feet, and in normal walking the two steps within each walking cycle tend to be of equal length.

Walking speed depends upon the length and rapidity of the steps, or the cadence, which is commonly measured by the number of steps per minute. For estimation purposes, normal cadence is about 115 steps per minute or just under 2 steps each second. Our studies reveal that as a result of taking shorter but not slower steps, normal people beyond the age of 60 walk at a slower velocity than younger people. The average velocity of normal women is slightly less than that of normal men in the same age range. Although the cadence of women is slightly faster than that of men, women take shorter steps, both in terms of actual step length and percent of body stature.

In normal locomotion, forward movement of the head is remarkably uniform, although there are periods of slight acceleration and deceleration within the walking cycle. The lateral and vertical pathways of the head of normal adults, as depicted in Figure 9–2, are smooth and sinusoidal in configuration and low in amplitude (averaging between 4 and 6 cm, depending on sex, age, and walking speed). The head and trunk reach their highest and most lateral positions during single-limb support and their lowest and most central positions during double-limb support.

The pelvis rotates through two small excursions of anterior and posterior tilting within each walking cycle. The periods of maximum anterior tilting occur shortly before heel-strike when the trunk is inclined forward, and the periods of maximum posterior tilting occur when the trunk assumes a more erect position over its single supporting base.

Normally the hip is flexed at heel-strike, and extends as the trunk moves forward over the supporting limb. At contralateral heel-strike, the hip reverses into flexion and continues to flex throughout most of the swing phase. The average total ex-

PATHWAYS OF THE HEAD AND NECK

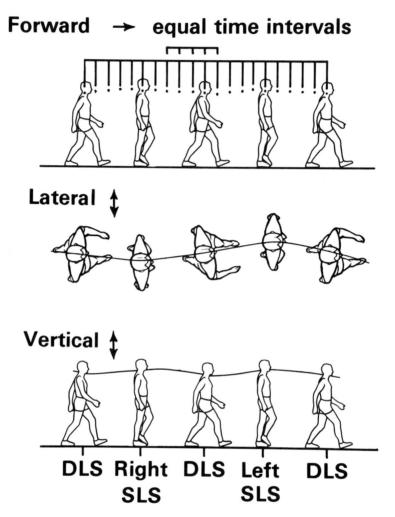

Forward → equal time intervals

Lateral ↕

Vertical ↕

DLS Right DLS Left DLS
      SLS   SLS

DLS = Double-limb support   SLS = Single-limb support

Fig. 9–2. Illustration depicting pathways of the head and trunk during normal locomotion. The forward motion of the head is uniform and the lateral and vertical pathways are smooth and sinusoidal.

cursion of hip flexion-extension during a walking cycle is approximately 40 to 50 degrees, depending upon the physical characteristics of the subject and the walking speed.

During normal walking, the knee is extended at heel-strike, then flexes slightly as weight is transferred onto the limb. The knee then extends as the trunk moves forward over the supporting limb. Before the end of the stance phase, the knee begins to flex, reaching its maximum flexion of approximately 65 to 75 degrees early in the swing phase, and knee extension then advances the foot forward for the next heel-strike.

After heel-strike, the ankle plantarflexes as the forepart of the foot descends to the floor. The ankle dorsiflexes early in stance as the leg segment rotates forward over the fixed foot and then plantarflexes, elevating the heel and shifting the weight to the forepart of the foot throughout the remaining stance phase. During the swing phase, ankle dorsiflexion provides foot-floor clearance.

The patterns of hip, knee, and ankle flexion-extension during normal walking are graphically depicted by the shaded areas in Figures 9-6 and 9-7.

During normal walking, the pelvis and thorax rotate simultaneously in opposite directions in the transverse plane. The thorax rotates forward on the side of the forward-swinging upper limb, while the pelvis rotates forward on the side of the forward-swinging lower limb. The forward and backward movement of the upper limb coincides with the forward and backward movement of the contralateral lower limb.

CLINICAL EXAMPLES

Following are the gait evaluations of five patients with hip disability. Each evaluation is structured so as to comment on those aspects of gait which are most commonly distorted in the patient with hip disability. With practice, the clinician can learn to recognize the following types of abnormalities: slow velocity; subnormal step lengths and cadences which contribute to slow velocity; lack of symmetry in successive step lengths; excessive motion or lack of smoothness in the forward, lateral, and vertical pathways of the head and trunk; lack of symmetry in the single-limb-support phases of the right and left limbs; and abnormal motion of individual body segments, particularly of the pelvis and thigh on the side of the disability. After reading this chapter, the clinician should be able to use such observations to begin to deduce the major source of the patient's difficulty in ambulating. Following the gait evaluations of the cases to be presented, their histories, physical findings, and quantitative data from the gait evaluations will be given, and the cases will be discussed.

Gait Evaluations

<u>Case 1.</u> This 67-year-old man walked with a cane in his right hand, and applied force to the cane during the left stance phase. When he walked without support, he appeared fairly disabled, with a slower than normal velocity, a pronounced limp, and more out-toeing on the left than on the right. On closer observation, it could be seen that his slow velocity resulted from taking shorter than normal steps and from a cadence that was subnormal. The limp consisted of a pronounced lateral lurch of his head and trunk toward the left during the left single-limb-support phase, and a rapid swing phase of the right lower limb resulting in a short period of single-limb support on the left. This made forward motion at the head irregular, and he appeared to speed up as he passed over the left foot and slow down as he passed over the right foot. The patient used very little hip flexion-extension on the left. The excursion of anterior-posterior pelvic tilting was not excessive, but the pelvis always seemed to be tilted anteriorly a little more than normal. The patient did not have symmetrical arm swing: the upper limb on the left side was in excessive extension at the shoulder and

excessive flexion at the elbow. No other abnormalities were observed. He was unable to increase his walking speed when asked to do so.

Case 2. This 40-year-old man used a cane in his right hand and applied apparently high force loads to the cane during the left stance phase. The types of gait abnormalities observed when he walked without support were similar to those in Case 1, but more pronounced. He was unlike Case 1 in that his successive steps were unequal in length, the shorter step occurring when the right foot was ahead, and he did not have striking differences between the out-toeing angles of his feet.

Case 3. This 59-year-old woman appeared to have a normal gait at first glance. However, careful observation revealed some abnormalities. Although her speed appeared to be near normal, as did her cadence, she seemed to have a subnormal step length when the left foot was forward. She had more out-toeing on the left than on the right. No asymmetry in timing of the left and right stance or swing phases was observed. Her forward and lateral motion of the head appeared to be normal. From the left side view, very little use of left hip motion was observed, and her pelvic tilting in the anterior-posterior directions seemed excessive. She was able to increase her speed only slightly when asked to walk faster.

Case 4. This 25-year-old man walked without any assistive device. His gait appeared to be of normal speed, symmetry in step lengths and timing, and smoothness of motion at the head. On closer observation, however, no motion was observed at the right hip. The right thigh and pelvis appeared to move as a unit so that, as the thigh swung forward, the pelvis tilted posteriorly, and as the thigh moved backward, the pelvis tilted anteriorly. The anterior-posterior pelvic tilting was considerably in excess of normal. The patient also appeared to rotate his pelvis excessively in the transverse plane. When the right foot was moved forward to take a step, the right pelvis rotated forward, and when the right foot was in back, the right pelvis was rotated back. He was able to walk considerably faster when asked to do so.

Case 5. This 53-year-old man did not use any assistive device. He walked slowly and had a pronounced limp. On closer observation, his cadence seemed normal, but his successive step lengths were both subnormal, particularly that step when the left foot was forward. He had normal side-to-side head motion. His limp consisted of irregular forward motion, as observed at the head and trunk. The left swing phase was rapid as compared to the right, and he appeared to hurry as he passed over the right foot. His pelvis and right thigh appeared to move as a unit, and no hip flexion-extension was observed on the right. His anterior-posterior pelvic tilting and his pelvic rotation in the transverse plane appeared to be excessive. When asked to walk fast, he walked at speeds similar to the free-speed of normal men.

Histories and Physical Findings

The 67-year-old patient in Case 1, with no history of previous injury or hip disease, had his onset of left hip pain 12 years prior to gait study. During this period, he used a cane and was treated with intraarticular injections of local anesthetic and steroids, which provided temporary relief of pain; he also took indomethacin. At the time of our gait study he had moderate hip pain on ambulation, minimal shortening of the left lower limb, and no orthopedic disability other than that of the left hip. Roentgenographic findings were typical of degenerative arthritis of the hip (Fig. 9–3A). The patient's hip range of motion was as follows:

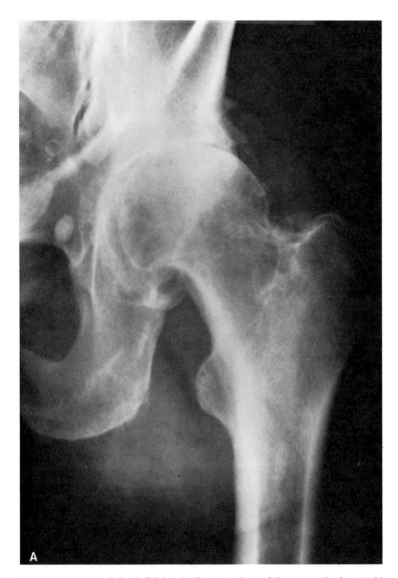

Fig. 9-3. Roentgenograms of the left hips in Cases 1, 2, and 3, respectively. (A) Narrowing of the joint space and sclerosis of the subchondral bone in the femoral head and acetabulum. *(Figure continues on p. 181.)*

	Painful Left Hip (in Degrees)	*Right Hip (in Degrees)*
Flexion	110	130
Extension	lacks 25	lacks 15
Abduction	5	25
Adduction	5	15
Inward rotation	−10	20
Outward rotation	40	20

The 40-year-old man in Case 2 had left hip pain for 3 years following a femoral neck fracture. The fracture, treated by reduction and internal fixation, subsequently healed, but avascular necrosis developed. The internal fixation device had been removed prior to gait study. The patient used a cane because of hip pain on ambulation. He had shortening of the left lower extremity. Roentgenograms of the left hip are illustrated in Figure 9–3B. He had no other orthopedic disability. The patient's hip range of motion was as follows:

	Painful Left Hip (in Degrees)	Right Hip (in Degrees)
Flexion	110	125
Extension	lacks 30	lacks 15
Abduction	5	20
Adduction	15	10
Inward rotation	10	15
Outward rotation	20	20

The 59-year-old woman in Case 3 was studied 4 years after a diagnosis of degenerative arthritis of the left hip was made. At the time of study, she had severely limited hip motion but very little hip pain and did not require external support for walking. Her primary complaint was that lack of hip motion interfered with her ability to walk fast and to perform activities such as kneeling, getting in and out of the bathtub, rising from low chairs, and so forth. The roentgenographic features of the left hip were those of advanced degenerative arthritis (Fig. 9–3C). She had no limb length inequality and no orthopedic disability other than that of the left hip. Her hip range of motion was as follows:

	Left Hip (in Degrees)	Right Hip (in Degrees)
Flexion	55	125
Extension	lacks 45	lacks 15
Abduction	0	20
Adduction	0	20
Inward rotation	−5	30
Outward rotation	15	20

The 25-year-old man in Case 4 had sustained a dislocated right hip from an automobile accident at the age of 18. Avascular necrosis with pain developed, and an intraarticular hip arthrodesis with a subtrochanteric osteotomy was performed 3 years following the accident. The arthrodesis became solid, but the osteotomy failed to unite; 9 months later, a bone graft with internal fixation was required to obtain union of the osteotomy. Examination at the time of gait testing revealed a solid asymptomatic fusion of the hip in a position of 45 degrees flexion, 0 degrees abduction-adduction, and 15 degrees outward rotation; and normal range of motion in all other joints of the lower extremities.

The 53-year-old man in Case 5 was treated for a painful right hip as a teenager, and did well until age 35 when his hip pain recurred and gradually became more severe. When the patient was 45 a hip arthrodesis was performed, and he was placed in a hip spica for approximately 12 months. At the time of gait testing he complained

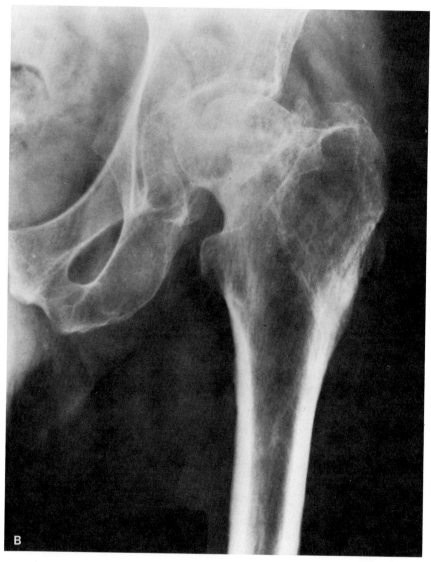

Fig. 9–3B. The femoral neck fracture has healed, but the superior-lateral segment of the femoral head was avascular and had collapsed, leaving a step-like offset in the articular surface. (*Figure continues on next page.*)

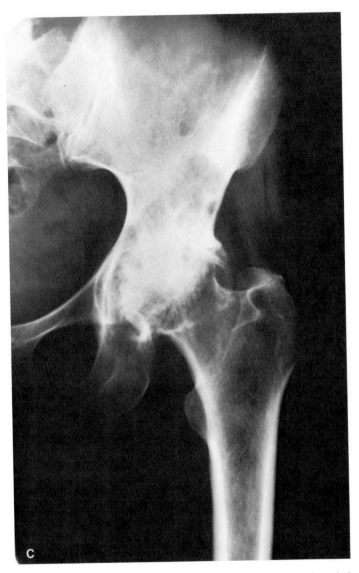

Fig. 9–3C. Almost complete absence of the joint space, subchondral sclerosis in the femoral head and acetabulum, and peripheral osteophyte formation.

of tiring with prolonged walking and of stiffness of the right knee. The patient had shortening of the right lower limb. His right hip was fused in a position of 40 degrees flexion, 15 degrees adduction, and 5 degrees outward rotation. Flexion of the right knee was limited to 45 degrees, but all other joints, other than this knee and the fused hip, had normal range of motion.

Discussion of Clinical Examples

The gait evaluation measurements in Cases 1 to 5 are compared with normal values in table 9–1. In Cases 1 and 2, the gait evaluations give no clues to the etiology of the hip problems. Case 1 had osteoarthritis and Case 2 had avascular necrosis. For both patients, however, pain was the main factor causing their difficulties in ambulation. Both patients displayed most of the gait abnormalities characteristic of the antalgic limp of patients with hip pain,[16] described in the five subsequent paragraphs.

Patients with hip pain walk slower than normal because of subnormal step lengths and subnormal cadences. Although the patient in Case 1 had fairly equal successive step lengths, the patient in Case 2 was more typical of patients with hip pain in that he took a longer step when the painful limb was directed forward.

In patients with hip pain, the stance phase on the painful limb is usually shorter than that on the sound limb. In addition, the swing phase on the sound limb is usually shorter than that on the painful limb, thus decreasing the period of single-limb support on the painful side. The bar graphs in Figure 9–4 illustrate the asymmetry in timing of the patients in Cases 1 and 2.

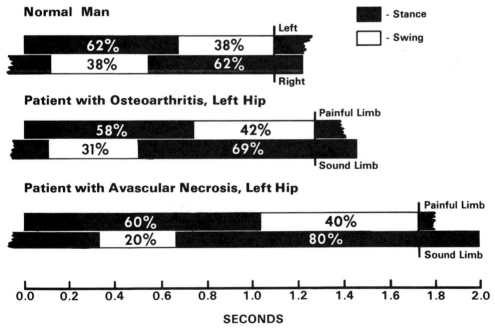

Fig. 9–4. Timing bars illustrate the durations of the stance and swing phases of both limbs of a normal man and in Cases 1 and 2, with unilateral hip pain. During free-speed walking both patients had longer-than-normal cycle durations and a subnormal period of single-limb support on the painful limb, as indicated by the short swing phase on the sound limb.

TABLE 9-1. GAIT COMPONENTS OF PATIENTS AND NORMAL INDIVIDUALS

Gait Component	Normal Men* Mean ± 1 SD	Case 1 Male, age 67, lt. hip pain	Case 2 Male, age 40, lt. hip pain
Fast Walking			
Velocity (m/min)	131 ± 15	50†	45†
Free-Speed Walking			
Velocity (m/min)	91 ± 12	50†	32†
Cycle duration (sec)	1.06 ± 0.09	1.28†	1.73†
Cadence (steps/min)	113 ± 9	94†	69†
Step length: Right to left (cm)	78 ± 6	52†	54†
Step length: Left to right (cm)	78 ± 6	54†	41†
Right stance (sec)	0.65 ± 0.07	0.87†	1.34†
Left stance (sec)	0.65 ± 0.07	0.74	1.03†
Right swing (sec)	0.41 ± 0.04	0.40	0.33
Left swing (sec)	0.41 ± 0.04	0.54†	0.70†
Lateral motion of the head (cm)	5.9 ± 1.7	12†	18†
Vertical motion of the head (cm)			
During right stance	4.8 ± 1.1	4.7	4.7
During left stance	4.8 ± 1.1	2.0†	1.5†
Velocity index (mean and range)	1.00 (0.97 – 1.04)	0.78†	0.53†
Hip flexion-extension used (deg)	48 ± 5	lt. 24† rt. 57	lt. 10† rt. 66†
Anterior-posterior pelvic tilting (deg)	7.1 ± 2.4	9.5	12.5
Transverse pelvic rotation (deg)	12 ± 4	10	8

Irregularities in vertical, lateral, and forward pathways of the head are characteristic of patients with hip pain. They have uneven successive vertical excursions of the head with subnormal vertical excursions during the stance phase on the painful limb. This irregularity is more easily observed if the patient also has structural or functional shortening of the disabled limb. Lateral lurching of the head and trunk toward the painful side during painful stance is common, as is irregular forward progression. The degree of irregularity in forward progression is measured by calculating a velocity index (Fig. 9–5). In normal subjects, this index is a ratio of the forward velocity during right single-limb support to the forward velocity during left single-limb support. For normal men, the velocity index averages 1.00, indicating uniform forward motion during the successive single-limb-support periods. In patients with hip pain, the index is a ratio of the forward velocity during sound single-limb support to the forward velocity during painful single-limb support. The low velocity indices in Cases 1 and 2 indicate faster forward motion during single-limb support on the painful side than on the sound side (Table 9–1).

In contrast to normal arm swing, the upper limbs (particularly the one on the painful side) are usually pulled up and back into excessive shoulder extension and elbow flexion. In addition, many patients with hip pain abduct their arm on the painful side during the stance phase on the painful limb.

Patients with hip pain use considerably less hip motion than normal on the painful side; this is observed even if the patient has no marked limitation in available hip

Normal Women**	Case 3 Female, age 59, limited lt. hip motion	Normal Men*	Case 4 Male, age 25, rt. hip fusion	Case 5 Male, age 53, rt. hip fusion
Mean ± 1 SD		Mean ± 1 SD		
113 ± 14	74†	131 ± 15	120	82†
74 ± 9	60	91 ± 12	93	58†
1.03 ± 0.08	1.07	1.06 ± 0.09	1.05	1.05
117 ± 9	112	113 ± 9	114	114
62 ± 5	56	78 ± 6	81	41†
62 ± 5	49†	78 ± 6	82	61†
0.64 ± 0.06	0.67	0.65 ± 0.07	0.60	0.60
0.64 ± 0.06	0.67	0.65 ± 0.07	0.60	0.70
0.39 ± 0.03	0.40	0.41 ± 0.04	0.45	0.45
0.39 ± 0.03	0.40	0.41 ± 0.03	0.45	0.35
4 ± 1.1	5.9	5.9 ± 1.7	6.7	4.4
4.1 ± 0.9	2.7	4.8 ± 1.1	4.3	3.9
4.1 ± 0.9	3.9	4.8 ± 1.1	5.7	4.7
1.00 (0.94–1.09)	0.95	1.00 (0.97–1.04)	0.95	0.68†
40 ± 4	lt. 10† rt. 45	48 ± 5	lt. 64† rt. 0	lt. 53 rt. 0
5.5 ± 1.3	13†	7.1 ± 2.4	17†	16†
10 ± 3	12	12 ± 4	21†	18

*See reference number 12.
**See reference number 15.
†Outside normal range of variability defined by two standard deviations above and below the normal mean.

flexion-extension, as was the case in Cases 1 and 2. Their patterns of hip flexion-extension during a walking cycle are plotted, with the range of normal variability, in Fig. 9–6. In addition, patients with hip pain may use greater than average anterior-posterior pelvic tilting, and also excessive flexion-extension of the sound hip. These are not considered pain-avoidance maneuvers *per se;* they compensate for limited use of motion of the painful hip, as will be discussed.

The primary disability in Case 3 was severely limited hip motion (not pain), and it is therefore not surprising that there were no abnormalities in the patient's gait indicative of pain-avoidance maneuvers (Table 9–1). She used excessive anterior-posterior pelvic tilting to augment the restricted flexion-extension of the disabled hip, and this maneuver is called a "compensatory mechanism."

The patients in Cases 4 and 5, who had hip fusions, were chosen for presentation because one walked extremely well and the other did not (Table 9–1); both also illustrate the use of other compensatory mechanisms typical of patients with limited or absent hip motion; they used excessive flexion-extension of the normal hip, excessive transverse rotation of the pelvis, and excessive anterior-posterior pelvic tilting in their walking. The patient in Case 4 appeared to have a remarkably normal gait in terms of its smoothness, rhythm, and speed. In contrast, the patient in Case 5 walked slowly because of an apparent inability to make steps of normal length, particularly

METHOD OF CALCULATING VELOCITY INDEX

NORMAL MAN

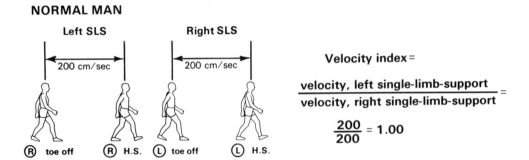

Velocity index =

$$\frac{\text{velocity, left single-limb-support}}{\text{velocity, right single-limb-support}} =$$

$$\frac{200}{200} = 1.00$$

PATIENT WITH RIGHT HIP PAIN

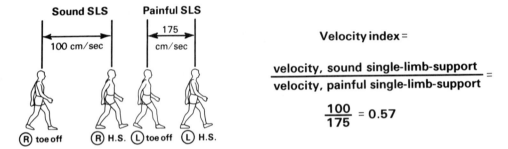

Velocity index =

$$\frac{\text{velocity, sound single-limb-support}}{\text{velocity, painful single-limb-support}} =$$

$$\frac{100}{175} = 0.57$$

Fig. 9–5. Schematic representation of the method used to calculate velocity index or uniformity in forward progression. SLS = duration of single-limb support.

when the sound limb was directed forward. In addition, the latter patient had irregularities in successive stance and swing times, and a marked stop-start limp, as indicated by his low velocity index.

The differences in quality of gait in Cases 4 and 5 probably related to the fact that the patient in Case 4 used compensatory maneuvers of greater amplitude than the patient in Case 5. (The last three gait measurements in the table are compensatory maneuvers.) Another compensatory maneuver of both men was continuous flexion of the knee on the side of fusion during the entire stance phase. This compensatory maneuver was also more pronounced in Case 4 (Fig. 9–7). The patterns of ankle motion for the two men were both abnormal when the limb on the side of the fusion was directed backward at the instant of toe-off, but the patient in Case 4 had excessive plantar flexion, which helps explain his relatively longer step length, while the patient in Case 5 had no plantar flexion (Fig. 9–7).

MECHANICAL CONSIDERATIONS OF GAIT DEVIATIONS

Two of the most important mechanical considerations in understanding the pathologic gait of patients with hip disorders are mechanical energy expenditure and hip joint loading.

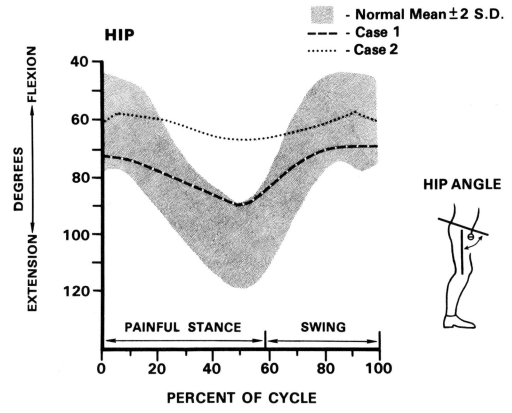

Fig. 9–6. Patterns of sagittal rotation of the hip during free-speed walking for the patients in Cases 1 and 2 with hip pain, as compared to the normal range of variability (shaded area), defined by two standard deviations above and below the mean for normal men.[14] Hip flexion-extension was measured as the anterior angle between the lateral pelvic target and a target parallel to the mechanical axis of the thigh.

Mechanical Energy Expenditure

As pointed out by Saunders, Inman and Eberhart,[20] the wheel provides the most economical means for translation of a body in a straight line with the least expenditure of mechanical energy. However, the most economical means of translation for bipedal locomotion is a smooth, rhythmic gait which results in gradual deflections of the center of gravity of the body through sinusoidal pathways of low amplitude. Our measurements of the pathways of the head of normal subjects during locomotion demonstrate remarkably smooth sinusoidal pathways of low amplitude, similar to the pathways of the center of gravity calculated for a normal man by Braune and Fischer.[2] Abrupt changes in speed or direction and excessive excursions of the head and trunk are frequently seen in patients with lower limb disability. These changes are particularly costly since they require abnormal accelerative and decelerative forces to produce and restrain the motion.

The increase in mechanical energy expenditure caused by excessive lateral motion is demonstrated by the following simplified example. Figure 9–8 shows the hypothetical lateral pathways of the center of gravity of a normal man and a patient with

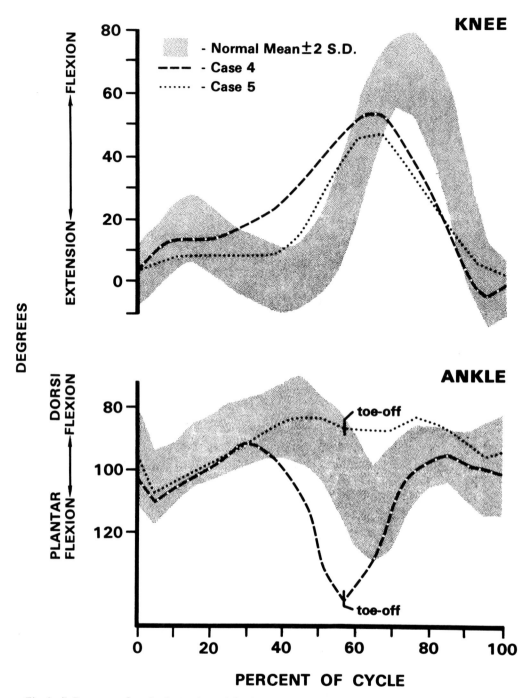

Fig. 9–7. Patterns of sagittal rotation of the knee and ankle on the side of fusion during free-speed walking in Cases 4 and 5, with unilateral surgical fusion of the hip, compared to the normal range of variability (shaded area) defined by two standard deviations above and below the mean for normal men.[12] Knee flexion-extension was measured as the angle between an inferior projection of the thigh target and a target parallel to the mechanical axis of the leg. Ankle dorsiflexion and plantarflexion were measured from the anterior angle between the leg target and the lower border of the heel of the shoe on the lateral side.

LATERAL PATHWAY OF THE CENTER OF GRAVITY

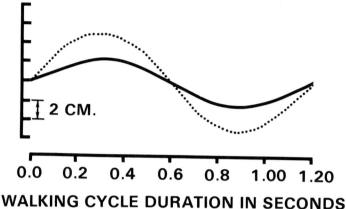

WALKING CYCLE DURATION IN SECONDS

Fig. 9–8. Hypothetical case showing lateral displacement of the center of gravity for a normal man and for a patient with a lateral lurch.

bilateral hip pain. In this example, the pathways of the center of gravity are assumed to be sine waves of the same frequency, but with different amplitudes, and the men are assumed to be walking at the same forward velocity. In a walking cycle, the center of gravity during the patient's gait moves a greater distance than it does during normal gait. It can be shown through manipulation of the equations for a sine wave that the lateral velocity of the center of gravity is proportional to the amplitude of its oscillation. Applying the formula for kinetic energy:

$$KE = 1/2mv^2$$

gives a rough idea of the increase in mechanical energy required because of the excessive lateral motion. The kinetic energy is proportional to the square of the velocity of the center of gravity in the lateral direction. Thus, the patient whose center of gravity oscillates laterally with approximately twice the normal amplitude will require about four times as much mechanical energy.

Vertical Hip Joint Loading

Walking is a dynamic act and, as such, a complete analysis of the loading on the hip joint would include the duration and amplitude of the forces acting in all three coordinate planes. For simplification, this analysis is limited to vertical hip joint loading under static conditions.

Vertical hip joint loading is affected by many factors, such as superimposed body weight and the position of the mass center of this weight. Additional compressive

FACTORS AFFECTING HIP JOINT LOAD

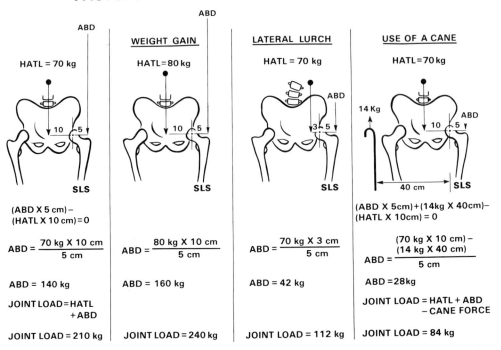

Fig. 9–9. Simplified diagrams illustrating some of the factors which affect vertical hip joint loading during single-limb support (SLS). The anatomic center of the hip joint is considered the fulcrum of the lever system. The calculations assume static conditions, in which counterclockwise moments about the hip must equal clockwise moments about the hip. The ABD vector indicates the vertical component of the hip abductor-muscle force. The resultant action line of the hip abductor muscles is estimated as 71 degrees from the horizontal[7, 21]; therefore the vertical component is proportional to the sine of 71 degrees. The HATL weight is the weight of the head, arms, trunk, and unsupported lower extremity.

forces will always be imposed by contraction of the muscles that function to produce or control hip motion, and by tension of fibrous tissue, particularly the anterior capsule when the hip joint is extended. Furthermore, forces that accelerate and decelerate the vertical motion of the center of gravity of the body during locomotion are transmitted through the hip joint, thereby affecting the hip joint load; in fact, the major muscle activity and periods of peak vertical hip joint loading during the stance phase coincide with periods when the vertical forces under the foot, as measured with force plates, are in excess of the body weight.

During walking the hips are at a mechanical disadvantage with respect to joint loading because they are positioned lateral to the vertical projection of the center of gravity of the body. During each single-limb-support phase of normal walking, the center of gravity of the head, arms, trunk, and unsupported lower limb (HATL weight) tends to cause the pelvis to rotate downward on the unsupported side, with the fulcrum of motion at the hip on the supporting side. This adduction of the supporting hip is normally controlled by a lengthening contraction of the hip abductor muscles on that side.

Some factors that affect the vertical hip joint load during single-limb support are shown in the simplified diagrams in Figure 9–9.[1, 3, 5, 7, 18] The diagrams assume conditions of equilibrium in which the sum of the clockwise and counterclockwise moments about the pivot of the supporting hip equals zero. The action line of the resultant force of the hip abductor muscles is estimated to be 71 degrees, and this force can be resolved into horizontal and vertical components.[7] The vertical component of the hip abductor-muscle force necessary to control hip adduction on the supporting side can be calculated if the following are known: the distance from the hip joint center to the point of attachment of the hip abductor muscles on the greater trochanter, and the moment caused by the HATL weight (the product of the HATL weight and the distance from the hip joint center to the vertical projection of the mass center of the HATL weight). Once the abductor-muscle force is known, the joint load may be calculated as the algebraic sum of the superimposed body weight plus the compressive forces of muscular contraction.

The hypothetical vertical hip joint load of a normal man during single-limb support is demonstrated on the left in Figure 9–9. His body weight is 82.4 kg, and his HATL weight is 70 kg, since a lower limb weighs approximately 15 percent of the body weight. The vertical component of the hip abductor-muscle force is 140 kg, and the total joint load is 210 kg, or approximately 2.5 times the body weight. This ratio underestimates the joint load because the activity of many other muscles, which span the hip joint and contract early and late in the stance phase, are not considered. Nonetheless, this ratio is similar to that measured in patients with transducers implanted in Moore hip prostheses.[19]

The deleterious effect of weight gain on a patient with hip pain is well recognized by orthopedic surgeons. If, under the previously described static conditions, the patient's HATL weight increased by 10 kg, his hip joint load would increase by 30 kg (Fig. 9–9).

Lateral lurching decreases the hip joint load (Fig. 9–9). If a patient with a HATL weight of 70 kg lurches laterally toward the painful side during single-limb support, so that the lever arm of the HATL weight decreases to 3 cm, the amount of the vertical component of the hip abductor-muscle force required by the patient to control the descent of the pelvis on the unsupported side would decrease from the original 140 to 42 kg, and the joint load would accordingly decrease from 210 to 112 kg.

Another means of reducing the hip joint load is by applying force to a cane in the hand opposite the painful hip (contralateral hand) during the painful stance phase.[1] In the final example in Figure 9–9, the patient applies 14 kg of force to a cane positioned 40 cm from the painful hip. A study using force-recording canes indicated that 14 kg is not an unusual amount of force to be applied to a cane by a patient with a hip joint disability.[13] It can be seen that the counterclockwise moment of the HATL weight is counteracted by the clockwise moment of the cane force plus the clockwise moment of the hip abductor-muscle force. Under these conditions, 28 kg of hip abductor-muscle force is generated and the joint load is decreased to 84 kg.

This theoretical conclusion is supported by some preliminary electromyographic testing of patients with unilateral hip pain. We have found that the electromyographic activity of the hip abductor muscles is decreased, but not absent, during walking with a cane in the contralateral hand.

A nomogram for a simplified method of determining the vertical component of the hip abductor-muscle force is given in Figure 9–10.

In addition to reducing the hip joint load, the use of a cane in the contralateral hand increases upright stability during painful single-limb support by increasing the size of the base of support (the area between the supportive foot and the application

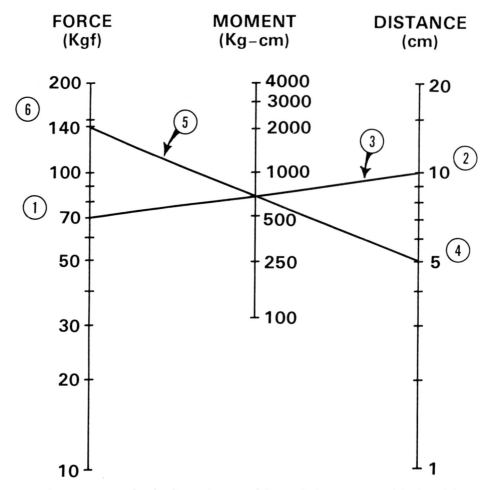

Fig. 9–10. A nomogram for the determination of the vertical component of the hip abductor-muscle force for the first example diagrammed in Figure 9–9. The following procedure is indicated on the nomogram by the circled numbers which correspond to the respective steps in the procedure.

1. Locate the value of the HATL weight on the force scale.
2. Locate the value of the perpendicular distance between the line of the HATL weight and the hip joint center on the distance scale.
3. Connect these two points with a straight line.
4. Locate the value of the perpendicular distance between the vertical component of the ABD force and the hip joint center on the distance scale.
5. Draw a line between this point and the point on the moment scale intersected by the first line and extend it to the force scale.
6. Read the value of the vertical component of the hip abductor-muscle force from the force scale where the second line intersects it.

of the cane to the floor). In contrast, the use of a cane in the ipsilateral hand during painful single-limb support substantially decreases the size of the base of support and shifts the base laterally. This lateral shift of the base of support is usually accompanied by an increase in lateral lurch in an attempt to get the line of gravity within or at least close to the base of support. The lateral lurching measured for the patient in Case 2, with hip pain, is illustrated in Figure 9–11. His lateral head motion was with-

LATERAL LURCH LIMP OF PATIENT WITH LEFT HIP PAIN

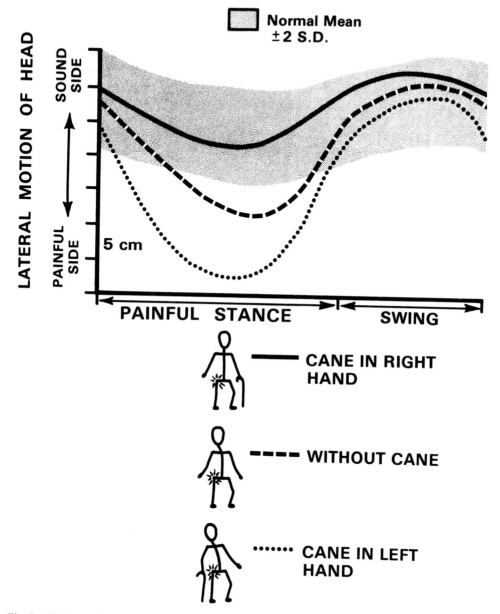

Fig. 9–11. Lateral pathways of the head of the patient in Case 2, who had left hip pain, compared to the normal range of variability, indicated by the shaded area.[12] The patient had a lurch toward the painful side during painful stance when he walked without a cane, and the lurch was accentuated when he used the cane in the left hand. His head motion was within normal limits when the cane was used in the hand opposite the pain.

in normal limits when he used a cane in the contralateral hand, but when he used the cane on the painful side, his lateral lurch was worse than when no cane was used.

Antalgic or Pain-Avoidance Maneuvers of Patients with Hip Pain

Since pain cannot be measured, it is important for the physician to recognize antalgic maneuvers, because they are objective indicators of the extent to which a patient's pain interferes with function. Antalgic maneuvers in patients with hip arthropathy are primarily attempts to decrease the effect of gravitational and muscular forces and moments acting at the hip joint. Accordingly, patients reduce hip pain during walking by decreasing both the amplitude of the joint load and the duration of heavy loading of the joint.

Calculations from mathematical models indicate that during walking the periods of peak loading of the hip joint occur early and late in the stance phase, with lesser but substantial loading during mid-stance.[17, 21] The joint load during the swing phase is much lower and is primarily the result of contraction of muscles spanning the hip joint,[19] which function to overcome the distraction force of the unsupported limb and to accelerate and decelerate the swing of the limb.

During the stance phase, three maneuvers are used by patients with hip pain to reduce the amplitude of the joint load. Early in the stance phase, when the hip is flexed, the gluteus maximus, hamstrings, and other hip extensor muscles normally contract to extend the hip and advance the trunk to a relatively erect position over the forward-placed foot. In patients with hip pain, this erect posture usually is not fully achieved during painful stance, resulting in uneven, successive vertical excursions of the trunk, with a lower vertical excursion during the painful as compared with the sound stance phase. This may be an attempt to reduce the joint load by reducing the contractile forces of the hip extensor muscles. During mid-stance on the painful limb, patients with hip pain typically reduce the amplitude of the joint load by lurching laterally toward the painful side. In some patients, this lateral shift of the center of gravity of the suprafemoral mass is accomplished by abducting their upper limb on the painful side during painful stance. Third, in contrast to normal subjects — who use almost full hip extension to direct the limb backward during the late stance phase — patients with hip pain reduce the amplitude of the hip joint load by limiting their hip extension. The period of high joint load that normally occurs during the late stance phase corresponds to a phase of contractile force of the muscles that initiate hip flexion preparatory to the swing phase. By avoiding hip extension, the patient does not require as much accelerative muscle force to initiate hip flexion as a normal subject. In addition, he avoids the joint compression that would result from tension of the anterior capsule. This restricted extension of the painful hip usually results in unequal successive step lengths, with the shorter step occurring when the painful limb is directed backward.

Other antalgic maneuvers function to reduce the duration of heavy hip joint loading. The characteristically shorter swing phase of the sound limb as compared with the painful limb is an obvious means of reducing the single-limb-support loading time on the painful hip. In addition, patients with painful hips consistently move forward faster during single-limb support on their painful limb than on their sound limb, representing an attempt to get body weight off the painful limb as quickly as possible. This results in a "stop-start" type of limp which is readily observable. The velocity index, which quantitates this type of limp, is a particularly sensitive indicator of the degree of patient disability.

The antalgic maneuvers typical of patients with hip pain have the beneficial effect of reducing the hip joint load, but the resultant irregularities in the forward, vertical, and lateral trajectories of the suprafemoral mass have the detrimental effect of increasing mechanical energy expenditure.

Compensatory Maneuvers of Patients with Hip Fusion

Loss of hip joint motion severely impairs the ability of a patient to make a step in a normal manner during walking. In patients with hip joint fusion and those with severely limited hip motion, compensatory maneuvers are those abnormal or excessive movements that attempt to supply the equivalent of the lost hip motion. As demonstrated in Cases 4 and 5, the principal means of compensation for total absence of hip joint motion during walking are excessive anterior-posterior tilting of the pelvis, excessive transverse rotation as measured at the pelvis, excessive flexion-extension of the sound hip, and continuous flexion of the knee on the fused side throughout stance on the fused side.[6] At the end of stance, some patients also use excessive plantarflexion of the ankle on the side of the fusion. Reference to Figure 9–12 will help visualize the use of these compensatory maneuvers, which are explained in detail in the following paragraphs.

The abnormally large excursion of anterior-posterior pelvic tilting produced by lumbar spine motion is perhaps the most striking compensatory mechanism used to lengthen the step in patients with hip fusion. The average total excursion of anterior-posterior pelvic tilting used by patients with hip fusion is 15 degrees during free-speed walking and 18 degrees during fast walking—about twice the normal average. During the stance phase, the pronounced excursion of anterior tilting occurs when the hip normally extends to direct the limb backward. During the swing phase, posterior tilting of the pelvis toward a more level position partially compensates for inability to flex the hip.

In patients with hip fusion, the excessive transverse rotation measured at the pelvis also serves as a step-lengthening mechanism. The width of the pelvis may be considered a radius that pivots around the longitudinal axes of the joints of the supportive limb and describes an arc as it moves forward on the side of the forward-swinging lower limb. The amount of transverse rotation normally occurring at the hip and knee during walking has been reported as being between 5 and 10 degrees.[4, 8-10] Obviously, in patients with hip fusion, no rotation can occur at the fused hip. Since the foot rarely pivots on the floor, the transverse rotation measured at the pelvis during stance on the fused side must be the result of rotation occurring at joints distal to the hip.

In order for patients with hip fusion to substantially advance the body forward during stance on the fused side, the knee on the fused side must continue to flex, thereby compensating for the inability of the patient to extend the fused hip (Fig. 9–12, left). The fused-to-sound limb step length would be even shorter without this knee flexion, and the ability of the limb on the fused side to propel the body forward would probably be decreased.

The excessive flexion-extension excursion of the sound hip also tends to lengthen the steps. Since the forward swing of the limb on the fused side demands that the

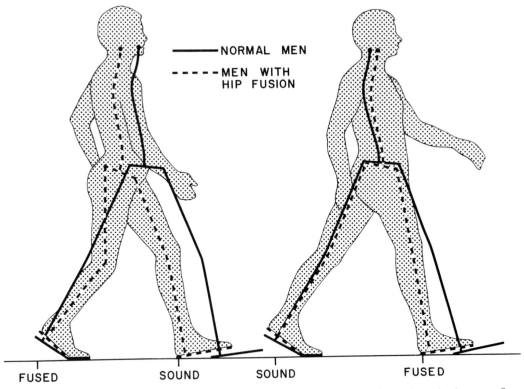

Fig. 9–12. Stick figures showing the positions of the lower limbs and trunk at the instant of heel-strike in patients with unilateral hip fusion compared with normal men. Angular positions of the lower limbs have been drawn to accentuate the differences between the patients and the normal men.

pelvis tilt posteriorly, toward a relatively level position, the patient, in order to make a substantial step, must be able to extend the sound hip considerably when the fused limb is directed forward. Therefore, a flexion contracture of the contralateral hip is particularly undesirable in the patient with a hip arthrodesis.

When the patient with a hip fusion maximizes the use of compensatory mechanisms, the prognosis for achievement of the speed and smoothness of a normal gait is enhanced. The ability to use compensatory mechanisms has a tendency to relate to some physical characteristics of patients. Age is a significant factor in the walking of patients with hip fusion because, among other factors, younger patients are usually more supple. Disability or limited motion of those joints that normally attempt to compensate for the loss of hip motion (lumbar spine, sound hip, and knee and ankle on the side of the fusion) will also limit the prognosis for a satisfactory gait. Other factors that relate to the walking ability of patients with hip fusion are the position of fusion and the height of the lift used to correct limb length inequality.[6]

Several of these factors help explain the differences in gait of the patients in Cases 4 and 5. The patient in Case 4 was 25 years of age, had a supple lumbar spine, and made excellent use of compensatory mechanisms. He achieved a gait of normal velocity, with equal successive step lengths, equal successive stance and swing times, and a remarkably uniform forward progression. The patient in Case 5 was 53 years

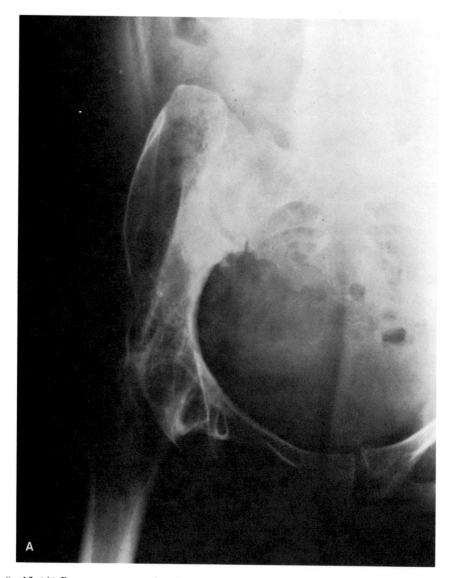

Fig. 9–13. (A) Roentgenogram showing secondary changes in the sacroiliac joint in a woman 15 years after successful hip fusion. *(Figure continues on next page.)*

of age, had shortening of the limb on the side of fusion, and limited flexion of the knee on the side of fusion. His compensatory motions were of lesser amplitude than those of the patient in Case 4, and his gait was slow with many irregularities.

The use of compensatory maneuvers appears beneficial in reducing mechanical energy expenditure requirements and providing a relatively faster gait. The stresses on the compensating joints, however, have not been identified, and may have a deleterious effect over a prolonged period of time. For example, a female who had a successful hip fusion at age 13 because of complications following a slipped capital femoral epiphysis, developed secondary changes in the sacroiliac joint on the fused side by age 28 (Fig. 9–13A). On the other hand, attempting to use compensatory

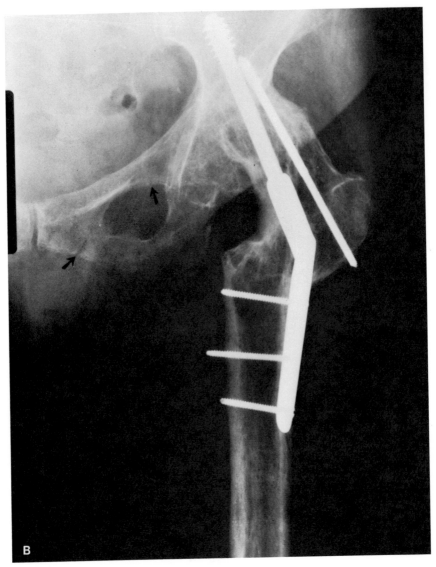

Fig. 9–13B. Roentgenogram showing stress fractures of the superior and inferior pubic rami of another woman 2 years after successful hip fusion.

motions, and not being able to, may also have a deleterious effect. For example, 2 years after a solid hip fusion, a 64-year-old female developed stress fractures of the superior and inferior pubic rami on the fusion side (Fig. 9–13B). It is possible that this woman's age and mild disability of the contralateral hip limited her ability to use compensatory mechanisms, and she may have developed the stress fractures secondary to this restriction.

Comparison of the Gaits of Patients with Hip Pain and Hip Fusion

Thus far we have discussed the gait of patients with hip pain, principally in terms of antalgic maneuvers, and the gait of patients with hip fusion in terms of compensatory maneuvers. It is more realistic, however, to emphasize that both types of deviant gait maneuvers are observed, although in different degrees, in patients with hip pain and those with hip fusion.[6, 16]

That compensatory mechanisms are used by patients with hip pain and by those with hip fusion should not be surprising, since most patients with hip pain also have limited hip motion or are reluctant to use the motion they have. As expected, however, compensatory maneuvers are of greater amplitude in patients with hip fusion than in those with hip pain. The four compensatory maneuvers described were: (a) excessive anterior-posterior pelvic tilting; (b) excessive transverse rotation of the pelvis; (c) excessive flexion-extension of the sound hip; and (d) continuous flexion of the knee on the side of fusion during the stance phase.

The maneuvers characteristic of patients with hip pain have been described in terms of their pain-avoidance function. These antalgic maneuvers include slow velocity, abnormal temporal components of gait, inequality in successive step lengths, stop-start lurching, lateral lurching, unevenness in successive vertical excursions of the head, and restricted use of hip joint motion. It must be recognized that these maneuvers are not exclusive to patients with hip pain; in fact, in patients with other disabilities, these same maneuvers may serve other functions or be indicative of non-painful conditions such as weakness, instability, or limited motion. In patients with hip fusion, few of these gait irregularities might be observed if the patient could use sufficient compensatory motions. In turn, the ability to compensate relates to physical characteristics such as age, position of fusion, height of shoe lift required, and the condition of other joints.

In patients with hip joint pain, antalgic maneuvers predominate over compensatory maneuvers. Minimizing hip joint loading appears to be of primary importance, even at the expense of increasing mechanical energy expenditure. On the other hand, in patients with pain-free hips but restricted motion, the ultimate goal of the compensatory maneuvers is to take steps long enough to permit the patient to achieve functional walking speeds without compromising energy expenditure. Thus, patients with hip fusion generally achieve a faster, smoother gait and would be expected to have greater walking tolerance or endurance than patients with hip pain.

ACKNOWLEDGMENTS

The authors deeply appreciate the assistance of Susan B. Sepic, P.T. and Gena M. Gardner, P.T. in the preparation of this chapter. This work was supported by the Rehabilitative Engineering Research and Development Service of the Veterans Administration and by USPHS Grant No. 13854 from the National Institute of Arthritis, Metabolism and Digestive Diseases.

REFERENCES

1. Blount, W. P.: Don't throw away the cane. J. Bone Joint Surg., *38A:* 695, 1956.
2. Braune, C. W., and Fischer, O.: Der gang des menschen. I teil. Versuche unbelasten und belasten menschen. Abhandl Math Phys C1 Sächs Gesellsch Wissensch *21:* 153, 1895.
3. Denham, R. A.: Hip mechanics. J. Bone Joint Surg., *41B:* 550, 1959.
4. Eberhart, H. D., Inman, V. T., Saunders, J. B. DeC.M., Levens, A. S., Bresler, B., and McCowan, T. D.: Fundamental studies of human locomotion and other information re-

lating to the design of artificial limbs. A report to the National Research Council Committee on Artificial Limbs, University of California, Berkeley, 1947.

5. Frankel, V. H., and Burstein, A. H.: Orthopaedic Biomechanics. Lea & Febiger, Philadelphia, 1970.

6. Gore, D. R., Murray, M. P., Sepic, S. B., and Gardner, G. M.: Walking patterns of men with unilateral surgical hip fusion. J. Bone Joint Surg., *57A:* 759, 1975.

7. Inman, V. T.: Functional aspects of the abductor muscles of the hip. J. Bone Joint Surg., *29:* 607, 1947.

8. Johnston, R. C., and Smidt, G. L.: Measurement of hip joint motion during walking. J. Bone Joint Surg., 51A: 1083, 1969.

9. Kettlekamp, D. B., Johnson, R. J., Smidt, G. L., Chao, E. Y. S., and Walker, M.: An electrogoniometric study of knee motion in normal gait. J. Bone Joint Surg., *52A:* 775, 1970.

10. Mann, R. A., and Hagy, J. L.: The popliteus muscle. J. Bone Joint Surg., *59A:* 924, 1977.

11. Murray, M. P., Drought, A. B., and Kory, R. C.: Walking patterns of normal men. J. Bone Joint Surg., *46A:* 335, 1964.

12. Murray, M. P., Kory, R. C., Clarkson, B. H., and Sepic, S. B.: Comparison of free and fast speed walking patterns of normal men. Am. J. Phys. Med., *45:* 8, 1966.

13. Murray, M. P., Brewer, B. J., and Zuege, R. C.: Kinesiologic measurements of functional performance before and after McKee-Farrar total hip replacement. J. Bone Joint Surg., *54A:* 237, 1972.

14. Murray, M. P., Kory, R. C., and Clarkson, B. H.: Walking patterns in healthy old men. J. Gerontol., *24:* 169, 1969.

15. Murray, M. P., Kory, R. C., and Sepic, S. B.: Walking patterns of normal women. Arch. Phys. Med. Rehab., *51:* 637, 1970.

16. Murray, M. P., Gore, D. R., and Clarkson, B. H.: Walking patterns of patients with unieral hip pain due to osteoarthritis and avascular necrosis. J. Bone Joint Surg., *53A:*259, 1971.

17. Paul, J. P.: Approaches to design: Force actions transmitted by joints in the human body. Proc. R. Soc. Lond. (Biol.), *192:* 163, 1976.

18. Pauwels, F. (1935) Der Schenkelsbruch ein mechanisches Problem. Ferdinand Enke Verlag, Stuttgart, 1935.

19. Rydell, N. W.: Forces Acting on the Femoral Head-Prosthesis. Tryckeri AB Litotyp, Göteborg, 1966.

20. Saunders, J. B. DeC.M., Inman, V. T., and Eberhart, H. D.: The major determinants in normal and pathological gait. J. Bone Joint Surg. *35A:* 543, 1953.

21. Seireg, A., and Arvikar, R. J.: The prediction of muscular load sharing and joint forces in the lower extremities during walking. J. Biomechanics, *8:* 89, 1975.

10
Compression Plating of Fracture of the Femoral Midshaft

H.-D. Sauer, M.D.; K. H. Jungbluth, Ph.D., M.D.;
J. H. Dumbleton, Ph.D.

FRACTURES OF THE FEMORAL MIDSHAFT

Anatomy of the Femur

The shaft of the femur is a long, hollow cylinder of compact bone partly filled with cancellous trabeculae and roughly cylindrical in cross-section throughout most of its extent. At either extremity the cortices are thinner and the femur is enlarged and nearly quadrilateral in cross-section. The linea aspera, with its medial and lateral labia, is an elevated ridge extending along the posterior border of the femoral shaft. It serves for the attachment of muscles and in the neighborhood of such attachments provides a blood supply to the outer one-third of the cortex. In contrast, the remainder of the cortex is vascularized internally by a centrifugal blood supply from the medullary cavity.[5] The linea aspera continues proximally with the intertrochanteric line and divides distally to become continuous with the supracondylar ridges. During evolution, the femoral shaft developed a curved and twisted shape. The shaft is not straight but is bowed forward and outward, as well as twisted inward.

In understanding the biomechanics of the thigh, the demonstration of the morphology of the femoral shaft remains incomplete without mentioning the fascia of the thigh as well as the thigh muscles. The thigh muscles are encased in the fibrous fascia lata, which is continuous with the deep fascia of the leg. On the lateral surface of the thigh the fascia lata is thickened, and this thickened portion—which is of the utmost biomechanical importance—is called the iliotibial band. The tension of this band is regulated by the tensor fascia femuris according to the loading on the thigh. From the deep surface of the fascia lata, bands of fibrous tissue reach the femur and separate the muscles of the thigh by forming compartments which contain the anterior extensors and the posterior flexors and adductors.

Because of the complicated shape of the femur, the numerous muscle attachments, and the role of the soft tissues, the analysis of stresses in the femur is difficult. Measurements and calculations, however, have indicated that the loads applied to the femur during ambulation are large, and that stresses in the femur can be high.[6, 13, 16]

Midshaft Fractures of the Femoral Shaft

Most fractures at the midshaft of the femur are due to direct impact injuries. Traffic accidents, especially those involving motorcyclists, produce many femoral fractures.

Because the femur is buried deep in the thigh and is surrounded by vascular muscle, ischemia of the bone ends is rather rare; thus, delayed union, if found, is more likely to be caused by inadequate fixation. Mal-union, however, is relatively common, either because of angulation or the overriding (overlapping) of the fragments. Shortening can occur but is usually of less than 2 cm, so that a limp does not develop.[10]

Treatment of Femoral Midshaft Fractures

There are several methods for treating midshaft femoral fractures, including closed reduction with immobilization, medullary rod nailing, and fixation using a plate and screws.[2]

The choice of treatment for midshaft fractures requires experience and judgment, but it is fair to say that centers carrying out treatment will, by tradition, experience, or local requirements tend to favor one type of treatment over the others. This is not to say that a particular treatment regime will be used inappropriately, but rather that a group using a particular method of treatment will tend to use that method for a wider range of indications than will a group committed to a different technique. It may be pointed out that in many instances open reduction is now the conservative method of treatment.

Reduction may often be achieved by manipulation and traction. Failure to reduce the fracture by manipulation is often due to interposition of soft tissue between the bone ends, and is an indication for open reduction. Immobilization is achieved by traction, and either fixed or sliding traction is used. The average time to union is 12 to 16 weeks for an adult. Some form of support device should then be employed to protect the fracture during walking. During the healing period attention is directed toward regaining knee flexion. As mentioned above, delayed union is rare; it is treated through internal fixation and onlay grafting.

The interposition of tissue between the fragments of a fracture may cause failure of the closed reduction. Other causes of failure include displacement of fragments by reflex contracture of muscles, constant motion from inability to fix the fragments firmly, and the application of mechanical forces insufficient to obtain reduction. Open methods have the advantage of permitting the fragments to be reduced under direct vision, so as to give good alignment, and permitting use of the fixation device to prevent displacement.

Open reduction changes a closed fracture to an open fracture in a favorable environment. Tissues are stripped from bone, and there may be produced scar tissue which interferes with muscle action. Problems may be encountered with the device itself. However, in a good environment and with experience, internal fixation of midshaft femoral fractures has a high success rate. Where there is a choice between closed or open reduction, the decision may be made in favor of open reduction because of the length of time needed to promote healing by reduction and traction.

Once it has been decided to carry out an open reduction, there are two alternatives: medullary nailing or the use of a plate and screws.[12] For the midshaft of the

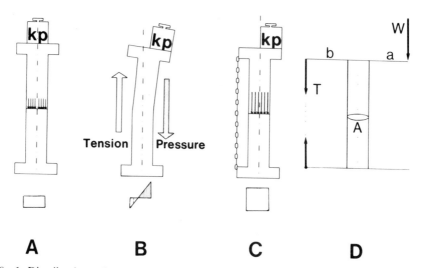

Fig. 10–1. Distribution of pressure and tension of an axial loaded column (A), non axially loaded column (B), and demonstration of the tension-band principle (C) according to Pauwels (D), schematic calculation of the stress.

femur, the treatment of choice is medullary fixation, since the fracture may be firmly fixed, permitting the mobilization of soft structures and adjacent joints to be carried out early and continued during healing. If the fix is stable, there may be no need for external support.

Medullary nailing is not suitable, however, in the case of long, oblique, or comminuted fractures, and for those fractures located at and distal to the junction of the middle and distal thirds of the femoral shaft. The placement of a plate for best advantage follows from a consideration of the loading at the femur. Figure 10–1 is a schematic view of the femur under a vertical load. The eccentricity of the loading tends to open a fracture on one side and to close it on the other. The plate must be placed on the tension side, which for the femur is the lateral side. If the plate were placed medially, on the compressive side, there would be little resistance to the tendency of the fracture to open.

Compression Plating of Femoral Midshaft Fractures

Plates come in many shapes and sizes. The evolution of plate configurations has taken place over many years, reflecting information gained by clinical usage and developments in engineering and material science. It was only recently, however, that it was realized that the application of compression would lead to more stable coaption of the bones, and healing by primary intention. The stability and consequent rigidity of the construct allows very little movement and little callus formation.

Animal studies with a transverse osteotomy model first showed that primary healing under compression was possible.[12] Other animal work, again using transverse osteotomy, indicated that a compression of over 100 kg diminished by only 50 percent over a 2 month period.[12] The use of compression applied to a fracture as part of the plating process was first advocated and developed in a systematic manner by the Swiss AO (Arbeitsgemeinschaft für Osteosynthese Fragen) group, but most orthopedic manufacturers now offer some form of compression plating system.

The use of a well-designed compression system does not obviate the need for surgical experience and judgment. Misuse of the method produces more poor results at the femoral shaft than at any other location. Mechanical errors include the use of plates that are too weak or too short, insertion of screws through only one cortex, and application of the plate with the fragments slightly distracted.

Apart from the complications mentioned above, difficulties may result from insufficient external support. Anderson[2] advocates suspension in a balanced splint for 12 to 14 days when a single plate is used for a femoral fracture, followed by a 1½ spica cast. The cast is discarded when bridging callus has developed, usually at 8 to 10 weeks. Walking with partial weight-bearing begins with crutches once the extremity can be controlled, and is gradually increased until the fracture has healed (4 to 5 months). When the fracture is fixed by two plates at 90 degrees to each other, as may be necessary in patients who will not tolerate a spica cast, no external support is required. In the final analysis, the degree of external support depends on the stability of the fixation and the degree of weight bearing. It must be remembered that no fixation device is strong enough to withstand, unaided, the full loading imposed during ambulation. Inadequate reduction of the fracture, so that the bone does not share load-bearing; lack of external support; or too vigorous activity can lead to fracture of the fixation device.

The very rigidity of the fixation that plays such an advantageous role in the early healing process can be a disadvantage in the long term because the plate takes too high a share of the load and bone remodelling takes place in response to the diminished bone loading. This is the so-called "stress protection" phenomenon. There are numerous reports of this effect,[1, 11, 19, 21] and it is recommended that the plate and screws be removed after 18 to 24 months.[12] Another reason for plate removal is the possibility of corrosion—either directly or because of fretting between the plate and underside of the screw heads; such an effect is more likely when the implant is made of stainless steel rather than other materials.

Stress protection is manifested by an increase in the diameter of the medullary cavity. It may be shown that the strength of the bone/fixation device combination is lower than that of the intact femur. Changes in the bone are most marked when two plates are used; this reflects the greater rigidity of the two-plate system (not to mention the increased damage caused to the bone in applying the second plate).

Since the bone has been weakened by remodelling, and the holes left by the removal of screws are particular sites of stress concentration, care should be taken to limit weight-bearing following plate removal. In the two-plate case, the removal should be done at two operations, 4 to 6 months apart, with cancellous bone grafts recommended at each operation.[12] Failure to take precautions can lead to refracture of the bone, usually through the sites of stress concentration.

Another complication involves fracture of the plate. Such fracture most commonly occurs when the plate is used for a femoral fracture. According to Steinemann[20] the overall incidence of failures is below 2 percent. For plate fixation of the tibia or femur the average time to implant failure is 4½ months. During this time it is estimated that the implant is subjected to some 100,000 load cycles. Thus plate fracture is due to repetitive loading of the device, and it is thus a fatigue failure. In almost all cases the cause of failure is a medial cortical defect which results in plate being subjected to excessive bending; the fracture if thus ultimately traced back to operative technique.

The complications, discussed above in some detail, were chosen not only because of their frequency of occurrence, but also because of their biomechanical origin. Other general complications, such as infection, do occur but have not been specifically discussed because infection, for example, can be a complication with any surgical procedure.

In summary, internal fixation of a femoral midshaft fracture has the advantage of permitting soft tissues to be exercised and joints mobilized at a far earlier stage than with techniques involving closed reduction and traction. The results, in the hands of a skilled surgeon in a good environment, are excellent. The cases hereafter presented have been chosen to illustrate the biomechanical features so far presented, and will illustrate the application of compression plating to femoral midshaft fractures.

CASE REPORTS

<u>Case 1.</u> An 8-year-old girl was heavily injured in a traffic accident. She suffered from very serious multiple injuries, with head injury. Besides a subcapital fracture of the humerus on the right side, both femoral shafts showed transverse fractures, and the skin of the thigh showed heavy contusions. The femoral fractures were immobilized by skeletal traction with Kirschner wires through the supracondylar region above the epiphysis. Owing to the severe head injury and constant cerebral convulsions during a period of nearly 3 months of unconsciousness, immobilization by skeletal traction failed absolutely. As soon as the contusions had healed it was decided to perform open reduction and internal fixation with a compression plate osteosynthesis. Because of the symmetric location of the injury, later differences in growth rate were not such a problem as usually would occur in operative management of fractures in childhood (Fig. 10–2A to G).

Compression plating was preferred to intramedullary nailing because of the possibility of damage to the proximal or distal epiphysis – or both – as well as to endosteal circulation by reaming of the medullary cavity. Complete healing, without any complications, was gained on both sides with compression plating. The fragments were reduced anatomically and compressed by means of adjusted pretension and angulation of the plate. Although the constant convulsions produced quite a considerable stress for the osteosynthesis, stability was absolute, as indicated by the absence of callus. Removal of the plates occurred after half a year. The x-ray demonstrates already slight signs of "stress protection" of the cortices, a common cause for refracture.

<u>Case 2.</u> The signs of "stress protection" and its consequences are demonstrated more distinctly in this case. At the age of 13, a boy suffered from short oblique fractures of both femoral midshafts. These fractures were treated with Lane plates (Fig. 10–3A to D). Ten years later the cortices of both femurs were seriously affected with a cancellous-like appearance. In consequence, a new trauma led to fracture directly above the plate on the left side. The Lane plates were removed and – because transverse and short oblique fractures of the midshaft are those of choice for nailing – the left femur was treated with Küntscher nailing.

The importance of an absolute integrity of the medial support for the stability of osteosynthesis has already been mentioned. This case demonstrates the failure of osteosynthesis owing to lack of medial support. It also demonstrates that although medial support was gained at the end of the operation, it could be lost due to necrobiosis of the fragments during the postoperative phase. In conclusion, it is absolutely

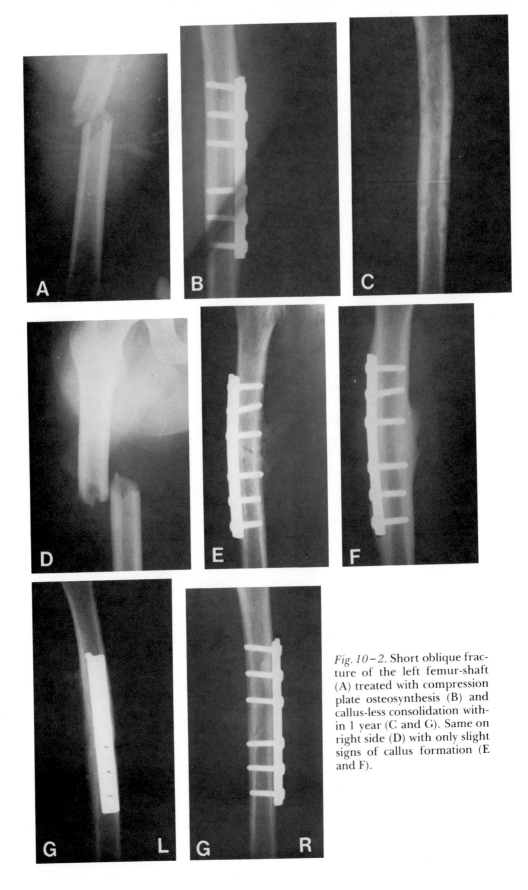

Fig. 10–2. Short oblique fracture of the left femur-shaft (A) treated with compression plate osteosynthesis (B) and callus-less consolidation within 1 year (C and G). Same on right side (D) with only slight signs of callus formation (E and F).

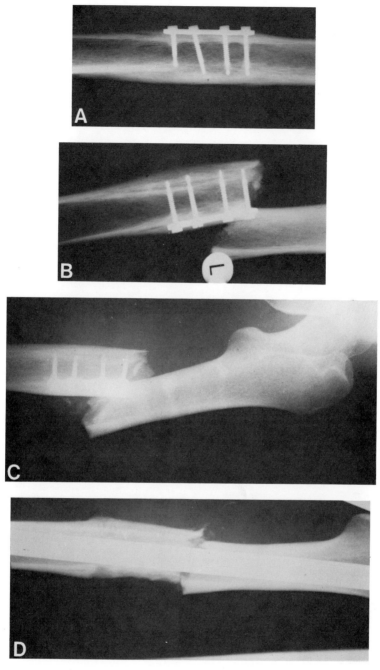

Fig. 10–3. Cancellous transformation of the cortices below and opposite Lane-plate osteo-syntheses of both femur-shafts due to "stress-protection" (A and B). Subsequently new trau-ma led to fracture above the plate on the left side (C). Treatment with Kuntscher nailing (D).

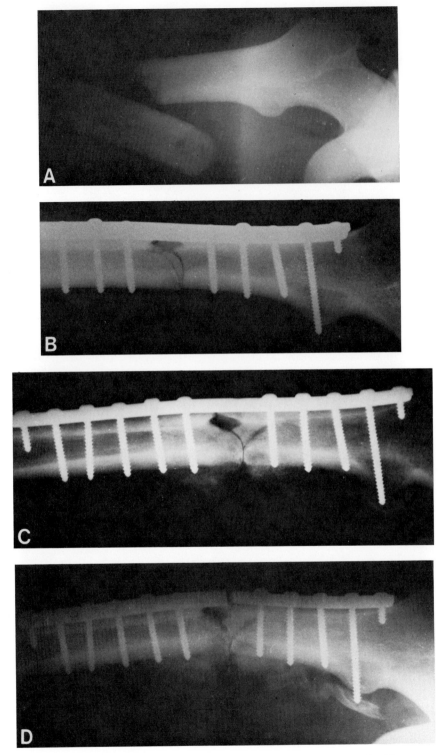

Fig. 10-4. Short oblique midshaft fracture of the right femur (A) treated with compression plate osteosynthesis (B). Necrobiosis of a small fragment led to loss of medial support (C). To-and-fro angulation of the plate due to absence of medial support results in metal fatigue (D).

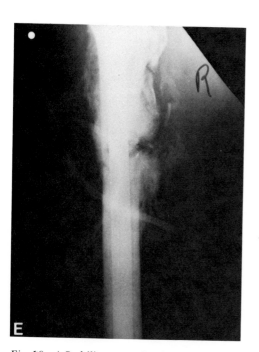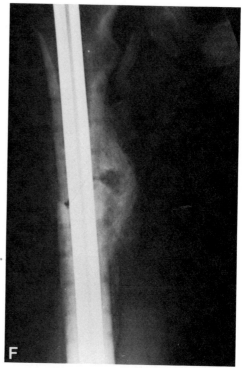

Fig. 10–4. Stability was gained by re-osteosynthesis by means of Kuntscher nailing (E and F).

necessary to perform autologous cancellous bone grafting in all cases in which necrosis of fragments is to be suspected because of serious trauma to soft tissues or open fractures, as well as by extensive separation of fragments from soft tissues.

Case 3. This young man, aged 21, was injured in a traffic accident. He suffered from a short, oblique closed fracture of the proximal shaft of the femur, with some small fragments. Although this fracture was quite suitable for nailing, it was managed by compression plate osteosynthesis *without* cancellous bone transplantation. Two months into the postoperative phase, the fragment that had been part of the medial support was necrotic. This little defect, hardly seen on x-ray, was of extraordinary biomechanical importance because it permitted to- and fro-angulation of the plate. In consequence, there was failure due to metal fatigue. Reoperation was necessary, and this time Küntscher nailing was performed. Three months later the cortices were remodelled (Fig. 10– 4A to F).

Especially in cases of comminuted fractures, medial support—despite exact reduction and static compression—is rarely satisfactorily achieved. All such cases should have the fixation protected by means of cancellous grafting, since the metal devices for compression plating (round hole plates as well as dynamic compression plates) generally do not (and should not!) have the stability of a prosthesis.

Open fractures are not as frequent in the femoral midshaft as in the shank; however, every open fracture is a very serious complication and presents the danger of osteomyelitis. Today it is generally accepted that faulty immobilization of fracture fragments in open fractures favors osteomyelitis. Consequently, osteosynthesis can diminish this danger, so far as absolute stability is achieved and guaranteed by addi-

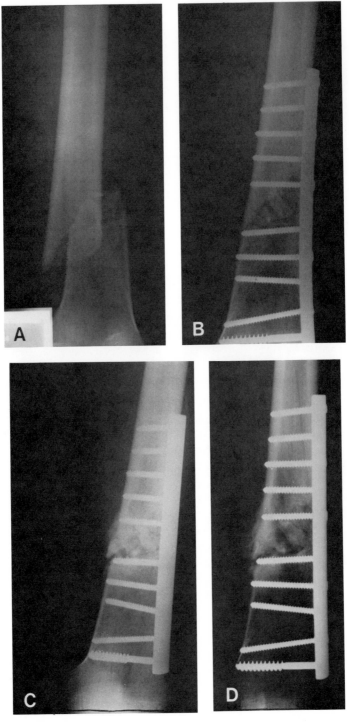

Fig. 10–5. Open comminuted fracture of femur-shaft sufficiently stabilized with compression plate osteosynthesis (A and B). Osteomyelitis developed (C and D).

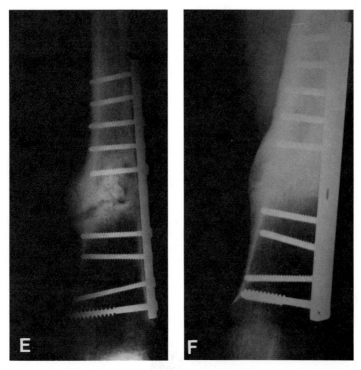

Fig. 10 – 5. Osteomyelitis was healed due to stability of the primary osteosynthesis after cancellous bone transplantation (E and F).

tional cancellous bone grafting. On the other hand, osteomyelitis can be treated only if the fracture is stabilized rigidly. It is absolutely wrong, from a biological as well as a biomechanical viewpoint, to remove sufficiently stabilizing implants in case of osteomyelitis.

Case 4. This case of a 20-year-old man demonstrates the above statement. This young man suffered a second-degree open comminuted femoral midshaft fracture. It was treated by compression plating as well as cancellous bone grafting. Nevertheless — although stability was achieved — osteomyelitis developed. Under antibiotic protection, and as soon as enough callus was produced to give further stability, reoperation was performed to remove all devitalized soft tissue and necrotic bone; another cancellous bone graft was then added. Under these conditions, and owing to the stability of the primary osteosynthesis, the osteomyelitis healed within 1 year. (Fig. 10 – 5A to F).

In the case of open fractures of the femoral shaft, compression plating must be preferred over intramedullary nailing because of the possibility of inflammation of the medullary cavity, whereas in plate osteosynthesis the infection is limited to the region of the fracture gap and plate. Another possibility for management, especially of third-degree open fractures, is external fixation, although there are some disadvantages to this because of the Steineman pins that must penetrate the thick layer of muscles. According to this viewpoint, compression plating is the method of choice for open fractures. Another definite indication for compression plating is the coincidence of a femoral midshaft fracture and injuries to the femoral vessels, nerves, or

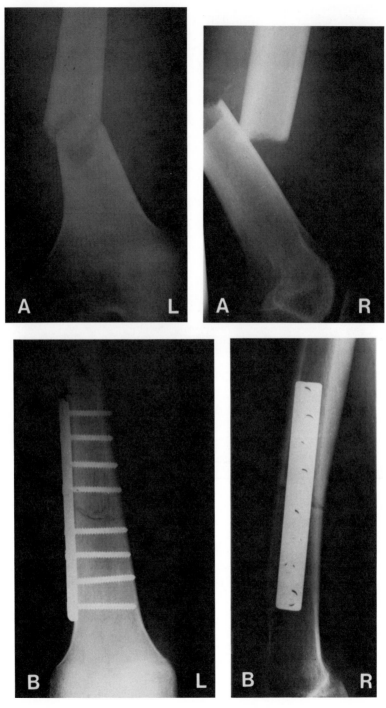

Fig. 10–6. Left side femoral fracture (A) treated with compression plate osteosynthesis (B). Delayed union resulted from unsatisfactory load relief on the injured leg.

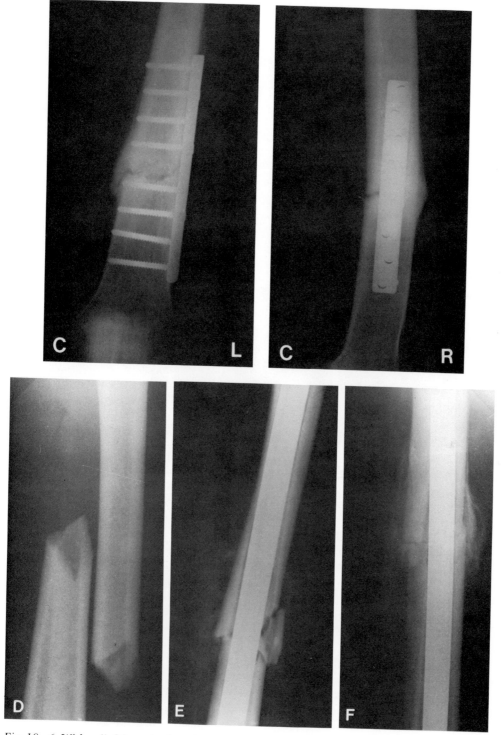

Fig. 10–6. With relief from body weight consolidation was gained by means of callus formation (C). The femoral fracture of the right side was nailed and showed regular consolidation (D to F).

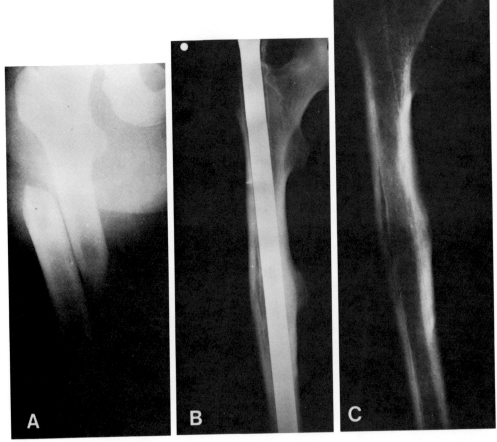

Fig. 10 – 7. Transverse and short oblique fractures are those of choice for nailing (A to C).

both. In such cases absolute stability is the prerequisite for the operative repair of vessels, nerves, or both.

One of the most extraordinary advantages of internal fixation is the possibility of immediate postoperative training of muscles and joints, permitting restoration as quickly as possible of both the anatomic situation and the total function of the injured leg. Some restrictions are absolutely necessary. Stability of the osteosynthesis is guaranteed only if the patient is cooperative and able to relieve his operated leg of body weight. Stability is endangered by additional bending moments due to weight-bearing. Loss of stability may lead to delayed union or nonunion. This is demonstrated in the next case.

Case 5. A young man suffered from femoral fractures on both sides, with rupture of the femoral artery on the left side. Osteosynthesis was performed on the left side, with compression plating and simultaneously venous grafting of the ruptured artery. The other side was treated by Küntscher nailing. Owing to the fact that both legs were injured, the patient loaded both legs. Consequently, on the left side, a delayed union resulted. By continuous x-ray control—necessary anyway in the case of compression plating—a persisting fracture gap was revealed. Instability was demonstrated by callus formed due to mechanical irritation. The patient was told to relieve

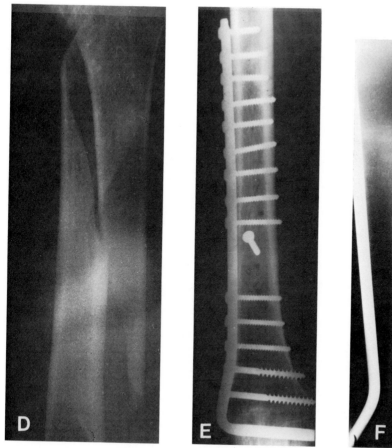
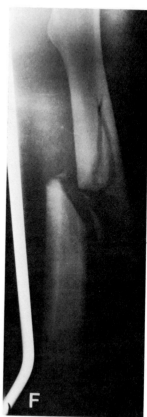

Fig. 10-7. Long oblique as well as comminuted and open fractures should be treated with compression plate osteosynthesis (D to I).

his left leg of body weight, and consolidation was achieved (Fig. 10-6A to F).

Patients most likely to be unreliable or noncooperative because of chronic alcoholism or mental disease should be excluded from compression plating.

Summarizing according to indication and contraindication, it must be said that the method of first choice for transverse and short oblique fractures of the midshaft of the femur is intramedullary osteosynthesis, with a Küntscher nail if the fracture is closed. In children and for open fractures as well as long oblique and comminuted fractures of the femur midshaft, compression plating according to the method, technique, and principles of AO is preferred (Fig. 10-7A to I).

MECHANICS OF MIDSHAFT COMPRESSION PLATING

Methods of Analysis

The system of forces acting at the femur, the shape of the femur itself, and the constraints of the soft tissue make the analysis of stresses in the femur a complex

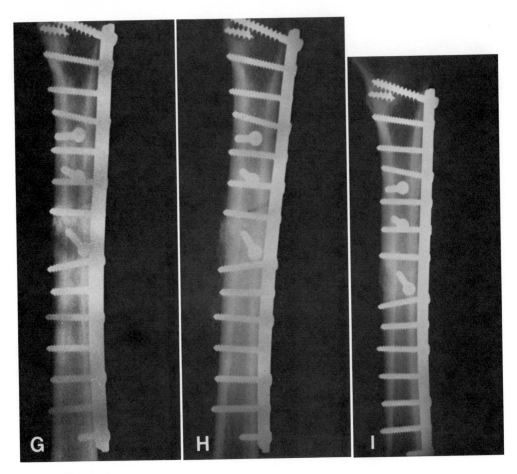

Fig. 10–7. (cont.)

task. The presence of a fracture and stabilization with an internal fixation device add to this complexity.

Biomechanics has only appeared in its own right during the last 15 years. Before that time, the fixation of fractures by plates and screws, hip nails, or intramedullary rods was well established and was based on experience rather than analysis. The application of biomechanics to orthopedic problems first concentrated on total hip replacement, and later on replacement of the tibiofemoral joint, since these procedures were gaining in popularity at the same time that biomechanics was introduced. It is only rather recently that attention has been directed to fracture fixation.

The analyses required to determine the stresses in the base and implant are no easier than they are for joint prostheses, and in some circumstances are more difficult. There have been many approaches to the analysis of internal fixation. Theoretical studies have included beam theory and finite element analysis,[9, 17, 15, 18, 22] while experimental studies[18, 3, 7, 8] have used strain gauges, load transducers, and photoelasticity measurements. The approach taken here is to illustrate basic principles by simple analysis; the results of more sophisticated studies will be presented for completeness.

Forces on the Femur and the Resulting Stresses

Some 20 muscles act about the hip joint, and the situation for the femoral shaft is no less complicated. Paul[13] has used a simplified muscle system for calculating the stresses in the femur; this is shown in Fig. 10–8. However, the first calculations of stress in the femur were carried out by Koch,[6] who treated the femur as a beam subjected to forces assumed to be much smaller than is now known to be the case for standing, walking, and running; however, this merely means that the results must be scaled appropriately. The results of Paul[13] indicate a maximum tensile stress in the femoral shaft of 5,100 lb/in² (35 MN/m²), and a maximum compressive stress of 5,-300 lb/in² (36.6 MN/m²). These values are high when compared to the tensile strength of cortical bone, which is about 12,000 lb/in² (83 MN/m²).

A review of calculations of stress in the femur has been done by Rybicki, Simonen and Weiss,[16] using the approximation that the femur could be treated as a beam, as well as computer-processed finite element analysis. Figure 10–9 shows the description used, in which the abductors and tensor fasciae latae – as well as the joint loading – were taken into account. Some of the values obtained are given in Table 10–1. Assuming that the weight of a 200 lb man, minus the weight of one leg, acts on the femoral head, the maximum tensile stress is only 1,680 lb/in² (11.6 MN/m²); however, equilibrium in the one-leg stance phase demands that the abductors be active in order to oppose the moment of the body weight about the fulcrum of the hip. The inclusion of an abductor force to produce equilibrium increases the resultant joint force, and the maximum tensile stress becomes 6,640 lb/in². However, if a tension of 200 lb is assumed in the tensor fascia latae, there is a reduction in this stress to 2850 lb/in². The assumed tension may be excessive, but a tension of at least 100 lb should be acting.

Schematically, the effect of actions other than joint loading is shown in Figure 10–1 D. The force F acting alone produces bending of the femur, with tension on one side and compression on the other. The moment produced by F may be resisted by the moment of the tension in the iliotibial band. Thus:

$$Tb = Fa \tag{1}$$

and

$$T = F \cdot \frac{a}{b} \tag{2}$$

The stress in the femur is now uniform compression over the entire cross-section. Although there is no bending, the stress has increased because the acting vertical force is now $(T + F)$ rather than F alone. The compressive stress is $(T + F)/A$, where A is the cross-sectional area. If the force generated in the iliotibial band cannot be large enough to satisfy Equation (2), there will still be bending in the shaft. However, the bending will not be so great as if the joint force F were acting alone.

The results presented in Table 10–1 indicate that rather high stresses are present in the femoral shaft, even for simple activities. Yet, allowing even for the approximations made in *all* of the calculations so far, it is clear that the femur is designed to withstand high loads. That fractures do not often occur in healthy bone during

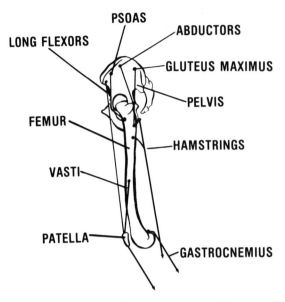

Fig. 10–8. Lateral view of the thigh showing simplified muscle system.

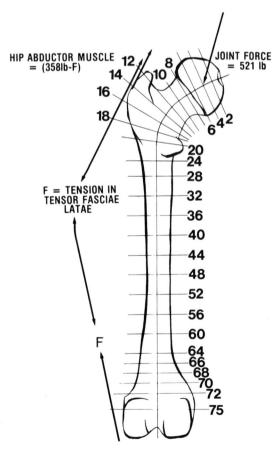

Fig. 10–9. Cross-section of the femur showing the joint and muscle loadings.

TABLE 10-1. VALUE OF THE MAXIMUM STRESS IN THE FEMORAL SHAFT FOR DIFFERENT LOADING CONDITIONS

Situation	Load (lbs)			Maximum Tensile Stress (lb/in^2)
	Joint Force	Abductors	TFL	
Axial load due to standing—no active muscles	169	–	–	1680
Axial load due to standing with active abductors	521	358	–	6640
Axial load due to standing with active abductors and tensor fascia latae	521	158	200	2850

(1 Kg = 9.81 N = 2.20 lbs).
(1 Kg/mm² = 9.81 MN/m² = 1423 lb/in²).
TFL = tensor fascia latae

normal activity indicates that perhaps there are other mechanisms for reducing the stress. Excessive loading due to trauma will lead to the strength being exceeded, and to bony fracture. Conversely, protection of the bone from the stresses normally encountered will lead to the remodelling found with rigid internal fixation.

There have been many published studies dealing with stress protection.[1, 11, 19, 21] Most of the reports have dealt with animal models in which implants—usually plates and screws—are fixed to intact bones. Although not a duplication of the clinical situation, the results are qualitatively similar and show that fixation produces a loss of cortical bone mainly caused by endosteal resorption with enlargement of the medullary cavity. Osteoporosis is usually not present in these studies, unless the bone is severely traumatized or osteotomized. The resorption is reversible following removal of the device.

Criteria for Fixation

There are several criteria for fixation:
1. The device must allow a stable reconstruction so that bone healing occurs freely.
2. The stresses in the device should not be so great as to cause permanent deformation or fracture.
3. The stresses in the bone must be low enough to allow healing to occur, but not so low as to produce stress protection remodelling.

Devices used for fracture fixation always represent a compromise. Figure 10-10 shows two ways in which the problem has been approached. Rigid fixation (Fig. 10-10A) gives a relatively low stress in the bone, so that healing can take place. The rigidity of the fixation is manifested by the relative lack of callus accompanying the healing process. As the fracture heals, the stress in the bone may increase somewhat but never approaches the normal stress in an intact bone. Flexible fixation (Fig. 10-10B), as obtained with a low modulus device such as a polymer plate, produces higher stresses in bone but less rigidity in the early stages of healing. Successful heal-

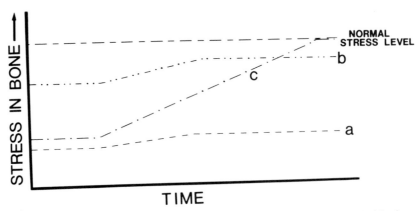

Fig. 10–10. Stress in bone for (A) rigid fixation, (B) flexible fixation, (C) ideal case.

ing of a fracture with flexible fixation is always accompanied by abundant callus formation, reflecting the movement allowed at the fracture site.

If flexible fixation is successful, it is preferable because the stress protection effects will be minor or absent; however, movement at the fracture site may result in delayed union or nonunion. The subject of the amount of movement allowed at the fracture site has recently been addressed by Perren,[14] who used the concept of strain tolerance of the tissues. Table 10–2 lists the properties of the tissues involved in fracture healing. The precursor tissues at the fracture gap can withstand a greater movement than can the tissues that appear later. However, the precursor tissues have far lower strength, and functional loads would far surpass the capacity of these initial tissues. Thus on the one hand the ability of the early tissue to withstand movement would seem to favor a flexible form of fixation, whereas on the other hand the existence of such movement would result in stresses far exceeding the strength of the tissue. Hence, the early development of the healing process depends on the fracture site being protected from the stresses imposed upon the bone, and this favors a rigid form of fixation. Presumably a flexible fixation will succeed, but only if the stress at the fracture site is externally influenced, i.e. by imposition of greater restrictions on weight bearing than would be required with a rigid appliance — which, in effect, produces stress limitation internally owing to its greater load carrying capacity.

Perhaps an ideal situation is one in which the fixation device allows for a rigid fixation in the early stages of healing, then gradually permits the bone to assume full

TABLE 10–2. STRENGTH AND MODULUS OF TISSUES
INVOLVED IN FRACTURE HEALING

Tissue	Tensile Strength (Kp/mm²)	Tensile Modulus (Kp/mm²)	Elongation to Break (percent)
Granulation	0.01	0.005	100
Cartilage	1.5	50	10
Lamellar bone	13	2000	2

TABLE 10-3. MECHANICAL CHARACTERISTICS OF
PLATES AND THE FEMORAL DIAPHYSIS

Structure	Tension/Compression		Bending	
	Stiffness $(kg \cdot mm^2)$	Strength $(kg \cdot mm^{-2})$	Stiffness $(kg \cdot mm^2)$	Strength $(kg \cdot mm^{-2})$
Compression (AO/ASIF type)				
narrow*	700,000	2100‡	570,000	450
broad*	1,300,000	3800‡	2,100,000	1900
narrow (DCP)*	750,000	—	610,000	650
Conventional plate				
narrow**	—	—	520,000	300
medium**	—	2900‡	1,300,000	
heavy**	—	3300‡	5,200,000	2400
Femur: diaphysis†	600,000	4500	50,000,000	25,000

*Cold-worked stainless steel: E = 19,000 kg mm^{-2}, yield stress = 80 kg. mm^{-2}.
**Annealed stainless steel: yield stress 40 kg · mm^{-2}. (the modulus is the same as for cold-worked stainless steel).
†For bone: E = 1,700 kg · mm^{-2} and the ultimate strength is 13 kg · mm^{-2}.
‡Measured breaking strength.

load bearing when healing and remodelling are complete. This is shown schematically in Figure 10–10 C. Efforts are underway to produce such a fixation device by use of resorbable materials such as polylactic acid with suitable reinforcing fibers.

In the above discussion, the terms stiffness and strength have been used. A fixation device on the femoral shaft is subjected to compression, bending, and torsional forces. This is a combined loading situation, but it is usual to examine the effects of the different types of load separately, and most studies neglect the effects of torsion as a first approximation. The quantities of interest are, therefore: (a) the compressive (tensile strength) and stiffness; and (b) the bending strength and stiffness. These quantities ought to be determined for the fixation device and for the bone, so that we may compare their respective rigidities and strengths.

The compressive (or tensile) strength is that stress necessary to cause fracture with no bending involved. The bending strength is that moment necessary to cause fracture, and corresponds to the stress at the tensile surface becoming equal to the tensile strength of the material.

Rigidity involves a resistance to deformation. For the case of compression (or tension) the rigidity is proportional to EA, where E is the Young's modulus (tensile modulus) and A the cross-sectional area. In bending, the rigidity is proportional to EI, where E is the modulus and I the moment of area. The moment of area reflects the distribution of material over the cross-section. When bending takes place the concave surface is in compression and the convex surface in tension (see Fig. 10–1). Between these surfaces there is a plane (the neutral plane) at which there is no tension or compression. The moment of area refers to the distribution of material with respect to the neutral plane; the greater the distance of the material from the neutral plane, the greater the contribution of the material to the moment of area. A femur has a great resistance to bending because it has a hollow cross-section, and the material (bone) is thus distributed well away from the neutral plane which runs down the length of the bone.

Table 10–3 lists the strength and rigidity of the femur and of typical fixation

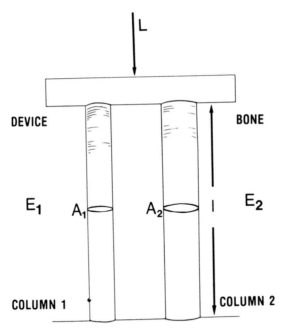

Fig. 10–11. Simple model for showing the way in which the load is divided between the fixation device and the bone for axial loading.

plates.[20] It will be seen that for tension/compression (as found in axial loading) the stiffness and strength of the devices are comparable to those of the femur. In bending, however, the stiffness and strength of the femur are about 25 times that of typical fixation plates.

The way in which the load is split between the device and the bone for axial loading may be considered by using a simple model (Fig. 10–11). The plate and bone are considered as two columns, which support the load L such that the change in length of each column is the same. The load is shared so that column 1 has load L_1 and column 2 load L_2, where:

$$L_1 + L_2 = L \tag{3}$$

If the stress is σ, the strain ϵ and the modulus E, then

$$\sigma_1 = \epsilon E_1 \tag{4}$$

where $\sigma_1 = L_1/A_1$, and A_1 is the cross-sectional area of column 1. Thus:

$$L_1 = \epsilon (E_1 A_1) \tag{5}$$

Similarly for column 2

$$L_2 = \epsilon (E_2 A_2) \tag{6}$$

and using Equation (3) gives:

$$L_1 = \frac{\left(\dfrac{A_1 E_1}{A_2 E_2}\right) \cdot L}{1 + \left(\dfrac{A_1 E_1}{A_2 E_2}\right)} \tag{7}$$

and

$$L_2 = \frac{L}{1 + \left(\dfrac{A_1 E_1}{A_2 E_2}\right)} \tag{8}$$

From Table 10–3, it is seen that the rigidity of the plate is about the same as that of bone and, hence, $(A_1 E_1) = (A_2 E_2)$. The load is therefore equally divided between the plate and the femur if there is no bending.

Table 10–4 shows how the load in the bone changes with change in the relative stiffness of implant and bone. It may be noted that a narrow DCP (direct compression plate) of the same dimensions, but made of polyacetal polymer, would give a rigidity ratio of about 0.02 and a load in the bone of about 98 percent of the applied load.

Note that the approach may equally well be used to give the load in the appliance, and hence the stress, by dividing by the cross-sectional area. The stress may be compared with the yield strength of the material of the device to determine whether permanent deformation can occur. If the stress exceeds the breaking stress, then the device will break.

The situation in bending is more difficult to analyze than in the case of axial loading. One approach is to consider the bone and plate combination as a composite beam. This gives a measure of how the plate influences the rigidity of the bone-plate combination.

TABLE 10–4. LOAD TAKEN
BY BONE AS A FUNCTION
OF RELATIVE STIFFNESS
OF PLATE AND BONE FOR
THE CASE OF
AXIAL LOADING

Fraction of Load on Bone (L_2/L)	Ratio of Rigidity $(A_1 E_1/A_2 E_2)$
0.33	2
0.5	1
0.67	0.5
0.91	0.1
0.95	0.05
0.99	0.01

It should be noted that the influence of the plate in bending is greater than would be assumed from Table 10–3. This is because Table 10–3 gives the bending characteristics of isolated plates. When a plate is applied to a bone the bending takes place about the neutral plane of the bone-plate combination, and not about the neutral plane of the plate itself (which is the case when an isolated plate is bent). Since the neutral plane lies in the medullary canal of the bone, the moment of area of the plate is effectively increased, because the distance of concern is that from the plate to the neutral plane of the combination, and is therefore approximately equal to the radius of the bone rather than half the thickness of the plate. For bending, Steinemann[20] has given the following:

$$\text{Maximum curvature} \simeq \frac{\sigma_{max}}{E^{2/3}} \cdot g \cdot \left(\frac{2W}{3S}\right)^{1/3} \tag{9}$$

where σ_{max} is the maximum stress allowable, g the stress concentration factor because of holes in the plate, W the moment of resistance, and S the stiffness. Note that for a rectangular plate of width w and thickness t:

$$A = wt \tag{10}$$

$$I = wt^3/12 \tag{11}$$

$$W = wt^2/6 \tag{12}$$

An effective way to increase the rigidity is to increase the thickness of the plate, since I increases as the cube of t (but note that the maximum curvature only changes as 1/t). Changing the material properties will change the ratio ($\sigma_{max}/E^{2/3}$). Table 10–5 illustrates the values of this ratio for some materials. From the table it will be seen that wrought cobalt-base alloy and titanium alloys are especially favored; ceramics have no promise; and polyacetal–despite the good ratio–would require a large implant (due to its low modulus), which is unacceptable on anatomic grounds.

TABLE 10–5. MATERIAL CRITERIA FOR FRACTURE FIXATION IMPLANTS

Material	σ $(kg \cdot mm^{-2})$	E $(kg \cdot mm^{-2})$	σ/E	$\sigma/E^{2/3}$ $(kg^{+1/3} \cdot mm^{+2/3})$
Bone	13[1]	1,700	0.0077	0.091
Cold-worked stainless steel	40[2]	19,000	0.0021	0.056
Cast cobalt base alloy	30[2]	21,000	0.0014	0.039
Wrought cobalt base alloy	50[2]	21,000	0.0024	0.066
Titanium and alloys	30–70[2]	11,000	0.0027–64	0.061–.140
Alumina	35[3]	39,000	0.0009	0.030
Polyacetal	7[1]	300	0.023	0.156

[1]Ultimate strength
[2]Fatigue strength
[3]Ultimate strength in bending or tension

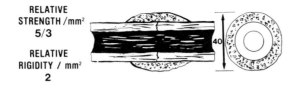

RELATIVE
STRENGTH /mm²
5/3

RELATIVE
RIGIDITY / mm²
2

RELATIVE
STRENGTH / mm²
1

RELATIVE
RIGIDITY / mm²
1

RELATIVE
STRENGTH / mm²
1/2

RELATIVE
RIGIDITY / mm²
1/4

Fig. 10–12. The femoral midshaft with 4 mm thickness of either endosteal or periosteal callus.

The criteria of strength and rigidity may be applied to callus stability. Perren[14] has compared the efficiency of endosteal, interfragmentary, and periosteal callus. To compensate for the different areas of cross-section under repair, the rigidity and strength were compared per unit area. The results are shown in Figure 10–12. Only the callus is presumed to have rigidity. Each case is considered as an annulus of callus. The rigidity is proportional to the moment of area, which for a hollow tube is:

$$I = \frac{\pi}{64}(d_1^4 - d_2^4) \tag{13}$$

where d_1 is the outside diameter and d_2 the inside diameter of the callus. Each value must be normalized by dividing by the area, which is:

$$A = \frac{\pi}{4}(d_1^2 - d_2^2) \tag{14}$$

so that equal areas are considered.

Thus the rigidity of the callus is proportional to:

$$F = (d_1^4 - d_2^4)/(d_1^2 - d_2^2) \tag{15}$$

Substituting the values $d_1 = 40$, $d_2 = 32$; $d_1 = 32$; and $d_2 = 16$, $d_1 = 8$ mm for the three cases considered gives the relative rigidities shown. Provided that the quality of the repair tissue remains the same, endosteal callus is much less efficient and resistant than interfragmentary callus, which in turn is less efficient than periosteal callus.

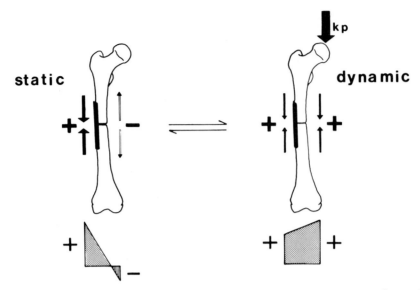

Fig. 10–13. Scheme of pressure distribution of femoral midshaft fracture under static compression with compression plating as well as dynamic compression according to the tension band principle.

The findings are a reflection of the increase in the moment of area as the callus is distributed further away from the center line of the bone.

Role of Compression

In the compression technique of fixation, the plate is placed in tension so that a compressive force is applied to the bone as the plate attempts to return to its original length. Figure 10–13 illustrates the stress distribution across the femur with the plate alone and with a dynamic load applied. As long as the static preload is greater than the dynamic traction, the fracture surfaces will be held together.

Frictional forces at the fracture surfaces will also resist the tendency for sliding movement in shear or torsion. This is especially valuable because fixation devices do not efficiently resist torsion. The friction force equals the product of the friction coefficient and the normal force. Typical values of the friction coefficient (μ) are in the range 0.2 to 0.4.[14] Figure 10–14 illustrates how torsion, as represented by an applied couple C, is resisted by the frictional couple (μFr), where F is the compressive force acting over the perpendicular osteotomy cut taken in this example.

One feature that should be mentioned is that asymmetry would be achieved by using two plates, one opposite the other, on the lateral and medial sides; but, apart from the difficulty in placement, the greater rigidity will lead to more severe remodelling of the bone. A more uniform compression may be achieved by prebending the plate before application (Fig. 10–15).

Medial Bone Support

The importance of medial bone support has been emphasized in the foregoing discussion, since the lack of medial support puts excessive bending stresses on the

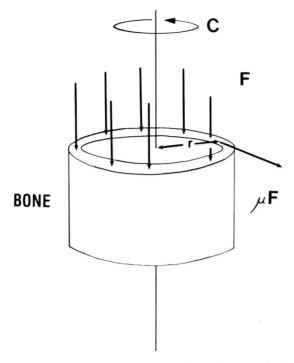

Fig. 10–14. Resistance of torsion by the frictional forces between the bone fragments.

fixation device. From Table 10–3 it was seen that the rigidity and strength of the appliance and of the intact bone were comparable. Thus the device does not suffer appreciably from added tensile load when the portion of the load taken by the bone is reduced. Failure of the device will more likely occur in bending since the bending strength, as measured by the moment needed to produce fracture, is much lower than that of the intact bone (Table 10–3).

In actual fact, plates rarely break by direct overloading, since the bone does carry part of the applied load. However, the continued backward and forward flexing of the plate produces a small crack which propagates to cause fracture. This mode of failure is known as fatigue. Laboratory experiments may be done to determine the fatigue properties of plates by applying a varying bending moment to flex the plate back and forth. A graph may be drawn of the applied moment versus the number of load cycles to failure. A typical graph for a stainless steel (AO) plate is shown in Figure 10–16.

With a medial instability, fracture of the plate generally occurs at about 100,000 (10^5) cycles, which indicates a bending moment of about 500 kg · mm, corresponding to a bending radius of about 1.2 m. This could only be due to a massive instability at the fracture site—a movement at the opposite cortex of about 2 mm. Note that Figure 10–16 indicates that under a bending moment of 250 kg · mm there will be no failure, whatever the number of cycles. This moment is much lower than the moment needed to cause breakage in one load application (see Table 10–3), and indicates that the fatigue strength of a material is lower than the breaking strength (usually by more than a factor of 2).

An increase in plate strength of a factor of 2 would make the plate "unbreakable" even under such circumstances, but it is uncertain whether this would be of great

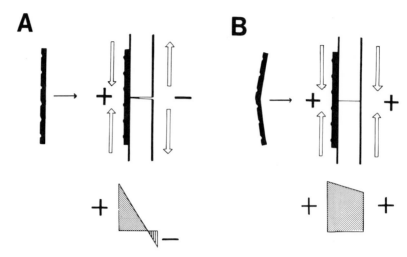

Fig. 10–15. Scheme of pressure distribution in the fracture gap under a pretensed linear compression plate (A). While the fracture gap opposite the plate is opened, preadjusted angulation of the plate (B) leads to high positive pressure over the whole cross section of the gap.

benefit, because the healing process could be inhibited by such large movements. It may be remarked that it *is* possible to increase the strength of the plate without increasing the stiffness, by choosing a higher strength alloy, having the same modulus, and keeping the dimensions of the device fixed.

Refracture Following Plate Removal

Refracture of the bone following removal of the implant is a recognized possible complication unless steps are taken to limit weight bearing for a reasonable time period. Remodelling of the cortices during healing leads to a bone of lower strength, but

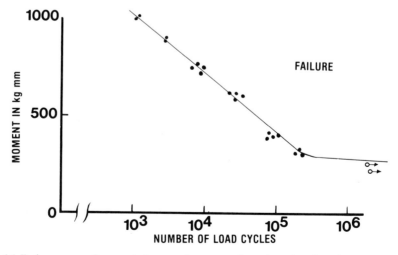

Fig. 10–16. Fatigue curve for a stainless steel compression plate showing the number of cycles to failure for a given bending moment.

there are also discontinuities because of removal of the screws. Discontinuities in structures under load give rise to neighboring concentrations of stress, and to a weakening effect much greater than would be expected from the decrease in section. This effect was shown 20 years ago by Bechtol,[4] who studied the strength of bone in bending, with and without drill holes. It was found that the smallest drill hole gave a strength reduction of 40 percent, and that the strength reduction was relatively insensitive to hole size. Thus the bone, following removal of plate and screws, must be protected from excessive stress so that post-healing cortical remodeling may *return* the bone to normal dimensions, and so that the screw holes will fill and their stress concentration effect will be eliminated.

Detailed Analysis of Plate Fixation

The simple calculations employed here are reasonably satisfactory for finding the stresses in the plate and in the bone away from the plate. The problem of the stresses in the screws themselves, and in the contact region between the plate and bone, cannot be solved in such a simple fashion, and it is necessary to turn to more sophisticated analysis.[3, 15, 17, 18, 22] The stresses acting across fractures of different configurations have also been determined.[15, 17]

SUMMARY

Fracture fixation using a plate and screws has the benefit of allowing early mobilization of joints and soft tissue. The results are satisfactory, however, only if attention is paid to certain principles which have biomechanical foundations. Compression should be used since this gives a more stable fixation. The plate must be placed on that side of the bone under tension. Care must be taken to see that the bone shares the load; in the case of the femoral shaft there should be no medial defect. Removal of the device is recommended after fracture healing because of cortical remodelling due to stress protection. These principles have been illustrated by clinical cases, and the underlying biomechanics have been presented.

REFERENCES

1. Akeson, W. H., Woo, S. L-Y, Rutherford, L., Coutts, R. D., Gonsalves, M. and Amiel, D.: The effects of rigidity of internal fixation plates on long bone remodelling. Acta Orthop. Scand., 47:241–249, 1976.

2. Anderson, L. D.: Fractures. In Campbell's Operative Orthopaedics, Vol. 1. C. V. Mosby Company, St. Louis, 1971.

3. Askew, M. J., Mow, V. C., Wirth, C. R., and Campbell, C. J.: Analysis of the intraosseous stress field due to compression plating. J. Biomech. *8*:203–212, 1975.

4. Bechtol, C. O., Ferguson, A. B., and Laing, P. G.: Metals and engineering in bone and joint surgery. The Williams and Wilkins Company, Baltimore, 1959.

5. Brookes, M.: The Blood Supply of Bone. Butterworths, London, 1971.

6. Koch, J. C.: The laws of bone architecture. Am. J. Anat., *21*:177–298, 1917.

7. Laurence, M., Freeman, M. A. R., and Swanson, S. A. V.: Engineering considerations in the internal fixation of fractures of the tibial shaft. J. Bone Joint Surg., *51B*:754–768, 1969.

8. Lindahl, O.: The rigidity of fracture immobilization with plates. Acta Orthop. Scand., *38*:101–114, 1967.

9. Minns, R. J., Bremble, G. R., and Campbell, J.: A biomechanical study of internal fixation of the tibial shaft. J. Biomech., *10*:569–579, 1977.

10. Monk, C. J. E.: Orthopaedics for Undergraduates. Oxford University Press, London, 1976.

11. Moyen, B. J. L., Lahey, P. J., Weinberg, E. H., and Harris, W. H.: Effects on intact femora of dogs of the application and removal of metal plates. J. Bone Joint Surg., *60A*: 940–947, 1978.

12. Müller, M. E., Allgöwer, M., and Willenegger, H.: Manual of Internal Fixation. Springer Verlag, New York, 1970.

13. Paul, J. P.: Load actions on the human femur in walking and some resultant stresses. *Exp. Mech.:*1–5, 1971.

14. Perren, S. M.: Physical and biological aspects of fracture healing with special reference to internal fixation. Clin. Orthop., *138*:175–196, 1979.

15. Rybicki, E. F. and Simonen, F. A.: Mechanics of oblique fracture fixation using a finite-element model. J. Biomech., *10*:141–148, 1977.

16. Rybicki, E. F., Simonen, F. A., and Weis, E. B., Jr.: On the mathematical analysis of stress in the human femur. J. Biomech., *5*:203–215, 1972.

17. Rybicki, E. F., Simonen, R. A., Mills, E. J., Hassler, C. R., Scoles, P., Milne, D., and Weis, E. B.: Mathematical and experimental studies on the mechanics of plated transverse fractures. J. Biomech., *7*:377–387, 1974.

18. Simon, B. R., Woo, S. L-Y, Stanley, G. M., Olmstead, S. R., McCarty, M. P., Jemmott, G. F., and Akeson, W. H.: Evaluation of one, two, and three-dimensional finite element and experimental models of internal fixation plates. J. Biomech., *10*:79–86, 1977.

19. Slatis, P., Karaharju, E., Holmström, T., Ahonen, J., and Paavolainen, P.: Structural changes in intact tubular bone after application of rigid plates with and without compression. J. Bone Joint Surg., *60A*:516–522, 1978.

20. Steinemann, S. G.: Characteristics of an ideal implant material for stable fixation. Presented at the Conference on Internal Fixation, Ottawa, Canada, May 1979.

21. Strömberg, L., and Dalen, N.: Atrophy of cortical bone caused by rigid internal fixation plates. Acta Orthop. Scand., *49*:448–456, 1978.

22. Woo, S. L-Y., Simon, B. R., Akeson, W. H., and McCarty, M. P.: An interdisciplinary approach to evaluate the effect of internal fixation plate on long bone remodelling. J. Biomech., *10*:87–95, 1977.

11
Proximal Tibial Osteotomy for Correction of Varus Deformity of the Knee

D. B. Kettelkamp, M.D., M.S.; E. Y. S. Chao, Ph.D.

INTRODUCTION

This chapter will consider the use of proximal tibial osteotomy for degenerative arthritis of the knee, and the biomechanics related to it. Since the early 1960s, many authors have proposed tibial osteotomy as a method of relieving pain and improving function for knees with unicompartmental osteoarthritis producing varus or valgus deformity.[2, 4, 5, 7, 8, 10, 14, 22, 26] The purpose of osteotomy is to decrease the stress on the affected side of the joint.

With the advent of knee replacement prostheses, hemiarthroplasty has been used as an alternative to proximal tibial osteotomy. Reports of the results of hemiarthroplasty for degenerative genu varum and genu valgum indicate mixed results.[9, 19, 23] The fundamental problems and goals in the treatment of degenerative genu varum and genu valgum by osteotomy and hemiarthroplasty are clinically and mechanically similar.

DESCRIPTION OF THE DISEASE

Degenerative Genu Varum

Degenerative genu varum usually develops in a knee without specific antecedent abnormalities, such as medial meniscectomy[11] or femoral or tibial shaft fractures with residual varus angulation,[18] which may antedate the development of degenerative genu varum by many years.

Previous meniscectomy increases the stress (force per unit area) on the medial plateau because of removal of the weight-bearing function of the meniscus.[16, 25] Healed fractures with varus angulation also increase the stress on the medial plateau because the deformity results in the line of weight-bearing-force passing medial to the center of the knee.[18]

In most patients, no predisposing cause of increased stress on the medial plateau can be determined. It would be tempting to postulate that these individuals have had genu varum since childhood, with degenerative arthritis as the late sequela. However, no data exist to support this hypothesis. Another possibility for increased stress

231

would be that cartilage degeneration was the primary event. With cartilage loss and narrowing of the joint space, the stress would increase because of the resulting varus. This mechanism as the primary inciting cause of degenerative genu varum also cannot be substantiated, and seems improbable, since whatever produces the initial cartilage degeneration on the medial side should affect the lateral side as well, leading to three-compartment disease. Three-compartment degenerative arthritis of the knee occurs infrequently, and usually does not produce a varus or valgus deformity.[1]

Thus far we have noted two mechanical features of degenerative genu varum: increased stress on the medial plateau, and a shift of the weight-bearing line from the center of the knee to the medial side. Either can initiate medial compartment degeneration. Furthermore, either of these changes tends to potentiate the other. Medial shift of the weight-bearing line increases tension in the lateral supporting structures (capsule, lateral collateral ligament, and iliotibial band), and this further increases the stress on the medial plateau.[12]

The body attempts to compensate for the increased stress by increasing the bone volume (density) under the medial plateau.[18] Wearing away of the cartilage and bone increases the area of contact, thus tending to decrease stress. However, the angulation deformity increases because of tissue loss, increasing the stress on the medial plateau. Depending on the magnitude of the stress and the body's ability to compensate, the deformity may either increase in severity or equilibrium may occur.[19]

Degenerative Genu Valgum

The pathologic biomechanical sequence of events in degenerative genu valgum is similar to that in degenerative genu varum. There are, however, some differences. In our experience, degenerative genu valgum occurs more frequently in women, perhaps related to the greater normal valgus tibio-femoral angle. Lateral subluxation or dislocation of the patella, which frequently is associated with degenerative genu valgum, may further increase the force transmitted through the lateral compartment. Meniscectomy, because of the larger meniscus contact area, leads to more frequent and severe degenerative changes on the lateral side.[11] Also, degeneration proceeds with greater certainty in degenerative genu valgum than with its varus counterpart.[19]

Finally, the pattern of bone attrition varies between valgus or varus deformities. In the varus deformity the greatest bone loss will occur on the tibial side, hence a proximal tibial osteotomy can correct the deformity while still maintaining the joint surface parallel to the ground. And with degenerative valgus, the greatest bone loss occurs from the lateral femoral condyle, which makes proximal tibial osteotomy less applicable because the joint surfaces cannot be made parallel to the ground. When the joint surfaces are not parallel, a medial thrust results, with frequent persistance or recurrence of symptoms.[22]

METHODS OF TREATMENT

Degenerative Genu Varum

We used two types of proximal tibial osteotomy in the following cases. Because many readers may not be familiar with both techniques, they will be briefly described.

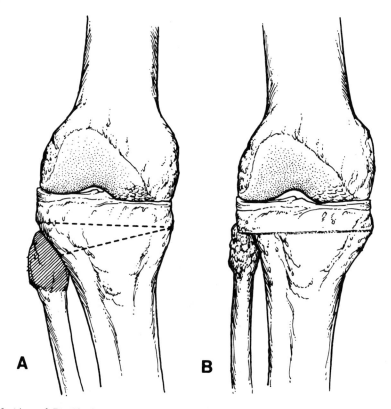

Fig. 11-1. (A and B) Closing wedge osteotomy. The wedge, based laterally for a varus deformity, is removed proximal to the patellar tendon insertion. The fibular head may be resected, hollowed, and crushed, or the fibula osteotomized.

Closing Wedge Osteotomy

The closing wedge osteotomy, popularized by Coventry,[4, 5] is the type most widely used in this country (Fig. 11-1). Correction is usually calculated from the tibio-femoral angle on weight-bearing roentgenograms. The optimum correction is to greater than a 5 degree valgus tibiofemoral angle,[14] and preferably to 10 degrees plus or minus a degree or two.[5] Wedge width at the base approximates 1 degree per millimeter up to 10 degrees, and then somewhat less. For a varus deformity the fibular head must be removed,[4, 5] hollowed, and buckled,[13] or osteotomized.

Barrel-Vault Osteotomy (Maquet)[18]

The barrel-vault osteotomy[18] has features of the dome osteotomy (Fig. 11-2A) — the use of compression fixation and anterior displacement of the tibial tubercle (distal tibia) (Fig. 11-2B). Correction is based on the mechanical axis of the leg — that is, on the angle made by a line from the center of the femoral head to the center of the knee, and a line through the center of the long axis of the tibia. The desired correction is the varus angle so measured, plus 2 to 4 degrees. A complete method for calculating the desired correction is given in the appendix to this chapter.

After resection of about 1 cm from the mid-third of the fibula, the curved cut for

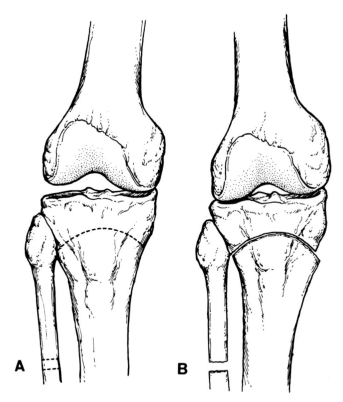

Fig. 11–2. (A and B) Barrel-vault osteotomy (Maquet). Correction is based on mechanical angle. The fibula is osteotomized in its middle third, and at least 1 cm of shaft is resected.

the tibia is outlined using a jig and a drill. The cut is completed with osteotomes. Heavy pins for the compression clamps are placed through the proximal and distal fragments under image intensification using a protractor jig. Correction and anterior displacement of 1 cm are obtained manually, and the compression clamps are applied.

BIOMECHANICS OF TIBIAL OSTEOTOMY

The effects of tibial ostoetomy can be discussed with regard to both clinical and biomechanical factors. The clinical factors consist of: (a) allowing damaged cartilage to regenerate;[2] (b) retardation of subchondral bone sclerosis and cyst formation; (c) prevention of further capsular-ligamentous laxity; (d) decreasing the clinical symptoms and pain; and (e) possible decompression of vascular pressure as a source of pain. The biomechanical factors include: (a) correction of knee-joint deformity in the anteroposterior plane; (b) realignment of the load-bearing axis; (c) increasing the joint stability (sometimes involving tightening of the contralateral ligament); and (d) altering the patello-femoral mechanics, which may be an important factor in causing symptomatic pain. All of these factors are intimately related. A clear understanding of the mechanics of the knee joint can be a great help in explaining the deformity associated with the joint deformity, and in analyzing the efficacy of the reconstruction.

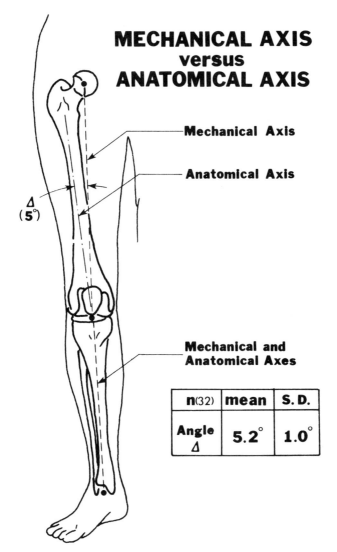

MECHANICAL AXIS
versus
ANATOMICAL AXIS

—**Mechanical Axis**

—**Anatomical Axis**

Δ
(**5°**)

—**Mechanical and
Anatomical Axes**

n(32)	mean	S.D.
Angle Δ	5.2°	1.0°

Fig. 11–3. Mechanical axis versus anatomic axis. The average difference based on 32 knees was 5.2 degrees (σ=10).

To study the quantitative mechanics of the knee joint in the frontal plane during normal standing, it is important to define the mechanical axes—in contrast to the anatomic axes—of the femur and tibia. As shown in Figure 11–3, the mechanical axis of the femur is the line joining the center of the femoral head to the midpoint of the intercondylar eminence; the anatomic axis of the femur is the bisector of the distal femoral shaft in the anteroposterior plane. The mechanical axis of the tibia is a line joining the midpoint of the intercondylar eminence and the centroid of the tibiofibular mortise; the anatomic axis of the tibia is defined in the same manner as that of the femur. A study of 32 normal anteroposterior x-ray films showed that the mechanical axis of the femur has a mean varus angle of 6 degrees (with a standard deviation of 1 degree), which matches closely with Steindler's finding.[24] In the same control group, the mechanical and anatomic axes of the tibia nearly coincide.

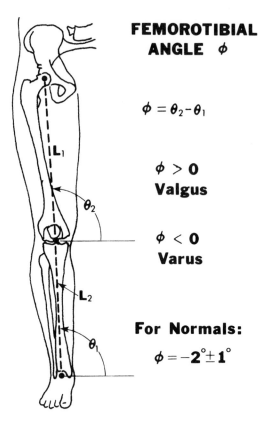

FEMOROTIBIAL ANGLE ϕ

$\phi = \theta_2 - \theta_1$

$\phi > 0$
Valgus

$\phi < 0$
Varus

For Normals:

$\phi = -2°\pm 1°$

Fig. 11–4. The femoro-tibial angle ϕ is the difference between the femoral orientation angle θ_2 and the tibial orientation angle θ_1. For normals the femoro-tibial angle averaged 2 degrees ($\sigma = 10$) of varus.

In defining knee joint deformity, mechanical axes are more reliable than anatomic axes, especially when there are other skeletal deformities away from the knee joint. The mechanical axes provide the precise description of load transmission through the knee joint. However, in certain clinical circumstances, anatomic axes would be more convenient to use. In such cases, a full appreciation of the residual difference between the mechanical and anatomic axes must be recognized for the accurate assessment of load transmission and subsequent osteotomy correction.

The angles of the mechanical axes of the femur and the tibia with respect to the horizontal line of the ground are used to define the knee-joint orientation in the frontal plane. As shown in Figure 11–4, θ_2 is the femoral orientation angle and θ_1 is the tibial orientation angle. The difference between these two angles ($\theta_2 - \theta_1$), designated as angle ϕ, is used to measure the varus or valgus deformity. If ϕ is larger than zero, the knee joint is defined in valgus; otherwise, it is said to be in varus. In the same group of normal knees studied, there was a residual varus angle of 2 degrees, with a standard deviation of ± 1 degree. In essence, a normal knee has a varus angle of about 2 degrees, as measured on the mechanical axes of the femur and tibia. If anatomic axes were used, a normal knee would have a 4 degree valgus angle. These data are limited to normal knees of young men and women. Additional data, includ-

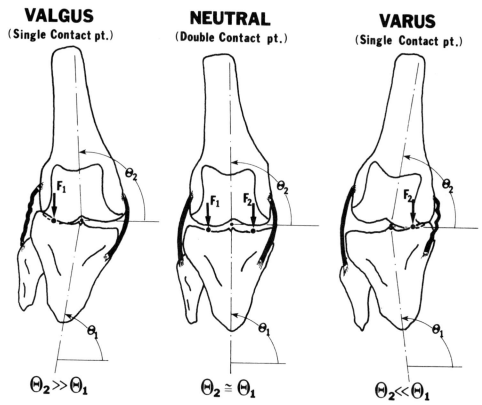

VALGUS (Single Contact pt.) **NEUTRAL** (Double Contact pt.) **VARUS** (Single Contact pt.)

$\theta_2 \gg \theta_1$ $\theta_2 \cong \theta_1$ $\theta_2 \ll \theta_1$

Fig. 11–5. In neutral position there is contact on both tibial plateaus, and the collateral ligaments are in neutral tension. In valgus deformity the force is only on the lateral plateau and tension is increased in the medial collateral ligament. The reverse situation occurs with varus deformity.

ing anthropomorphic, age, and sex factors, are required to establish a more reliable reference base.

Additional features associated with joint deformity can also be defined on the basis of plateau force distribution and laxity of collateral ligaments. As shown in Figure 11–5, the normal knee allows contact force on both the medial and lateral plateaus, while the collateral ligaments are in neutral length with minimal tension. In this case, $\theta_1 \cong \theta_2$. In either deformity, there is only one plateau subject to contact force, which causes excessive tension of the contralateral collateral ligament. A correction by upper tibial osteotomy, if done properly, can provide effective plateau force redistribution. Proper relief of collateral ligament tension can be achieved at the same time.

It is postulated that excessive knee deformity in the anteroposterior plane also alters patello-femoral mechanics, which may cause clinical symptoms. A high tibial osteotomy may correct the pathomechanics of the patella and eliminate one possible source of pain. As illustrated in Figure 11–6, a severe varus knee may cause a strong medial pull of the patella, thereby introducing localized high stress in the patellar cartilage. Valgus osteotomy would tend to relieve such stress by altering the dominant medial pull to the neutral or even to the lateral direction, which may alleviate pain.

EFFECT OF HIGH TIBIAL OSTEOTOMY TO PATELLOFEMORAL MECHANICS

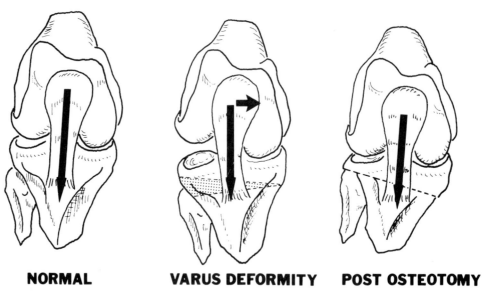

NORMAL   VARUS DEFORMITY   POST OSTEOTOMY

Fig. 11–6. In varus deformity there may be a medial displacing force on the patella, with increased stress on the medial patellar facet. This force may be corrected with realignment secondary to a valgus osteotomy.

DESCRIPTION OF CASES

Case 1. This 62-year-old man with genu varum had developed a dull ache in his right knee one year before presentation. Intraarticular steroid injections produced no benefit, and the pain had become constant with all weight bearing. He climbed and descended stairs with the affected knee extended. Knee stiffness was present on waking, after sitting, and after walking. On examination, the collateral and cruciate ligaments were found to be intact, there was no effusion, and flexion was present from 0 to 100 degrees. Weight-bearing anteroposterior roentgenograms showed narrowing of the medial joint space to greater than half the normal, and a tibio-femoral angle of 3 degrees of varus (Fig. 11–7A). Closing wedge proximal tibial osteotomy was performed, with correction to a tibiofemoral angle of 11 degrees of valgus (Fig. 11–7B). The osteotomy healed uneventfully. Four years later the patient has only minimal stiffness, usually associated with weather changes, and can walk any distance desired.

Case 2. This 72-year-old man gave a 10-year history of progressive bilateral medial knee pain. The right knee was the most symptomatic. Pain was present with every step, and occasionally woke the patient at night. He could not go up and down stairs and required arm support to rise and sit in a chair. Pain limited his walking to less than two blocks. On examination he was found to have severe bilateral degenerative genu varum. The varus deformity of the right knee could be positively improved, but not to a normal tibio-femoral angle. Flexion was 0 to 100 degrees and the ligaments were stable. Weight-bearing roentgenograms showed complete loss of the

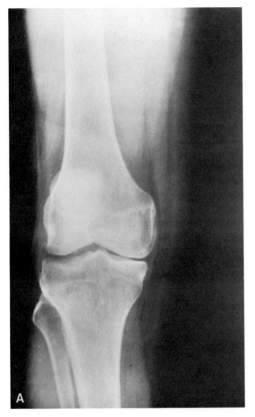
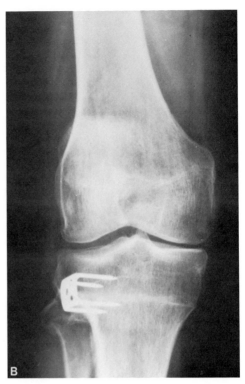

Fig. 11–7. Case 1 (A and B) Closing wedge osteotomy. (A) shows the pre-operative weight-bearing roentgenogram with a 40 degree varus tibio-femoral angle. (B) Four years after osteotomy, illustrates the correction to a 4 degree valgus tibiofemoral angle.

medial joint space, without subluxation of the right knee (Fig. 11–8A). The tibiofemoral angle measured 11 degrees of varus. The left knee showed loss of the medial joint space and marked lateral subluxation of the tibia, which contraindicates proximal tibial osteotomy. A valgus stress film of the right knee demonstrated a good lateral joint space, moderate medial-plateau bone loss, and a correction of the tibiofemoral angle of 1 degree of varus (Fig. 11–8B). A closing wedge proximal tibial osteotomy was performed, with correction to a tibio-femoral angle of 10 degrees of valgus (Fig. 11–8C). Partial weight bearing in a long leg cast was continued for 6 weeks postoperatively. At 4 months the patient was walking well with a cane, and at 10 months he could walk 3 miles without difficulty. This level of activity has been maintained through the 2 years since surgery. It should be noted that the improvement in the right knee has permitted this functional level despite severe degenerative disease in the left knee.

Case 3. This 58-year-old obese woman had originally developed pain in both knees—the left more severe than the right—3 years before being seen. The pain was present with all weight bearing, awakened the patient at night, and restricted her walking to several blocks only. The patient gave a history of essential hypertension and myasthensia gravis. Examination of the left knee revealed flexion from 5 to 120

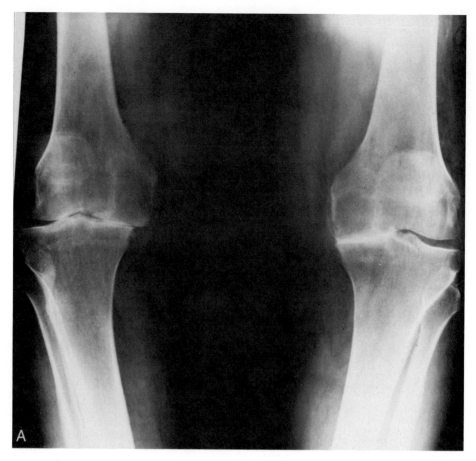

Fig. 11–8. Case 2 (A, B, and C) Closing wedge osteotomy. The preoperative weight-bearing roentgenogram in (A) shows loss of medial joint space and an 11 degree varus tibiofemoral angle on the most symptomatic and least involved knee.

degrees, and 8 degrees of varus, with stable ligaments. Roentgenograms confirmed an 8 degree varus tibio-femoral angle with complete loss of the medial joint space and minimal bone attrition from the medial tibial plateau (Fig. 11–9A). A closing wedge proximal tibial osteotomy was performed with correction to a tibiofemoral angle of 7 degrees of valgus (Fig. 11–9B). Postoperative phlebitis prophylaxis consisted of coumadin, with prothrombin levels at 40 to 60 percent of normal, because aspirin was prohibited by her medications for myasthenia gravis. The osteotomy healed slowly, requiring 10 months, probably secondary to the crystalline sodium warfarin (Coumadin). Correction was lost to a 0 degree tibio-femoral angle (Fig. 11–9C). A duo-condylar prosthesis was inserted 14 months after the original osteotomy because of continued medial joint pain (Fig. 11–9D).

Case 4. This 50-year-old man had developed increasing left knee pain over the preceding 3 years. He had medial joint-line pain with all weight-bearing. Examination revealed a varus knee with medial joint-line tenderness and patello-femoral crepitation with patellofemoral tenderness. Weight-bearing roentgenograms showed loss of the medial joint space (Fig. 11–10A), and a mechanical angle of 8 degrees

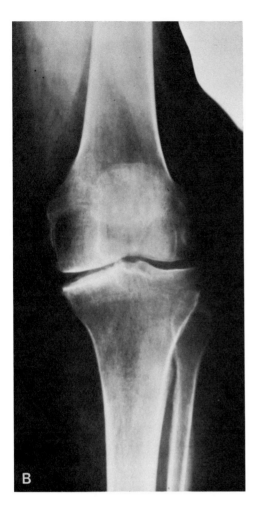

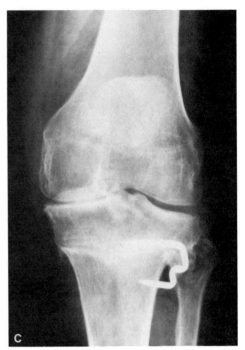

varus (Fig. 11–10A). A barrel-vault (Maquet) proximal tibial osteotomy was performed after fibular osteotomy (Fig. 11–10B), with correction to a mechanical angle of 2 degrees valgus (Fig. 11–10B). The distal tibia, including the tibial tubercle, was displaced anteriorly by approximately 5 millimeters (Fig. 11–10C). The patient started quadriceps exercises 1 day after surgery. Four days postoperatively the splint was removed, and touched weight-bearing with crutches began. The clamps were tightened weekly, and the pins and clamps were removed at 8 weeks postoperatively. At that time the patient had a full range of knee motion, good quadriceps control, and no effusion (Fig. 11–10D). At 6 months postosteotomy, there was no pain and an adequate non-protected activity level.

Case 5. This 64-year-old man, who had genu varum that was treated by barrel-vault osteotomy, gave a history of progressive right knee pain over the past 2 years. He now has night pain, and noted increasing bowlegged deformity. He was diabetic. Examination revealed a 20 degree knee flexion deformity, with a 10 degree tibio-femoral fixed varus deformity (Fig. 11–11A). The mechanical angle measured 11

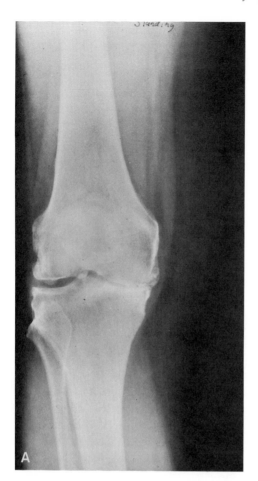

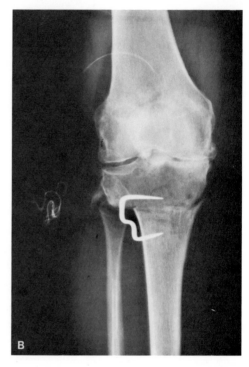

Fig. 11–9. Case 3 (A, B, C, and D) Closing wedge osteotomy. Weight-bearing preoperative roentgenogram (A) shows complete loss of medial joint space with minimal bone attrition from the medial tibial plateau and 8 degree varus tibio-femoral angle. At surgery (B), correction was obtained to a 7 degree valgus tibio-femoral angle. Healing was delayed, probably related to Coumadin prophylaxis for thrombophlebitis, and correction lost to a 0 degree tibio-femoral angle (C).

degrees (Fig. 11–11A). A barrel-vault osteotomy with slight anterior displacement was done (Fig. 11–11B and C). Correction was obtained to a mechanical angle of 1 degree valgus (Fig. 11–11B). This is less than the desired correction, which would be to 4 degrees mechanical valgus. Pins and clamps were removed at 8 weeks. At 3 months after osteotomy, the patient had flexion from 15 to 110 degrees, no pain, good quadriceps, and progressed to full weight bearing. The long-term results in this patient may be compromised by the residual flexion deformity of 15 degrees and by the correction of the mechanical angle to only 1 degree valgus.

BIOMECHANICAL ANALYSIS

The human knee joint must adapt to large forces and at the same time maintain a high degree of mobility and stability. These functional demands make it more sus-

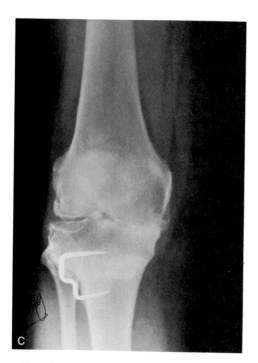
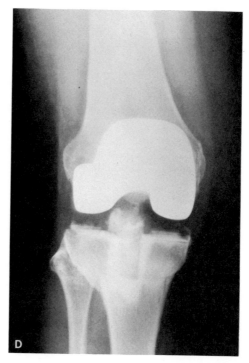

Fig. 11–9(D) A duo-condylar type implant was then used (D), because of continued pain.

ceptible to traumatic injury and degenerative joint disease. In any reconstructive procedure involving the knee, complex and demanding functions must be taken into account. In performing a tibial or femoral osteotomy, the following biomechanical factors must be carefully considered in order to achieve long-term success.

1. Pathomechanics of genu varum and genu valgum deformities.
2. Tibial plateau contact-force redistribution.
3. Effects of other anatomic abnormalities of the knee joint.
4. Functional performance of the knee in gain and other activities.

This part of the chapter attempts to discuss these factors on the basis of theoretical concepts and clinical considerations. It is hoped that such a discussion can provide information helpful for establishing proper guidelines for patient selection, surgical technique, and long-term prognosis with regard to tibial osteotomy. The underlying principles included in the presentation apply equally well to high tibial osteotomy as to supracondylar osteotomy.

Although the ensuing analysis emphasizes the biomechanical aspects of proximal tibial osteotomy, the clinical advantages of the procedures are fully appreciated.[2, 5] It is important to recognize the potential clinical advantages of these surgical procedures so that their theoretical results can be put into the proper perspective of their clinical significance.

We have previously discussed the pathomechanics of the genu varum and genu valgum deformities, and possible resulting changes in the patello-femoral forces. These factors help to explain the abnormal mechanics associated with knee deformity. Plateau force redistribution subsequent to planned surgery can be analyzed to predict the effectiveness of osteotomy in correcting a deformity and relieving the pain. Other anatomic factors, although not included here, should also be kept in

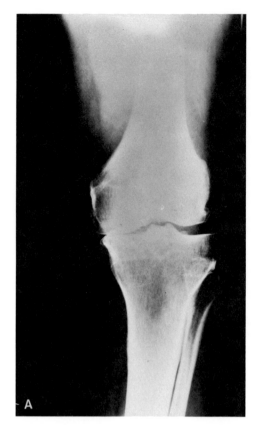
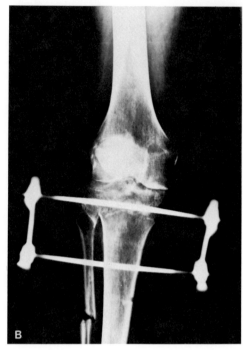

Fig. 11–10. Case 4 (A, B, C, D, and E) Barrel-vault osteotomy (Maquet). (A) Preoperative weight-bearing view mechanical angle of 8° varus. (B) Anterior-posterior roentgenogram.

mind, since they also contribute to the mechanics of knee-joint deformity. These factors consist of the joint contact area, femoral condylar geometry, cruciate ligament functional status, meniscal contribution to stability, and force equilibrium in the saggittal plane. For reasons of simplicity, the following analysis does not include these variables. However, this simplified analysis provides information useful in the preoperative planning of tibial osteotomy.

Plateau Force Distribution

In calculating the tibial plateau forces, the following basic criteria have been assumed:
1. The knee deformity is defined by the femoro-tibial angle ($\theta_2 - \theta_1$), measured on the mechanical axes of the femur and tibia.
2. The force analysis is static, and measured only in two dimensions[12, 17] in the anteroposterior plane with the patient in a normal standing posture.
3. The geometric data are measured based on a 6-foot-long film, with load bearing.
4. Only the plateau forces and the collateral ligament forces are considered in the modeling.
The force analysis is based on the free-body diagram shown in Figure 11–12. F_1

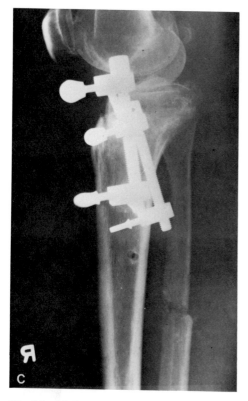
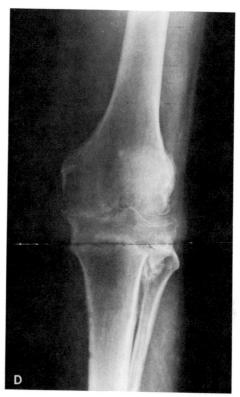

Fig. 11–10(C) Lateral roentgenogram after a barrel vault osteotomy. (D) Healed osteotomy. *(Figure continues on next page.)*

and F_2 are the lateral and medial plateau contact forces, applied at points A and B, respectively. P and Q represent the lateral and medial collateral ligamentous forces, with the orientation and point of application as illustrated. R is the ground reaction force, which has the value of half the subject's body weight. W is the weight of the leg plus the foot and shoe, which is assumed to be 6.2 percent of the body weight, according to Morrison.[21] L_1 and L_2 are the mechanical axes of the tibia and femur, respectively. All parameters used in this diagram were measured directly from the standing 6-foot film.

Static analysis was done on the basis of the following input data[12]:

1. Body weight.
2. Plateau width and surface curvature.
4. Tibial axis length and orientation.
4. Direction of collateral ligament pull.

The equilibrium equations for various deformity cases are derived. The solutions of these equations provide the plateau and collateral ligament forces, which are plotted against the femorotibial angle, ϕ. A typical force plot for a normal subject is shown in Figure 11–13. From this diagram, it was found that in standing, more force is applied to the medial plateaus, and that the collateral ligaments carry virtually no load. Figure 11–14 depicts an abnormal knee with an approximately 7 degree varus deformity in which all plateau force is on the medial compartment (F_1), with significant lateral collateral ligament force (P) present. For proper correction (with 4 degrees of over-correction), a wedge of approximately 10 degrees is to be excised.

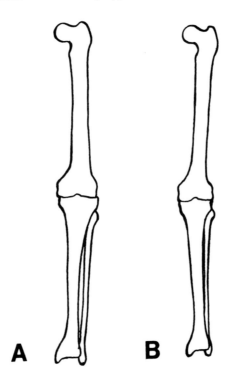

A   B

Fig. 11 – 10E(A) Outline of the femur and tibia preoperatively with a mechanical angle of 8 degrees varus. After the osteotomy, (E) (B), the mechanical angle was 2 degrees valgus.

To establish the normal plateau contact forces during standing, 26 apparently normal knees of 13 persons were analyzed. The results are presented in Figure 11 – 15, the lateral-medial contact force ratios are plotted against the femoro-tibial angle, ϕ $(\theta_2 - \theta_1)$. The region bounded by two dashed lines, which includes data from all the normal subjects, is defined as the normal range. The slant of the lines results from the idea that the variation of the femoro-tibial angle maintains a certain linear relationship with the contact force ratio. The lower limit of this normal range is full medial compartment weight bearing, with from 5 degrees varus to 3 degrees valgus angulation. The upper boundary of this region is full lateral plateau weight bearing, with from 3 degrees varus to 5 degrees valgus angulation. This provides a reliable statistical range for most normal subjects. This range serves as a reference base in evaluating plateau force redistribution after tibial osteotomy.

In a separate series of 10 preoperative and 41 postoperative patients, the plateau force distribution data were used to estimate the proper angular correction in osteotomy. The postoperative results for the 41 patients were divided into varus and valgus groups, based on the patients' preoperative histories.

Clinical assessments (to be described in the next section) – classified as good, acceptable, and poor – were also identified. Among the patients with varus knees whose postoperative results were classified as good, 85 percent had plateau distributions either within the normal range or in the range of slight overcorrection to valgus. The remaining 10 percent of the patients with good results had their knees undercorrected. The majority of undercorrected knees had deteriorated clinical

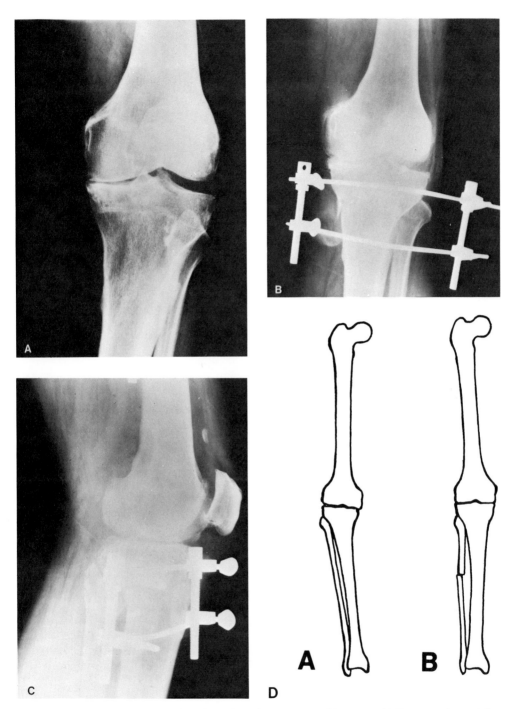

Fig. 11–11. Case 5 (A, B, C, and D) Barrel-vault osteotomy (Maquet). (A) Preoperative weight-bearing roentgenogram. (B) Anterior-posterior view; (C) Lateral view after barrel-vault osteotomy. (D) (A) Preoperative tibio-femoral alignment with a mechanical angle of 11 degrees varus. (D) (B) Postoperative alignment of 1 degree valgus (mechanical angle).

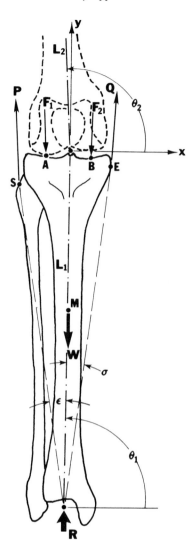

Fig. 11–12. Plateau force analysis is based on this free-body diagram with forces and measurements shown.

results in subsequent evaluations after a period of 2 years. Assessments of some of the patients who were corrected to neutral changed from good to acceptable, and some changed from good and acceptable to poor. This reflects the importance of overcorrection in knees with varus deformity. There were similar findings in knees with valgus deformity.

Patient Functional Evaluation

The patients included in clinical and biomechanical evaluations (functional gait and plateau force distribution) were separated as follows:

Control group: 13 persons with normal knees (26 knees).

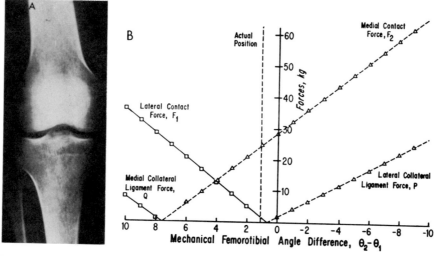

Fig. 11–13. Force distribution on the tibial plateaus calculated for this normal knee. (Kettelkamp, D. B., and Chao, E. Y. S.: A method for quantitative analysis of medial and lateral compression forces at the knee during standing. Clin. Orthop., *83*:202, 1972.)

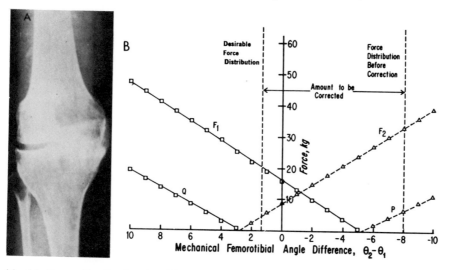

Fig. 11–14. Force distribution in this varus deformity was entirely on the medial plateau. (Kettelkamp, D. B., and Chao, E. Y. S.: A method for quantitative analysis of medial and lateral compression forces at the knee during standing. Clin. Orthop., *83*:202, 1972.)

Patient group: 25 men and 32 women, of whom 54 had unilateral involvement and 3 had bilateral involvement (a total of 60 knees in 57 patients).

Thirty-eight individuals in the latter group had degenerative disorders. Long-term results were available after 39 months and 63 months. The clinical assessment rating scale is defined as follows:

Good: No support required and no pain present with activity.
Acceptable: No support required but some pain present.
Poor: Support required and presistent pain present.

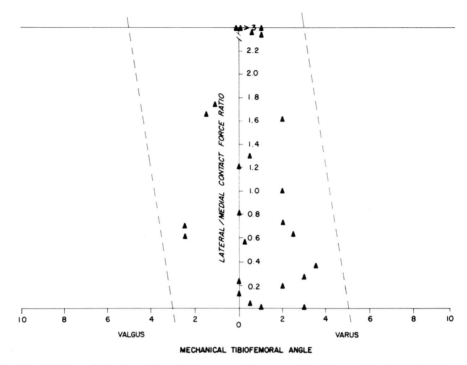

Fig. 11 – 15. Plot of the lateral-medial contract forces against the mechanical tibio-femoral angle for 13 normal knees.

The evaluation results for two consecutive follow-up series were:

Clinical Assessment	1972	1974
Good	32	24
Acceptable	16	16
Poor	12	20

A significant deterioration of results is clearly shown. When the results were examined with respect to the degree of overcorrection, an interesting finding emerged (Table 11 – 1). In knees with varus deformity, that also had overcorrections of 5 de-

TABLE 11 – 1. RESULTS OF OVERCORRECTION OF
VARUS DEFORMITY

Results	Less than 5° Overcorrection		5° Overcorrection	
	39 mo	*63 mo*	*39 mo*	*63 mo*
Good or acceptable	15 (75%)	9 (45%)	25 (89%)	23 (72%)
Poor	5 (25%)	11 (55%)	3 (11%)	5 (18%)

TABLE 11-2. RESULTS OF POSTOPERATIVE EVALUATION OF
CHANGES IN GAIT

Clinical Result	Swing-phase Flexion Extension	Stance-phase Flexion Extension	Abduction Adduction	Rotation	Cadence	Stride
Good	NS	Increased	NS	Decreased	Increased	Increased
Acceptable	NS	NS	NS	Decreased	Increased	Increased
Poor	NS	NS	NS	NS	NS	NS

grees or greater, the long-term results are more stable, reflecting the close correlation between long-lasting good results and overcorrection in valgus osteotomy (varus-deformed knees).

An electrogoniometer was used in evaluating level walking and activities of daily living. The evaluation procedure and the definition of gait parameters have been reported elsewhere.[6, 14, 15] The most significant gait parameters were found to be: (a) swing-phase flexion; (b) Stance-phase flexion; (c) range of abduction and adduction; (d) range of axial rotation; and (e) cadence and stride length.

A summary of gain evaluation results for different clinical classifications is presented in Table 11-2. A close correlation between gait evaluation results and the subjective clinical assessments was found.

DISCUSSION

In view of its well-established clinical advantages and the results of biomechanical analysis, tibial osteotomy is certainly a justified reconstructive procedure for the knee in a selected group of patients who have unicompartmental osteoarthritis with acceptable ligamentous stability and minimal bone loss. The results of long-term clinical and biomechanical gait evaluation further substantiate the efficacy of this procedure when it is properly performed. However, since the longevity of such a procedure in treating varus deformity is highly dependent on overcorrection, careful preoperative analysis of joint deformity, as determined on a 6-foot, weight-bearing long film, is essential.

An ideal correction for various degrees of varus or valgus angulation can be determined on the basis of the previously defined method. In the case of genu varus, overcorrection to 3 to 5 degrees of valgus angulation, based on mechanical axes, is recommended. Valgus deformity should be corrected only to neutral (zero degrees) on the basis of anatomic axes, because of the residual varus angle inherent in normal knee geometry. Since there is a varus thrust existing in normal walking, which tends to shift the plateau force medially during stance phase, slightly less overcorrection for valgus knees may be allowed.

Among the other potential advantages of high tibial osteotomy are that it alters patellofemoral mechanics and contributes to the elimination of pain. However, if rotatory deformity is involved, a three-dimensional, dome-type osteotomy[10] may be more beneficial, since it takes into account the change of patellar-tendon force direction, as well as the correction of rotation. Additional analysis in this respect is certainly warranted to expand the present scope of knee osteotomy. Collateral ligament tightening may also be an important factor, since from a purely mechanical standpoint, tightening the collateral ligaments is a good practice for enhancing joint stability after osteotomy. However, biologic tightening of the ligaments may occur spontaneously over a period of time, making surgical advancement of collateral ligaments unnecessary.

For excessive valgus deformity, particularly with an inclined femoral condylar line, supracondylar osteotomy may be more desirable and effective in shifting the plateau force medially and maintaining an adequate joint line. If the osteotomy site is more proximal, the lower leg will be brought closer to the body-center line, which would provide the required medial load transfer without significantly altering the tibio-femoral orientation. The common problem of bony nonunion experienced in such cases may be avoided by applying the closing wedge technique and by the use of external fixation devices. However, if the patient has an abnormal hip on the same side, particularly with pronounced adduction, proper correction at the hip is essential in order to maintain the effectiveness of the osteotomy performed at the knee level.

It may seem misleading and improper to correlate clinical evaluation results with biomechanical force and gait analyses, since there are many other factors that could affect the outcome of osteotomy. These factors are largely anatomic and pathologic. They consist of the scope of soft tissue reconstruction, vascular status, nerve impingement and tension, and so forth, which can only be assessed subjectively by clinical means. However, the present results show a strong correlation between biomechanical evaluation and clinical observation. Valuable information can be extracted from objective analysis of the patient's functional status. For example, if a patient begins to have symptoms of functional deterioration, it can easily be detected by biomechanical evaluation procedures using force analysis and the level-walking examination as the main criteria. These results should suggest the proper therapeutic course for the patient. Improvement in surgical technique can also be studied on the basis of retrospective and prospective protocols involving biomechanical analyses.

The question of whether high tibial osteotomy could be used as prophylactically in younger patients with excessive varus deformity but without clinical symptoms of pain or instability is interesting but must be carefully reviewed. Such a question cannot be answered easily without considering many medical and socioeconomic factors. The age of the individual is important and should be considered in the light of the current development of joint replacement. Many experienced orthopedic surgeons believe that if a relatively young patient is asymptomatic—regardless of the patient's knee-joint deformity—tibial osteotomy is contraindicated. For older patients demanding a high level of activity, such a corrective procedure may be indicated to retard cartilage and subchondral bone degeneration. Such decisions must be made on the basis of the individual merits of each case.

Patients with pronounced knee deformity but no symptoms should be examined frequently. Any slight complaint of pain or discomfort may be an early warning sign of pathologic changes as a result of uneven loading at the plateaus. In such a circumstance, proper therapeutic precautions may be recommended, and a high tibial osteotomy at a later time should be considered as a strong possibility. It must be kept in mind, however, that if a joint is allowed to deteriorate, the articular surface and ligamentous structure may experience excessive and irreversible damage that can make any reconstructive procedure difficult and less effective. In any event, tibial osteotomy will remain an effective surgical procedure, and must be carefully considered before the application of more radical surgery such as total or partial joint replacement arthroplasty.

CONCLUSIONS

The following conclusions may be drawn from the results of the clinical and biomechanical studies presented in this chapter. Application of the principles stated should be carefully weighed against other clinical and surgical factors in each individual case:

1. Biomechanical principles justify tibial osteotomy as a surgical procedure for redistributing knee plateau forces.
2. In determining the proper wedge angle, the *mechanical* axes rather than the *anatomic* axes should be used for the femur and tibia. If only the anatomic axes are available, careful correction for the residual valgus angle of the femur should be considered.
3. A slight overcorrection, of approximately 3 to 5 degrees is recommended for the varus knee; the patient's preoperative condition should be taken into account in implementing this principle.
4. For moderate valgus deformity, the osteotomy should bring the anatomic axes for the tibia and femur into line, which produces a slight varus angle for the load-transmitting mechanical axes because of the residual valgus angle of the anatomic axis of the femur.
5. Supracondylar osteotomy may be indicated for excessive valgus deformity with an inclined plateau line, but a secure means of bony fixation without causing loss of knee joint motion should be used. If the hip is abnormally adducted, correction at the hip is essential to ensure the desired postoperative result after osteotomy at the knee joint.
6. For rheumatoid arthritis patients with flexion contracture and rotatory deformity in addition to genu varum, a dome-type osteotomy may be more effective in providing three-dimensional correction.
7. Careful patient selection through preoperative analysis can improve the operation results greatly, thereby enhancing long-term success.
8. Objective evaluation methods should be applied to selected patients postoperatively in order to determine the prognosis of tibial osteotomy.

Both clinical and biomechanical evaluation results indicate that high tibial osteotomy is acceptable—perhaps the procedure of choice—for certain groups of patients with degenerative unicompartmental disease. The many clinical advantages inherent in this reconstructive surgery make it appealing, even in the wake of the present enthusiasm for total knee arthroplasty. Consequently, knee osteotomy should be seriously considered before a more radical operation, such as hemiarthroplasty or total knee replacement, is considered. Even if high tibial osteotomy provides only a limited number of years of pain relief and functional service, it is still a safer method than other reconstructive procedures. With more study and experience, knee osteotomy may become a prophylactic procedure for eliminating the pathologic changes in joint tissues caused by early varus or valgus deformity.

APPENDIX

Calculation of Desired Correction

There are two methods of calculating the desired correction for a proximal tibial osteotomy: on the basis of the anatomic tibio-femoral angle or on the basis of the mechanical tibio-femoral angle. The more accurate method—using the mechanical angle—is described here.

First obtain weight-bearing anterior-posterior view roentgenograms of the patient's entire lower extremities (femoral head to ankle). The x-ray machine should be centered at the knees. When the patient is too large to get the entire extremity on a single large film, three films can be taken and joined together later with clear tape. If this is done, markers must be placed on the thigh and calf so that the films can be cut and taped in such a way that the extremity length is correct.

The mechanical axes are then marked on the film—that is, one line from the center of the femoral head to the center of the knee and a second line from the center of the ankle through the center of the knee. The angle formed by the intersection of these lines at the knee is the mechanical angle. If we assume a varus deformity with a mechanical angle of 8 degrees, then the desired correction would be 8 degrees plus from 3 to 5 degrees,[3] or from plus 2 to 4 degrees.[17] If we compromise at 4 degrees then the total correction should be 12 degrees.

Because the mechanical angle cannot be measured radiographically in the operating room, the correction based on the mechanical angle must be converted to the anatomic angle. To do this, the uncorrected distal femur and proximal tibia should be traced on a separate piece of paper. This is either cut out, the correction made, and then traced on a second paper, or the femur and upper tibia are traced on paper, the correction made, and the tracing of the distal tibia is completed. Since the anatomic angle is the angle formed by the intersection of a line down the shaft of the femur and up the shaft of the tibia, it can be measured from the corrected tracing. This angle can then be used on the operating room postcorrection roentgenogram to be sure of having the desired correction.

This method of calculation of correction and conversion to anatomic angle is the only way to avoid potentially serious error when there is a deformity in either the femur or tibia outside of the knee area.

Coventry[5] has stated that 1 mm of the base of the wedge equals about 1 degree of correction, up to about 10 mm. We agree with this observation. Another method of measurement is to insert two small Steinmann pins with image control and then measure the angle directly.

REFERENCES

1. Ahlback, S.: Osteoarthrosis of the knee. A radiographic investigation. Acta. Radiol., (Supp.):277, 1968.
2. Baeur, G. C., Insall, J., and Tomihisa, K.: Tibial Osteotomy in Gonarthrosis (Osteoarthritis of the Knee). J. Bone Joint Surg., *51A*: 1545, 1969.
3. Chao, E. Y. S.: Biomechanics of high tibial osteotomy. In: Symposium on Reconstructive Surgery of the Knee. pp 143–160. American Academy of Orthopedic Surgery. C. V. Mosby Co. St. Louis, 1978.

4. Coventry, M. B.: Osteotomy of the upper portion of the tibia for degenerative arthritis of the knee. A preliminary report. J. Bone Joint Surg., *47A*:984, 1965.

5. Coventry, M. B.: Osteotomy about the knee for degenerative and rheumatoid arthritis. Indications, operative techniques, and results. J. Bone Joint Surg., *55A*:23, 1973.

6. Gýory, A. N., Chao, E. Y., and Stauffer, R. N.: Functional evaluation of normal and pathologica knees during gait. Arch. Phys. Med. Rehab., *57*:571, 1976.

7. Harris, W. R., and Kostuik, J. P. High tibial osteotomy for osteoarthritis of the knee. J. Bone Joint Surg., *52A*:330, 1970.

8. Insall, J., Shoji, H., and Mayer, V.: High tibial osteotomy. A five-year evaluation. J. Bone Joint Surg., *56A*:1397, 1974.

9. Insall, J., and Walker, P.: Unicondylar knee replacement. Clin. Orthop., *120*:83, 1976.

10. Jackson, J. P., and Waugh, W.: Tibial osteotomy for osteoarthritis of the knee. J. Bone Joint Surg., *43B*:746, 1961.

11. Johnson, R. J., Kettelkamp, D. B., Clark, W., and Leaverton, P.: Factors affecting late meniscectomy results. J. Bone Joint Surg., *56A*:719, 1974.

12. Kettelkamp, D. B., and Chao, E. Y. S.: A method for quantitative analysis of medial and lateral compression forces at the knee during standing. Clin. Orthop., *83*:202, 1972.

13. Kettelkamp, D. B.: Proximal tibial osteotomy. Clin. Orthop., *103*:46, 1974.

14. Kettelkamp, D. B., Wenger, D. R., Chao, E. Y. S., and Thompson, L.: Results of proximal tibial osteotomy. The effects of tibiofemoral angle, stance-phase flexion-extension, and medial phateau force. J. Bone Joint Surg., *58A*:952, 1976.

15. Kettelkamp, D. B., Johnson, R. J., Smidt, G. L., Chao, E. Y., and Walker, M.: An electrogoniometric study of knee motion in normal gait. J. Bone Joint Surg., *52A*:775, 1970.

16. Kruise, W. R., Pope, M. H., Johnson, R. J., and Wilder, D. G.: Mechanical changes in the knee after menisectomy. J. Bone Joint Surg., *58A*:599, 1976.

17. Maquet, P., Simonet, J., and de Marchin, P.: Biomécanique du genou et gonarthrose. In: Symposium les gonarthroses d'Origine Statique. Rev. Chir. Orthop. *53*:111, 1967.

18. Maquet, P. G. L.: Biomechanics of the Knee. Springer-Verlag, New York, 1976.

19. Marmor, L.: Single compartment replacement with the Marmor modular knee. Orthop. Rev., *6*:81, 1977.

20. Miller, R., Kettelkamp, D. B., Lambenthal, K. N., Karagioras, A., and Smidt, G. L.: Quantitative Correlations in Degenerative Arthritis of the Knee. J. Bone Joint Surg., *55A*:956, 1973.

21. Morrison, J. B.: The Forces Transmitted by the Human Knee Joint. Thesis, University of Strathcyde, Glasgow, Scotland, 1967.

22. Shoji, H., and Insall, J.: High tibial osteotomy for osteoarthritis of the knee with valgus deformity. J. Bone Joint Surg., *55A*:963, 1973.

23. Solnick, M. D., Bryan, R. L., and Peterson, L. F. A.: Unicompartmental polycentric knee arthroplasty. Clin. Orthop., *112*:208, 1975.

24. Steindler, A.: Kinesiology of the Human Body. p. 331. Springfield, Ill., Charles C Thomas, 1955.

25. Walker, P., and Erkman, M. J.: The role of the meniscus in force transmission across the knee. Clin. Orthop., *109*:184, 1975.

26. Wardle, E. N.: Osteotomy of the tibia and fibula. Surg. Gynecol. Obstet., *115*:61, 1962.

12

Total Replacement of the Osteoarthritic Knee with a Semiconstrained Device

P. A. Lotke, M.D.

INTRODUCTION

Osteoarthritis of the Knee Joint

Osteoarthritis is one of the most common causes of disability in the elderly, and the knee joint is second only to the hip as the most commonly affected joint. Although the etiology of osteoarthritis is unknown, the biomechanical factors contributing to arthritis can be clearly demonstrated within the knee. No joint in the human body must bear such varied stress as the knee. It must absorb high dynamic loads and attenuate impulsive shock loads. It must provide a wide range of motion while maintaining stability, as well as producing forces capable of propelling the body forward. In spite of such a complex and demanding function, there is very little intrinsic bony stability about the knee joint. Its entire structural integrity is maintained by ligaments and soft tissue structures. Additionally, it lies between two long bones which produce large stresses through long lever arms. Therefore, it is not surprising that the knee joint is one of the body's musculoskeletal units that is most vulnerable to external stress and to the development of the degenerative changes in osteoarthritis.

Since the knee joint is such a common site for the development of osteoarthritis, a brief background on this disease is appropriate here. Primary osteoarthritis, also called degenerative joint disease (DJD), is characterized by the progressive degeneration of cartilage, thickening of subchondral bone, remodeling of bone, and formation of marginal spurs and large cysts. Although metabolic and enzymatic factors are involved in the joint degeneration in DJD, the process is best thought of as a final common pathway of mechanical deterioration of the joint, which may occur from any severe imbalance of the stresses applied to the joint and the resultant inability of the joint structures to resist those stresses. The development of osteoarthritis occurs insidiously and slowly over a period of decades. Initially, the clinical manifestations are mild pain, stiffness, and swelling about the joint. With time and further deterioration of the joint, increasing effusions, synovial inflammation, and destructive changes become more apparent. The disease—which initially starts as a focally destructive process within the joint—gradually expands to include the entire joint.

Microscopic and biochemical events occur within the deteriorating articular cartilage. Human articular cartilage is comprised of two major components: collagen and

256

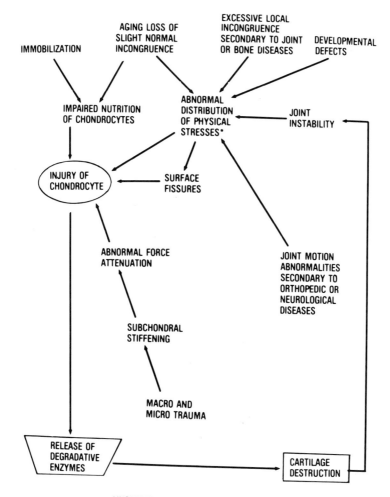

Fig. 12–1. Hypothetical effects of biomechanical factors in promoting osteoarthritis.

proteoglycan. Both are extensively damaged mechanically, culminating in the complete loss of both materials and exposure of eburnated bone. Chondrocytes are originally dispersed within the extracellular collagen-proteoglycan matrix. Focal areas of chondrocyte destruction, as well as focal areas of cellular proliferation and attempts at repair of the matrix, begin to appear. Initially, chondrocyte metabolic activity increases significantly, apparently to compensate for increasing damage and degradation of the chondrocytes external milieu, but later their activity diminishes below normal. Histologically, the staining properties of the proteoglycans are gradually lost, fissures then develop within the articular surface, followed by the development of erosions within the articular surface that enlarge; eventually, all normal articular matrix is destroyed.

As the degenerative process proceeds, the joint cartilage and capsule, together with the subchondral bone, become less able to distribute the loads of weight bearing. The frictional forces across the joint increase, the compressibility and elasticity of the articular cartilage diminishes, and destructive effects occur within the joint,

Unconstrained

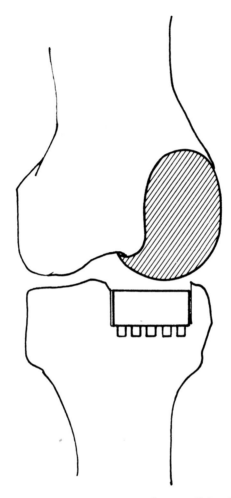

Fig. 12-2. Unconstrained knee replacement. May replace medial or lateral or both compartments. The flat tibial surface offers no restriction to motion, and the patient's own ligaments and surrounding soft tissues maintain stability.

reinforcing each other with increasing deleterious effect. A scheme developed by Howell et al.[4] demonstrates the importance of these mechanical effects (Fig. 12-1).

TOTAL KNEE REPLACEMENT: A TREATMENT FOR OSTEOARTHRITIS

Various medical and surgical treatments are available for osteoarthritis, but many patients reach a state in which the combination of pain, stiffness, deformity, and instability makes the affected joint practically unusable. Prior to this end stage of the disease, numerous therapeutic modalities are available to alleviate symptoms and perhaps slow the progression of the destruction. At first, mild antiinflammatory

Semi-constrained

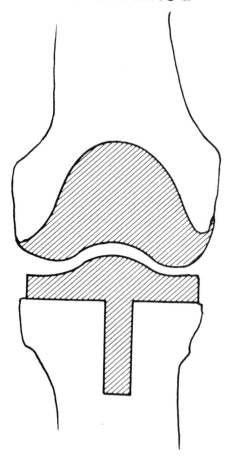

Fig. 12–3. Semiconstrained knee replacement. Various degrees of tibia conformity increase continuity and serve to stabilize the prosthesis.

medications, such as salicylates or indomethecin, may be prescribed; later, physical therapy, braces, and a cane or crutches will be required to reduce loads on the affected joint. The occasional use of intraarticular injections of cortisone derivatives can frequently be quite effective. But despite these treatments, and after decades of conservative care, there will be a group of patients whose knee joint is so severely affected that a total knee replacement may be the only treatment.

In general, there are three basic types of total knee replacements: the unconstrained; semiconstrained; and totally constrained or hinged device. The unconstrained device consists of a flat surface replacing the tibial plateau and a curved surface replacement covering the femoral condyles (Fig. 12–2). This curve-on-flat configuration offers no restraint to motion, and all support from this device is derived from the surrounding ligaments and soft tissue.

Semiconstrained devices comprise a variety of total knees consisting of tibial plateaus with concave grooves or cups into which a convex, curved femoral mold is fitted. The conformity and congruence of the femoral condyle to the molded tibial pla-

Hinge

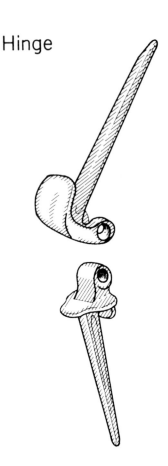

Fig. 12–4. Hinged knee replacement. These devices allow only one range of motion, flexion and extension, and require no ligament support from the patient. (Adapted from Howell, D. S., Sapolsky, A. I., and Pita, J. C.: The pathogenesis of osteoarthritis. Semin. Arthr. Rheum., 5:365–383, 1976.

teau varies from device to device and adds increasing constraint with increasing conformity. These devices are also nonarticulated, and their function depends on the integrity of the surrounding ligaments and soft tissues (Fig. 12–3).

The third type of total knee replacement is the totally constrained or hinge-type replacement (Fig. 12–4). These consist of long-stemmed devices inserted into the medullary canal of the femur and tibia and articulated together through a hinge mechanism. The axis of this hinge may be placed in variable locations depending upon its design, but the freedom of motion about the knee is confined to flexion and extension. Since the device is totally constrained, no ligaments are necessary to support this prosthesis. These devices are discussed in the next chapter.

REPLACEMENT OF THE KNEE: CLINICAL APPLICATION

In this chapter we will discuss exclusively the use of the semiconstrained knee prosthesis of the total condylar type.

One of the most commonly used semiconstrained knee prostheses in the United States is a total condylar type of total knee (Fig. 12–3). This prosthesis allows the knee to flex up to 105 degrees with a controlled amount of rotation in the axial and sagittal planes. The large conforming contact area between the femoral component and the tibial component reduces contact stresses and enhances the stability of the prosthesis. The femoral component is made of chrome-cobalt-molybdenum alloy or stainless steel, with highly polished articulating surfaces. The tibial piece is made of an ultrahigh-molecular weight polyethylene. The concave articulating tibial surfaces are slightly larger in diameter and slightly longer in the anteroposterior dimension than the femoral component. The central peg projects into the central position of the tibia and rests obliquely against the posterior tibial cortex. The components are "cemented," that is, imbedded in the bone with polymethyl methacrylate cement. This device has shown itself to be highly successful, although it may not represent the ultimate design in semiconstrained knee prostheses.

The goals of the total knee replacement are relief of pain, correction of deformity, and increase of function, in that order of priority. At present, these goals may be achieved for up to 5 years in 90 to 95 percent of patients. However, physicians must always be aware of the possibility of failure, and that there are a certain number of patients who will have some complications. Extreme care must be used to prevent these problems, and we must have available adequate salvage techniques.

The major short-term problem with all total knee replacement surgery is infection, which occurs in 0.5 to 1 percent of all cases. This frequently is noted in the perioperative period, but may also occur from 1 to 10 years later. Some of these infections may be treated with device removal, antibiotic therapy, and reimplantation, but most will require removal of the prosthesis and fusion of the knee. Therefore, one of the requirements in the design of a total knee is to avoid removal of so much bone as to prohibit adequate salvage, or a reasonable chance to obtain fusion of the knee.

The major long-term complication is mechanical loosening of the bone-cement interface. When this occurs it is associated with pain, deformity, and swelling about the knee. This may occur suddenly from trauma, or insidiously through excessive stresses applied to the device (see the section on biomechanics of total knee replacement). Correct surgical technique and proper patient selection may help to prevent this complication.

We describe below a series of cases in which semiconstrained total knee prostheses have been used. These cases will demonstrate the rewards versus the risks in total knee surgery. They serve as guides in discussing the results, in view of the biomechanical events which have contributed to their success or failure. Whether or not to perform surgery on any individual patient is a complex decision involving many factors including age, weight, disability, and previous procedures. In general, the most common reason for surgery is relief of pain. However, osteoarthritis is an insidious disease with gradual onset of pain, increasing intermittently over the decades. How then does the clinician decide upon the appropriate time for surgical intervention? Hopefully, these cases will give some insight into the clinical judgments for surgery.

CLINICAL CASES

Case 1. This patient was a 72-year-old female with longstanding osteoarthritis of the left knee. She had pain after activity and could hardly walk more than one block. A cane offered her some relief. She had tried numerous medications, including high-dose aspirin and other antiinflammatory medications, numerous intraarticular cortisone injections, and braces. She obtained varying degrees of relief, but continued to have pain and increasing disability. Her examination revealed bilateral genu varum deformities (bowed legs) of 15 degrees (Fig. 12–5). When she walked she had

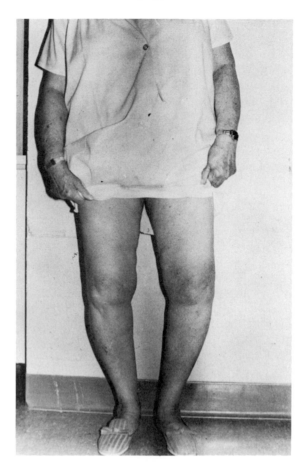

Fig. 12 – 5. Case 1: Preoperative photograph showing marked varus deformity of both knees.

a lateral "thrust" at heel strike. Her range of motion was limited by a 5 degree flexion contracture and further flexion of 110 degrees. There was crepitus about the joint, worse on the medial side, and tenderness along the medial border of the knee. There was also gross medial-lateral instability of approximately 15 degrees. The synovium was thickened and there was a mild effusion. The x-ray of her right knee (Fig. 12 – 6) shows changes typical of severe medial femorol-tibial osteoarthritis, with collapse of the medial compartment, mild subluxation, sclerosis, small cysts, moderately-sized osteophytes, and an 11 degree angular deformity (best seen in this standing anteroposterior view).

   This patient was an ideal candidate for total knee replacement. She had increasing disability and pain despite good medical management. She was in the older age group, active, and had a deformity which could be corrected by utilizing the patient's own ligamentous support. She had good bone support and no evidence of previous infections. The patient received a total condylar-type total knee (Fig. 12 – 7). Note that the tibial plateau sits horizontally on the bone support. There was a minimum amount of cement used for fixation, the femoral component is in 5 degrees of valgus, and the knee can extend fully. The forces on the tibial plateau are well distributed over a broad area on the plateau. The posterior tibial cortex, which has an oblique

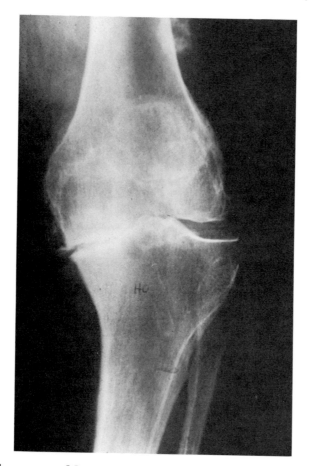

Fig. 12 – 6. Right knee, x-ray of Case 1. The medial compartment is completely obliterated and there is early medial subluxation of the femur on the tibia.

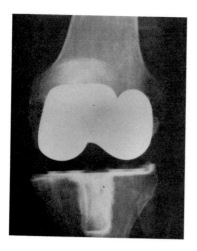

Fig. 12 – 7. X-ray of total knee replacement. Case 1. The femoral component is in good alignment and the polymethylmethacrylate cement appears well distributed around the base of the polyethylene tibial component, not visualized by radiographs.

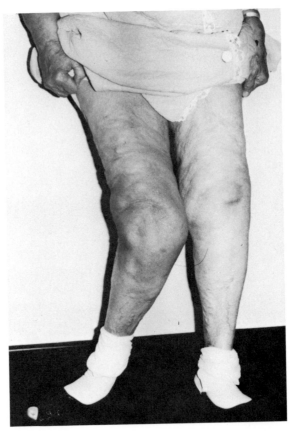

Fig. 12–8. Case 2, an obese female with a valgus deformity of the right knee. This has been increasing for many years.

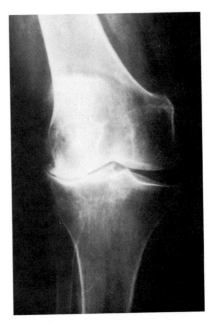

Fig. 12–9. X-rays of right knee from Case 2. Note similar findings on the lateral aspect of the knee as seen in the previous patient. She has collapse of the lateral compartment and subsequent angular deformity.

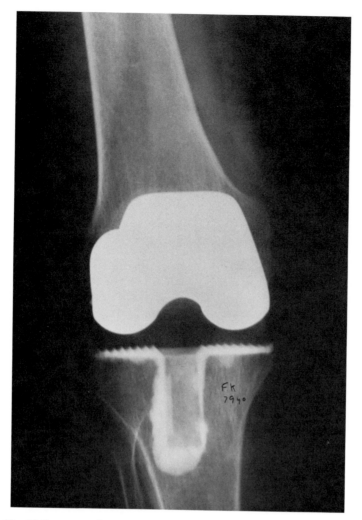

Fig. 12–10. Postoperative x-ray of case 2. This device is also well positioned.

configuration in the midsection, is accepting the vertical forces from the central peg of the tibial plateau. This patient did well and was doing well at her 4 year follow-up examination. She had no pain, could walk as far as her general health allowed, flexed to 100 degrees, and noted only minimal discomfort after a full day's activity.

Case 2. This was an obese 58-year-old female with increasing pain in the right knee and valgus (knock-knee) deformities (Fig. 12–8). The patient had also tried numerous medications and intraarticular cortisone injections with varying degrees of relief. Because of increasing pain, she could hardly walk. Her examination showed a 15 degrees valgus deformity with moderate synovitis and effusions. She had 10 degrees of medial-lateral instability, which could not be corrected to neutral. There was tenderness and crepitus over the lateral aspect of her knee. Her x-rays showed lateral compartment osteoarthritis with narrowing of the joint line, sclerosis, osteophytes, and a valgus deformity (Fig. 12–9).

This patient presented several problems in the decision to perform total knee sur-

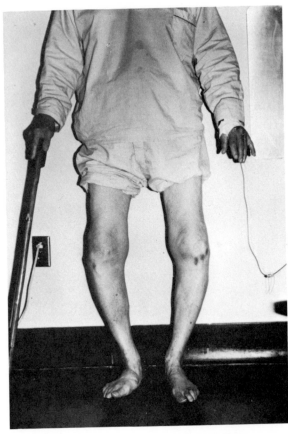

Fig. 12–11. Preoperative photograph of case 3. This patient has severe bilateral varus deformities.

gery. She was an obese, active female in the younger age group. The stresses she would apply to any prosthetic device would be greater and for a longer duration than would a frail, thin person. We knew that the joint reaction forces are 4 to 10 times body weight for everyday activities, but we did not know if any artificial knee device could withstand 20 years of these repetitive, heavy loads before it or the underlying bone failed. Other prostheses, such as the total hip replacement, will fail with time and overuse. However, if we did nothing, the patient would continue to have pain, be unable to walk, and be increasingly disabled. After due consideration and careful discussion with the patient, she had a total knee replacement. At surgery the ligaments on the lateral side of both knees were released in order to achieve proper alignment. Postoperatively, her prosthesis position is excellent (Fig. 12–10), and she has done well since surgery. At her last follow-up examination, 3 years after surgery, she had lost 50 lb, was continuing to work, was avoiding excess stress on her knees, and was returning for regular follow-up visits.

  Case 3. This patient was a 63-year-old male with longstanding medial-femoro-tibial osteoarthritis with severe varus deformity. He had inconsistently taken numerous antiinflammatory medications with varying results. He refused routinely to use a cane for cosmetic reasons, and had seen four different physicians in the previous 2

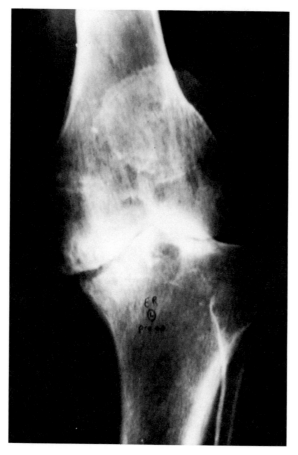

Fig. 12–12. Preoperative x-ray of case 3. He has severe deformity involving both compartments of the knee, resulting in significant bone loss from the medial tibial plateau.

years. His examination showed a 10 degree varus deformity with subluxation (Fig. 12–11). There was a marked synovitis and mild effusion. The knee had a range of motion from 10 to 90 degrees, with pain at the extremes of motion. There were 10 degrees of medial-lateral instability, and the varus deformity could not be corrected to neutral. The x-rays (Fig. 12–12) showed narrowing of the medial joint line, subluxation, sclerosis, and osteophytes.

The patient had a total knee replacement at another, unidentified hospital. The prosthesis was seated in varus on the tibial plateau (Fig. 12–13). The medial and lateral ligaments were not released to allow the knee to come into proper alignment, and the leg remained in varus. The patient did not cooperate in the postoperative period and never did well after surgery. He continued to have pain, varus deformity, recurrent effusions, and increasing disability when seen by us 1½ years after surgery. His x-rays showed the initial poor position of the device immediately after surgery (Fig. 12–13). With use, there were excessive stress concentrations which eventually caused failure of the underlying bone (Fig. 12–14). After the bone began to fail, it continued to fragment. The device deformed and the patient experienced continued pain, deformity, and significant loss of bone stock (Fig. 12–15).

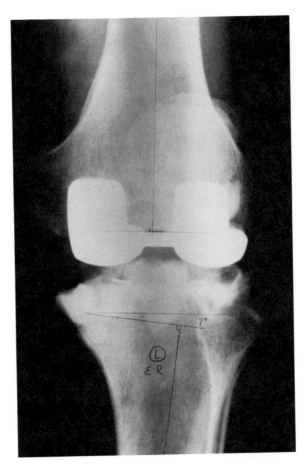

Fig. 12–13. Postoperative x-ray of earlier model total knee. It is positioned with the knee still in varus and there has been an attempt to make up the bone loss on the medial side of the tibia with cement.

The latter problems made further reconstruction more difficult. Several mechanical factors were operative in the failure of the device. First, the patient was heavy, and this excessively stressed the device. Next, the device was positioned in varus, increasing the stress on the medial compartment (Fig. 12–16), and third, this particular design of prosthesis allows for stress risers to develop beneath the medial and lateral tibial "spikes." All of these factors, plus other undetermined forces, led to failure of the device and the need for a secondary reconstructive procedure. The patient had a salvage procedure with a total condylar device. This was satisfactory, and at follow-up the patient remains pleased with his result.

Case 4. This is a 68-year-old female who sustained a lateral tibial plateau fracture in an automobile accident. Seven years later she developed lateral compartment osteoarthritis with pain, deformity, and disability which required a total knee replacement. Initially, she did well but in 4 to 6 months noted a progressive valgus deformity and increasing pain. She presented for evaluation with the x-rays shown in Figure 12–17.

This patient had sustained a rupture of the medial collateral ligament at the time

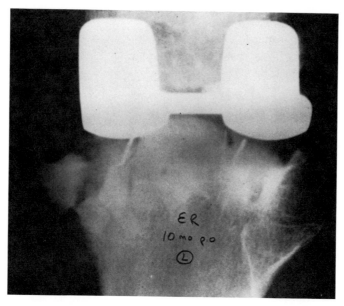

Fig. 12–14. Case 3, 10 months postoperatively on initial total knee replacement. The bone has begun to fracture beneath the device.

of her initial injury and lateral plateau fracture. This rupture had not been recognized, and the tensile forces on the medial collateral ligament gradually stretched the remaining portions of the ligament, creating a valgus deformity. As the deformity increased, the angle increased and the stresses on the ligament and prosthesis became greater and created increasing pain and deformity. This was treated by replacing the total knee and repairing the medial collateral ligament with a staple.

Biomechanically, it is important to appreciate the stresses applied to the ligament about the knee (Fig. 12–18). Any abduction-adduction force is translated to the collateral ligaments. These must be intact or repaired in order to prevent deformity. If the ligaments are not present, a constrained device, which is capable of compensating mechanically for the loss of these ligaments, must be selected.

BIOMECHANICS OF TOTAL KNEE REPLACEMENT

In order to better understand the clinical application of a total knee replacement, it is important that we review the biomechanical requirements which this prosthesis must satisfy. We shall discuss the knee in relation to the motions required for daily activities, and the forces acting across the joint surfaces.

Kinematic function—the range of motion of the knee joint in a variety of daily functions—has been determined.[5] It has been shown that the knee flexes and extends but that, in addition, there are large angular displacements into varus and valgus, as well as axial rotation. During normal walking, beginning at the stance phase, the knee is in extension with the maximum external rotation and maximum abduction of the tibia. Between heel-strike and flat-foot there is flexion, internal rotation, and adduction. From flat-foot to heel-off, a maximum stance-phase flexion of 20 degrees is achieved and extension begins. Usually, internal rotation continues and

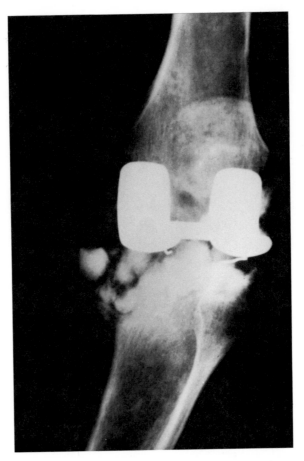

Fig. 12–15. Case 3, 2 years after surgery. The bone and cement have begun to fragment. The prosthesis is loose and deformed. The patient has increasing pain and deformity. He required revision.

adduction decreases. Between heel-off and toe-off, extension is completed and the flexion of swing-phase begins. Internal rotation usually increases, and adduction increases, to toe-off. In swing-phase, a maximum flexion of 67 degrees in maximum adduction and internal rotation occurs. As extension toward heel-strike occurs, the maximum external rotation and abduction coincide. The adduction-abduction range of motion averages 11 degrees and the internal-external rotation averages 13 degrees.

Activities other than walking require greater ranges of motion. (Table 12–1) Sitting requires up to 90 degrees of flexion, depending on the height of the individual and the chair. Walking up stairs requires 83 degrees, and bending to lift an object requires 71 degrees. In these motions the normal person needs an average of 17 degrees of adduction-abduction, and 16 degrees of internal-external rotation. When setting goals for replacement surgery it is important to attempt to achieve these ranges of motion, which are dictated to a degree by the muscles, capsular structure, and articulating surfaces of the joint.

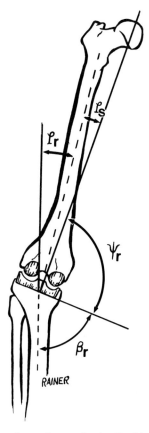

Fig. 12–16. Schematic representation of prosthesis. Position in varus with displaced tibial component.

KINETIC ANALYSIS

With kinetic information, it is possible by indirect methods to determine the forces in static or dynamic situations on the knee joint. In the static state all forces are in equilibrium and the body is at rest. Vector analysis of the static joint-contact force on the femoral tibial articular surface, by Frankel and Burstein,[2] has indicated a force of 4.1 times body weight in stair climbing. The dynamic state requires a more complicated analysis, such as carried out by Morrison.[7] He calculated the external forces acting upon the joint during walking. He used a walkway, an electromyograph, filmed the gait cycle, and measured the forces applied at each instant during the cycle. In Morrison's studies it was first necessary to find the accelerations about the knee and the external forces acting on the knee. He was then able to calculate the forces and identify the muscles that were contracting. He noted that the knee joint forces varied from two to four times body weight, with an average of three times body weight during level walking (Fig. 12–19).

Examination of the individual muscle sources acting across the knee indicated that the first peak stress occurs due to hamstring contraction at heel-strike; that the next peak occurs from quadriceps contraction at midstance; and that the third peak oc-

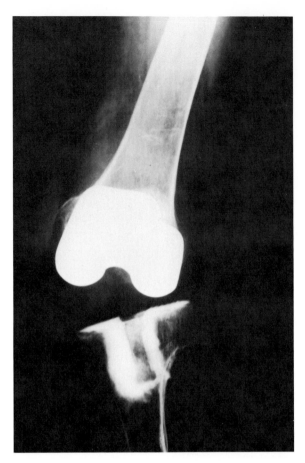

Fig. 12–17. Case 4 after a total condylar total knee replacement. The alignment at the tibial and femoral components is correct, but the knee is in severe valgus because of the loss of the medial collateral ligament. This required surgical repair.

curs from gastrocnemius function during heel-off (Fig. 12–20). The muscle forces increased significantly with activity such as walking up a ramp or up stairs. By measuring the other orthogonal axis, the forces on the cruciate and collateral ligaments were also calculated. In general, a mean maximum force of 74 lb was applied to the posterior cruciate ligament, which was approximately twice the forces applied to the anterior cruciate and medial collateral ligaments, respectively. It is of interest to note the ratio of the forces applied to these ligaments is the same as the overall strength of these ligaments, since it has been noted that the posterior cruciate ligament is twice as strong as the anterior cruciate and medial collateral ligaments.[8]

Morrison also measured the position of the center of the force on either condyle during the gait cycle.[7] It was noted that during the stance phase of walking, the center of pressure was positioned over the medial condyle. However, it shifted to different positions of the joint during the gait cycle (Fig. 12–21). From a mechanical point of view, in the normal knee, a greater portion of the force transmitted by the medial condyle, as opposed to the lateral condyle, would be structurally more favorable, since the medial condyle has a large bearing surface and compressive stresses on this particular surface would be lower. From a mechanical point of view in a pros-

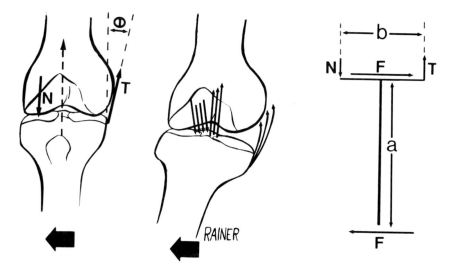

Fig. 12-18. There is a simple couple which resists abduction. The abduction is accompanied by medial-lateral translation which is resisted in part by the medial-lateral component of the contact force of the semiconstrained device and the medial collateral ligament. Absence of the ligament leads to malalignment of the knee and deformity, as noted in case 4.

thetic knee, the effects of these joint reaction forces, the ligament and muscle forces, and the shifting of the weight distribution during the gait cycle to various portions of the joint creates within the prosthesis tongues and stresses which may be important to the loosening of the prosthesis, as will be discussed later.

The instant center of rotation is another important kinetic concept in the knee.[3] It has been known for many years that the instant center of rotation in the knee is constantly changing, and it is possible to construct a pathway along which the instant center moves as the knee joint goes from flexion to extension (see Ch. 17). As the knee moves from flexion to extension the surface velocity can be measured on the condylar surface to the femur. The least resistance of motion will occur when the direction of the surface velocity at the contact point is tangent to the contact surface. This condition obtains when the instant center lies along a line which is perpendicular to the articular surfaces at their contact point (Fig. 12-22). If an instant center does not lie on a line perpendicular to the articular surface at the contact point, sliding may take place, but the motion will tend to separate or compress the joint surfaces, increasing the frictional compressive forces. As the development of osteoarthritis increases, certain structural changes occur in the joint; these include

TABLE 12-1. MOTION OF THE KNEE IN DAILY ACTIVITIES

	Flexion-Extension	*Swing-phase*	*Stance-phase*	*Abduction Adduction*	*Rotation*
Walking	0°	67°	21°	11°	13°
Stairs	83°	—	—	17°	16°
Sitting	90°	—	—	15°	14°
Lifting	71°	—	—	12°	13°

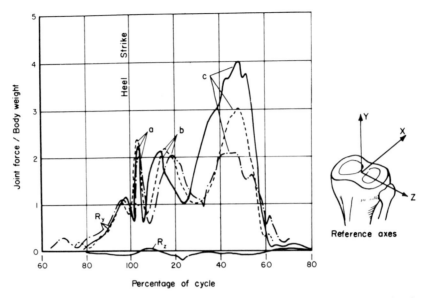

Fig. 12–19. Joint forces at knee during level walking. (Morrison, J. B.: The mechanics of the knee joint in relation to normal walking. J. Biomech. *3*:51–61, 1970.)

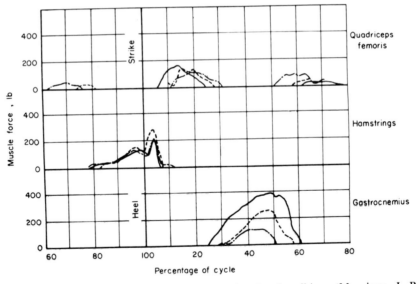

Fig. 12–20. Muscle forces acting across knee during level walking. (Morrison, J. B.: The mechanics of the knee joint in relation to normal walking. J. Biomech. *3*:51–61, 1970.)

flexion contractures and varus and valgus malalignment. The results of these deformities usually lead to a decreased tibio-femoral contact surface area for weight bearing, and to increased joint-surface forces. This maldistribution of forces serves to increase the destructive changes occurring in osteoarthritis and to hasten the total destruction within the joint.

When considering the development of a semiconstrained total knee replacement, we would, therefore, take into consideration all of the forces and motions acting on the joint. If due consideration is not given to these forces in the design of the pros-

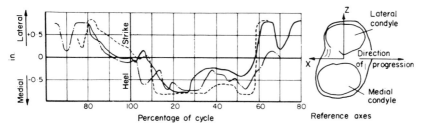

Fig. 12–21. Position of the center of pressure on the condyles during level walking. (Morrison, J. B.: The mechanics of the knee joint in relation to normal walking. J. Biomech. *3*:51–61, 1970.)

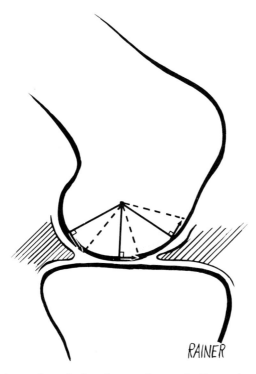

Fig. 12–22. The direction at the velocity of any point on the femoral component can be found by connecting that point to the instant center. The least resistance to motion will occur when the direction of the velocity at the contact surface point is tangent to the contact surface.

thesis, excessive stress will be applied to the prosthetic bone interface, and loosening may occur.

Chao and Mullen[1] have developed a theoretical analysis of the loosening forces about the knee joint. They state that the mechanical loosening of the tibial component may be related to an imbalance in the distribution of the contact forces on each component rather than to the magnitude of the forces (Figs. 12–23, 12–24, and 12–25). Chao and Mullen have used Morrison's data to compute the tibial loosening-moment resultant forces in the component throughout the stance-phase of walking. They have noted that anatomic limits have been considered in designing knee prostheses, but that no mechanical limits have been established for component orientation. Chao and Mullen note that slight variations in prosthetic orientation significantly alter the contact force distribution between the medial-lateral joint sur-

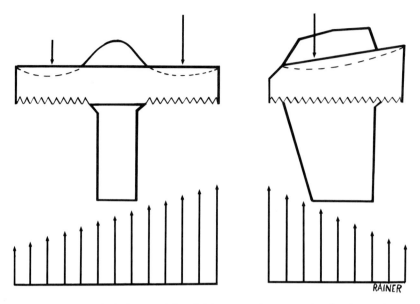

Fig. 12–23. An imbalance of the tibial contact force may lead to loosening.

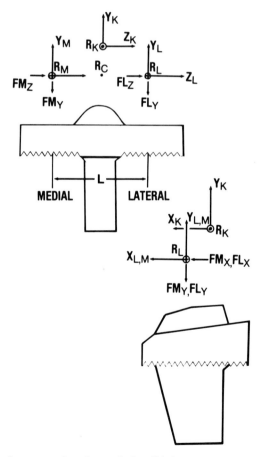

Fig. 12–24. Resultant forces passing through the tibial component. R_K, knee joint center. R_C, center for tibial component. R_M, medial plateau center, and R_L, lateral plateau center.

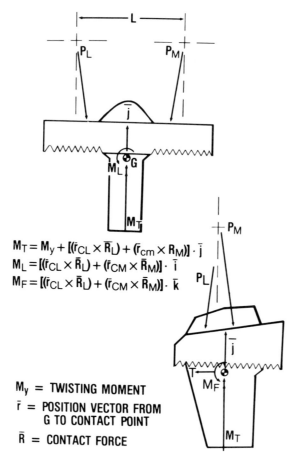

$$M_T = M_y + [(\bar{r}_{CL} \times \bar{R}_L) + (\bar{r}_{cm} \times R_M)] \cdot \bar{j}$$
$$M_L = [(\bar{r}_{CL} \times \bar{R}_L) + (\bar{r}_{CM} \times \bar{R}_M)] \cdot \bar{i}$$
$$M_F = [(\bar{r}_{CL} \times \bar{R}_L) + (\bar{r}_{CM} \times \bar{R}_M)] \cdot \bar{k}$$

M_y = TWISTING MOMENT
$\bar{r}$ = POSITION VECTOR FROM G TO CONTACT POINT
$\bar{R}$ = CONTACT FORCE

Fig. 12–25. Calculations can be made of the loosening moment with respect to the contact surface.

faces. Thus not only do the forces from normal walking create imbalances, but the malposition of the prosthesis increases these loosening forces. Therefore, unbalanced force distribution on the tibial component from exercise torque transmitted to the knee joint during the gait cycle, and due to variations in prosthetic orientation, will be a main mechanical factor contributing to failure and loosening in knee prostheses.

The clinical effects of prosthesis malposition have been clearly shown by Lotke and Ecker.[6] By reviewing a series of patients with total knee replacements, they were able to grade the results in relation to success with the patient, as well as to the position of the device. There was a statistically significant correlation between those patients who had a well-positioned prosthesis and those who had the best clinical results. In addition, there were fewer complications in those patients who had well-positioned prostheses. By analysis (Fig. 12–16), we can appreciate that a prosthesis that is positioned in varus (defined as $\beta_r > 90$ degrees), or a knee positioned in varus (defined as $\beta_r + \psi_r < 180$ degrees) will lead to increasing forces on the medial compartment of the knee both at the femorotibial prosthetic interface and at the medial tibial bone-cement interface. It can be appreciated that this situation is similar to that in Case 3, leading to the failure of a total knee replacement.

In summary, the biomechanical events that occur across the knee joint are important in the development of osteoarthritis of the knee and are important to the success or failure of a total knee replacement. Adequate knowledge and appreciation of these events is imperative for the successful use of a semiconstrained total knee replacement for the treatment of osteoarthritis of the knee.

REFERENCES

1. Chao, E. Y., and Mullen, J. O.: Theoretical and experimental analyses of the interface strength in geometric total knee replacement. Closed Loop, 6:3–16, 1976.
2. Frankel, V. H., and Burstein, A. H.: Orthopaedic Biomechanics. Lea & Febiger, Philadelphia, 1970.
3. Frankel, V. H., Burstein, A. H., and Brooks, D. B.: Biomechanics of internal derangement of the knee. J. Bone Joint Surg., 53A:945–962, 1971.
4. Howell, D. S., Sapolsky, A. I., and Pita, J. C.: The pathogenesis of osteoarthritis. Semin. Arthr. Rheum., 5:365–383, 1976.
5. Kettelkamp, D. B., Johnson, R. J., Smidt, G. L., Chao, E. Y., and Walker, M.: An electrogoniometric study of knee motion in normal gait. J. Bone Joint Surg., 52A:775–790, 1970.
6. Lotke, P. A., and Ecker, M. L.: Influence of positioning of prosthesis in total knee replacement. J. Bone Joint Surg. 59A:77–79, 1977.
7. Morrison, J. B.: The mechanics of the knee joint in relation to normal walking. J. Biomech. 3:51–61, 1970.
8. Noyes, F. R., Torvik, P. J., Hyde, W. B., and Delucas, J. L.: Biomechanics of ligament failure. II. J. Bone Joint Surg., 56A:1406–18, 1974.

13
Total Replacement of the Knee in Rheumatoid Arthritis

N. G. Gschwend, M.D.; U.P. Wyss, Ph.D.

THE PATHOLOGY OF THE KNEE IN RHEUMATOID ARTHRITIS

Knee involvement is very frequent in rheumatoid arthritis. A study of 300 patients with an average duration of rheumatoid arthritis of 10 years has revealed an incidence of 74 percent. Pain and swelling are the major symptoms manifested in the early stages of the disease. The principal site of involvement in this systemic disease is the synovial layer; this becomes several times thicker than normal, has a villous-polypoid appearance, and produces a joint effusion that is characteristic of rheumatoid arthritis.

Joint destruction takes place in several ways. Some of these are more biologic and others more mechanical in character. The destruction of the joint cartilage by developing pannus may, for example, be regarded as biologic. Arising first as a vascular membrane, the pannus overgrows parts of the joint surface, proceeding from the synovialis-covered marginal zones, until it finally covers the cartilage. As a result of this, blood vessels and cellular elements penetrate the normally avascular cartilage, also preventing nutrition of the cartilage by the synovial fluid of the joint space. Joint destruction as a result of overstretching of the capsulo-ligamentary elements of the joint through chronic effusion may, however, be regarded as partly mechanical. Since the joint fluid in this case is rich in so-called lysosomal enzymes and poor in normal nutrients—leading to the well-known softening of the cartilage (chondromalacia)—joint destruction through effusion shows a biologic aspect, too.

The pain arising from the processes just described is also of a mixed biologic-mechanical character. In this connection, mention should be made of the pain-induced "sparing" of the knee in a best-possible relaxed, mostly flexed position by patients themselves, and leading on the one hand to atrophy of single muscle groups and on the other to contracture of these muscles' antagonists. Flexion contractures—in cases of a weak quadriceps muscle and shortened hamstrings; or valgus—which above all is characteristic of juvenile polyarthritis—underlying among others a shortening of the lateral structures (ilio-tibial tract, biceps muscle of the thigh) and a weakening of the medial antagonists (muscles of the pes anserinus), serve as typical examples. Fixed external rotation of the lower leg, which in most cases is combined with flexion- and valgus-deformity, can be attributed to the same pathogenic factors.

The derangement in joint mechanics resulting from all of these fixed deformities is of a distinctly mechanical character. If essential elements of the joint structure, such as the menisci of the ligaments, are destroyed at the same time—a worse condition yet—joint disintegration proper follows in the course of time. It is above all the insufficiency of the cruciate ligaments that contributes to a breakdown in normal knee joint kinematics, which has a secondary effect on the collateral ligaments, provided that these were not already insufficient.

Localized overpressure in one part of the joint results in a gradual local loss of the joint cartilage, which has already been damaged biologically, and subjects the underlying and frequently already affected bone (steroid osteoporosis, inactivity-induced atrophy) to stresses which it can withstand for a limited time only. The collapse of one tibial plateau or femoral condyle very rarely results from an accident-like extraordinary occurrence; in most cases it is the result of a continual accumulation of "microcracks" over months or even years, or of the confluence of several erosions into a larger "crater." Once this has occurred, there follows an incessant, vicious circle, in which the cause and the effect influence each other reciprocally and detrimentally. Severe and frequently almost grotesque valgus knees, displaying as much as 70 degrees of angulation, are not at all rare in advanced polyarthritis. Less frequently observed are varus knees, which occur primarily in osteoarthritis and—in contrast to valgus knees, which are more connected with hyperextension—are more frequently associated with flexion contracture.

Parallel with the destruction of the tibio-femoral joint goes the destruction of the patello-femoral compartment. This can be attributed as much to biologic as to mechanical factors. The former include chondromalacia, which is predominantly attributable to the action of lysosomal enzymes; the latter include localized and unilateral overstresses observed also in osteoarthritis of the femoro-patellar joint, in consequence of a changed direction of pull of the extensor mechanism. The increased external rotation of the lower leg alters the so-called Q-angle—the angle between the patellar ligament and the pull direction of the quadriceps muscle. The same is to be expected in the case of the pathological valgus knee. The overstretching of the medial, and the contracture of the lateral retinacular ligaments leads to the fixation of this pathological condition. The fixed flexion deformity of the knee joint destroys, in its turn, the cartilage of the patello-femoral joint through overpressure. Here, too, frequent collapse of one or the other articular facet can be consequently observed, this being undoubtedly promoted in most cases by the weakening of the bone as a result of erosion, osteoporosis, or both.

In the case of progressive joint destruction, the clinical picture is more and more dominated by pain, increasing limitation of movement, loss of stability, and reduced ability to walk, with complete loss of ambulation in severe cases.

The purpose of every therapy is to prevent or arrest this extensive joint disintegration. This can be achieved casually by any measure capable of effectively eliminating the inflammatory swelling of the synovial membrane (synovitis) at the earliest possible stage—that is, before any of the critical elements of the joint structure (above all the articular cartilage and the ligaments) are so irreversibly damaged that the degenerative process continues, even after healing of the inflammatory component of the disease, ultimately leading to the final picture of arthritic destruction.

In cases in which drug therapy cannot eliminate the swelling within a useful period of time (in the first year after the onset of the disease), the synovial sheet is completely removed (synovectomy), either surgically or by the application of radioisotopes (β-ray emitters such as yttrium).

If the articular cartilage and the ligaments are satisfactorily preserved, corrective

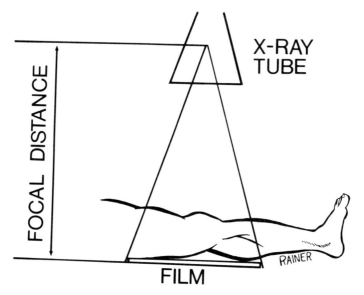

Fig. 13 – 1. Arrangement for x-ray photography.

osteotomies or operations designed to relax the soft tissues can save the joint for at least some years. Soft tissues operations – as for example severing or lengthening of the iliotibial tract and the biceps femoris muscle, or release of the posterior capsule and lengthening of the hamstrings – have proved effective above all in juvenile polyarthritis of the growing individual. The foregoing is true as well of osteotomies in the vicinity of the knee, which also have a chance of success provided that the inflammatory activity has largely subsided, and the matter comes only to one of mechanically saving a still sufficiently congruent joint from destruction through overpressure.

In cases in which the joint congruence has been irreversibly destroyed, whether by complete disintegration of the cartilage, extensive erosive changes, collapse of the articular surfaces, or through severe ligament insufficiency, only prosthetic replacement of the knee can save the function of the joint, or at least restore it partly. In principle, this is possible in various ways:

1. By partial or total prosthetic surface replacement (condylar prosthesis).
2. By hinge-like joints which can be constrained or nonconstrained, most of the latter having some degree of rotation and a physiologically migrating axis. Nonconstrained joints are therefore called "physiological hinges." The purpose of this chapter is to analyze the biomechanics of these hinges and their possible short-comings.

Since there are individual differences in the size and anatomy of humans, it is obvious that a single prosthesis size cannot be used for all patients.

The implantation of a prosthesis should be carefully planned on the basis of radiographs and the results of other examinations. In order to be able to determine the correct size of the prosthesis to be implanted, the orthopedic surgeon needs accurate radiographs. Experience has shown that too little attention has hitherto been paid to this major point. As illustrated in Figure 13 – 1, anteroposterior and lateral x-ray photographs can be taken with a magnification of 5 to 15 percent. The radiographic measurements must be then calculated back to actual data in order to determine the appropriate prosthesis size. Relevant tables for this can be obtained from x-ray equipment manufacturers.

The precautions to take in obtaining radiographs are:

1. Film focus distance: 120 ± 5 cm.
2. Film placed directly under the leg (smallest possible film-bone distance in order to prevent extreme magnifications).

TYPES OF KNEE PROSTHESES USED IN TREATING RHEUMATOID ARTHRITIS

Since all knee prostheses with a fixed axis do not have the same geometry, the biomechanics of the respective implants will thus also vary. The biomechanical differences are illustrated by comparing the Walldius and Shiers prosthesis with one having a traveling axis (GSB). The Guepar, Attenborough, and Sheehan prosthesis (not illustrated) will also be touched upon in this connection.

Design details have been intentionally omitted, and only major relationships are discussed. This should facilitate the choice of a prosthesis in rheumatoid arthritis. The biomechanics of the ligaments need not be discussed in connection with prostheses for the treatment of severe rheumatoid arthritis since the prostheses mentioned have been designed in accordance with the clinical picture of the disease, so that the ligaments are consequently not really required. The three prostheses studied and compared with each other will now be described briefly.

Shiers prosthesis

Experiments with the Shiers prosthesis were conducted as early as 1950, and the first prosthesis of this type was implanted in 1953. As shown by Figure 13–2, it consists of a femoral and a tibial component, which are bolted together after insertion into the respective medullary cavities so as to provide a fixed axis. Both components are cemented intramedullary.[16]

Walldius prosthesis

The Walldius prosthesis also consists of a femoral and a tibial component that are joined by means of a bolt after insertion (Fig. 13–3). The major feature of this prosthesis is that it may be anchored without acrylic cement. Clinical experience gathered with this design since 1957 has been reported in the literature.[19]

GSB prosthesis

The first prosthesis of the GSB design (Fig. 13–4) was implanted in 1972. Although it does not have a fixed axis, it should be compared with the above prostheses, because it can be used in cases having the same operative indications. The femoral component, with its central box, is slipped over the fin of the tibial component, whereby the prosthetic components are aligned with the axis of the femoral component, without, however, being linked rigidly.[4]

Fig. 13 – 2. Shiers total knee replacement.

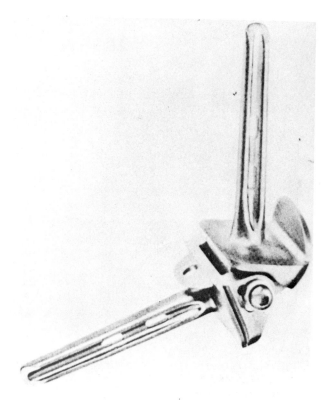

Fig. 13 – 3. Walldius total knee replacement.

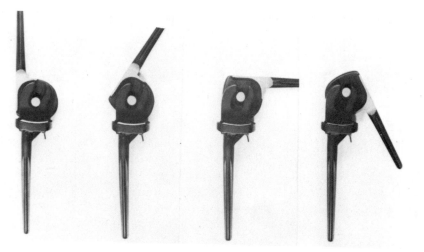

Fig. 13 – 4. GSB total knee replacement, showing motion during articulation.

SELECTION CRITERIA FOR A RA PROSTHESIS

1. The amount of resection required.
2. The position of the axis of rotation.
3. The bearing surface of the prosthesis on the bone.
4. Wear.

A number of other points could be added to the above list. It is, however, more practical to consider only the above four points, since they inherently encompass far more criteria. The position of the axis of rotation, for example, will influence the extensor mechanism with, on the whole, the patella and the knee muscles, the mode of fixation, the position and amount of resection, and so forth.

Amount of resection required

When selecting a prosthesis, the leading principle should be the minimally possible bone resection. Only this can ensure the chance of resorting to an alternative procedure—reoperation in the form of prosthetic replacement or, ultimately, arthrodesis—in case of complications. Unfortunately, a number of very sophisticated prosthesis designs are nowadays being frequently offered without adequate attention being given to this vital point. Every decision, however, eventually represents a compromise between mechanical function, mode of fixation, amount of bone resection, and some other parameters. The decision should not be taken too lightly, because it is, after all, the patient who will have to bear the consequences.

Figure 13–5 shows clearly that the GSB prosthesis requires minimal resection, or that only a few millimeters of bone must be removed. The prosthetic stems of all three types of device are anchored intramedullarly. Whereas large portions of the femoral and tibial condyles are removed in the case of the Shiers and Walldius knee hinges, the amount of bone resected with the GSB prosthesis is minimal. However, this does not imply that arthrodesis is impossible after removal of a Shiers of Walldius prosthesis; it is made considerably more difficult, though. The small amount of bone resection required with the GSB prosthesis has been achieved through a different design solution.

The remaining selection criteria will be discussed in the section on biomechanics of the knee in rheumatoid arthritis.

CLINICAL CASES OF TOTAL KNEE REPLACEMENT

Some cases given below are designed to demonstrate the abnormal stresses to which constrained and nonconstrained hinge joints, and consequently the bone-cement interface and the bone itself are ultimately subjected:

Case 1. This female patient, born in 1930, had rheumatoid arthritis from the age of 12 (Fig. 13–6). In the course of the disease, nearly all of the joints were involved. At the age of 37, the patient was nearly unable to walk. Therefore a Walldius hinge prosthesis was implanted in the left knee joint. When the patient was 43, the right knee joint was treated with a GSB knee arthroplasty. The patient now has a near normal gait despite some pain on the left side, where the Walldius prosthesis was implanted. X-rays show a definite loosening of the Walldius prosthesis, with bone resorption

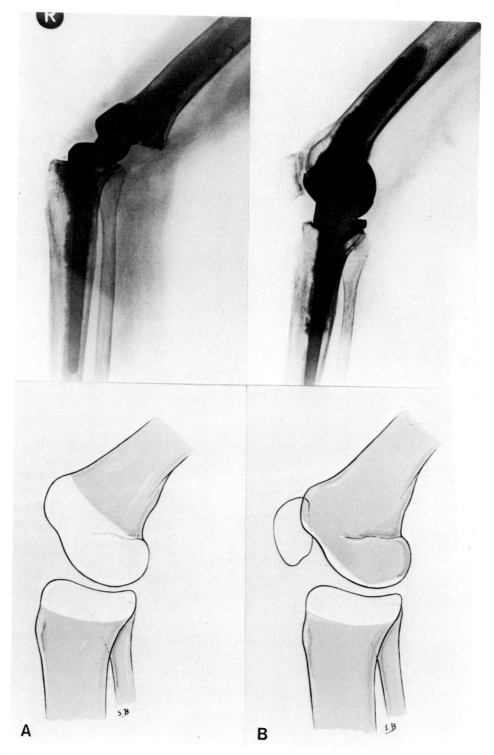

Fig. 13–5. Required resection of bone for insertion of Shiers prosthesis (A) and GSB Prosthesis (B).

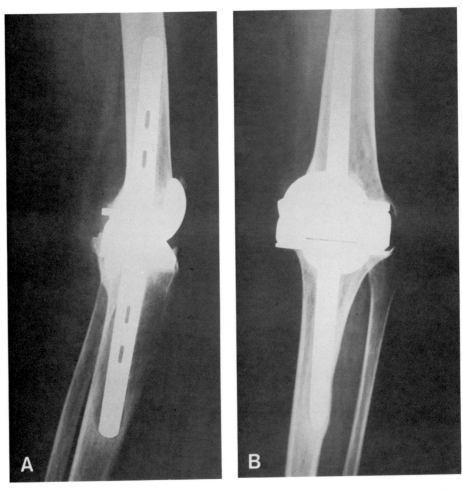

Fig. 13–6. Case 1: 11 years after insertion of Walldius prosthesis in left knee. (A) Lateral view; (B) Antero-posterior view.

around the stem, particularly on the tibial side, and an increased radiolucent line on both the tibial and femoral sides. The reason for the patient's suffering relatively little pain despite this loosening may be that no cement was used for fixation of the stems, and that the prosthesis rests on a larger area on the bone, thereby reducing the stress between metal and bone.[2]

Case 2. This female patient, born in 1903 and suffering from familiar diabetes mellitus, first complained about knee trouble at age 58 (Fig. 13–7). This was accompanied by recurrent swellings and by a steadily increasing varus deformity of the right and valgus deformity of the left knee. The simultaneous occurrence of analogous destructive changes in other joints, in particular the wrists, and the markedly increased sedimentation rate despite a negative latex agglutination test pointed to a case of atypical polyarthritis. The massive bone destruction, resembling a Charcot's joint with reduced protective sensibility, was striking. Apart from diabetes, a pertinent explanation for the bone loss was offered by the effective analgesic agents taken over a long period of time (indomethacin, and occasionally cortisone), which enabled the patient to walk despite the increasing instability and deformity of the knee joint. Shortly before the patient became unable to walk, the surgeons decided to perform an arthroplasty using the Shiers endoprosthesis available at that time, instead of attempting arthrodesis, which was otherwise the only alternative procedure to be considered. Arthroplasty was performed on both knees at an interval of 6 weeks, with healing of the wounds proceeding without any complications.

After an initially impressive success with regard to alleviation and relief of pain, and correction of the deformity and instability, there was a recurrence of effusions and pain in the years following, which resulted in a marked worsening of the patient's ability to walk. Radiography revealed sinking of the right prosthesis into the shaft of the femur and tibia, and simultaneously all relevant signs of loosening, with extensive radiolucency between the bone cement and the bone. There even occurred spontaneous fractures in the vicinity of the bearing surface of the prosthesis on the femur. After device removal, the attempt to achieve a compression arthrodesis according to the Charnley method failed, owing to poor bone surfaces and limited pos-

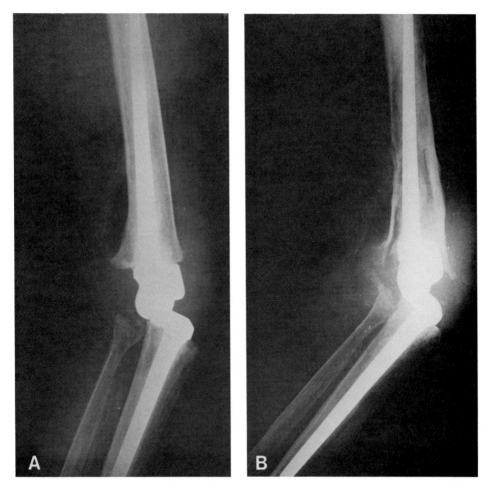

Fig. 13–7. Case 2: 7 years after insertion of Shiers Prostheses in both knees. (A) Lateral view, right; (B) Antero-posterior view, right. *(Figure continues on next page.)*

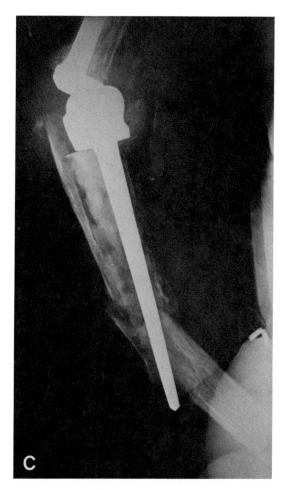

Fig. 13–7 (cont.). (C) fracture and angulation, shortly thereafter, right.

sibilities of fixing them solidly. The unstable fibrous ankylosis which resulted necessitated the use of a caliper splint for walking. The endoprosthesis on the opposite side was left in position despite clear signs of loosening since, owing to the patient's very poor general condition, another operation would have been too much for her. Consequently, spontaneous fractures occurred on the left side, both in the femur and the tibia. This case, with undoubtedly increased fragility and tendency to bone resorption, demonstrates in a particularly telling manner the detrimental forces that are transmitted by a constrained hinge joint to the anchorage (acrylic cement) and the bone.

Case 3. This man, born in 1933, suffered from the effects of juvenile polyarthritis that began when he was age 3 (Fig. 13–8). The knees and hips were the most involved joints. All four joints were fibrously ankylosed in flexion of 60 degrees. The steadily increasing pain associated with this condition has threatened the patient's ability to walk to such an extent that a salvage procedure was the only alternative to confinement to a wheelchair. The fixed flexion contracture of all four joints compelled the surgeons to first replace not only one joint, but also the other three joints,

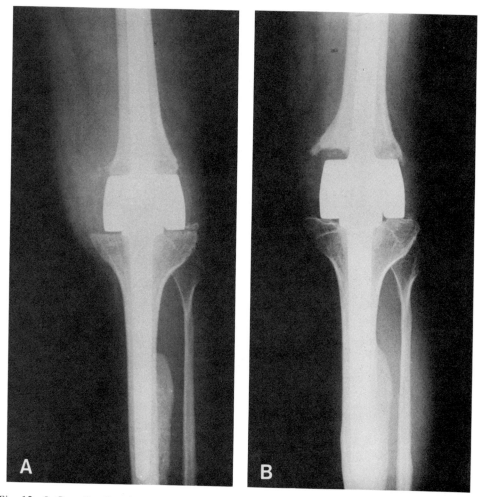

Fig. 13–8. Case 3, after insertion of Shiers prostheses in both knees. (A) Antero-posterior view, immediately postoperative; (B) antero-posterior view, 8 years postoperatively. Note fracture of proximal stem.

for only this could ensure the patient's ability to stand in the upright position. The left hip joint was replaced by a small Mueller prosthesis (metal ball head and metal socket with polyethylene studs) when the patient was aged 36. The right hip joint was operated on later the same year by implanting a prosthesis of the same type. The right and left knee joint were also replaced by Shiers prostheses in two procedures a month apart in the same year.

The postoperative course after knee replacement was uncomplicated. After 6 months, the patient was able to walk without using a stick, and to a large extent without limping. Considering the patient's relatively young age, and for the sake of reducing the loading of the artificial joints, the surgeons advised the patient, who had a predominantly sedentarial job, to use a walking stick for longer walks. The man was capable of walking virtually without any trouble, and of working full-time for 9 years. After more than 8 years of implantation, suddenly and without any known external cause, pain occurred in the man's left knee, making walking rapidly more

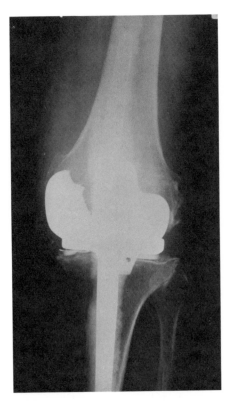

Fig. 13 – 9. Case 4, fracture of medial portion of GSB prosthesis, 3 years after insertion in left knee.

and more difficult. Radiography revealed a clear fracture of the prosthesis stem just above the prosthesis – that is in the transition zone from the portion of the firmly fixed stem in the bone to the badly embedded stem portion in the cement (situated close to the joint). The bone was partly resorbed here. The fractured prosthesis was subsequently replaced by a new Shiers prosthesis.

Case 4. This female patient, born in 1900, was first admitted to the clinic at age 72 because of increasing pain in both knees (Fig. 13 – 9). This reduced her ability to walk to 15 minutes, and troubled her even at night. Her worse, right knee was operated upon, and a St. George sledge prosthesis was implanted. The operation brought the patient relief from pain; however, when, 2 years later, the pain in her left knee became worse and threatened anew her ability to walk, we replaced the left knee joint with a GSB prosthesis. Here the flexion achieved was 120 to 0 degrees, exceeding the right side, with only 95 to 0 degrees. The patient was now able to walk satisfactorily without using a stick. Three years later, at age 77, on making a sudden movement while taking off her stocking, she suddenly felt a sharp pain in the left knee. This was so bad that she could not walk. Examination revealed swelling of the knee joint and a metallic sound during motion. Radiography showed a fracture of the medial femoral bearing surface of the GSB prosthesis. The broken component was replaced, and the patient has been completely without trouble since the operation.

Case 5. This female patient, born in 1931, had been suffering since age 23 from

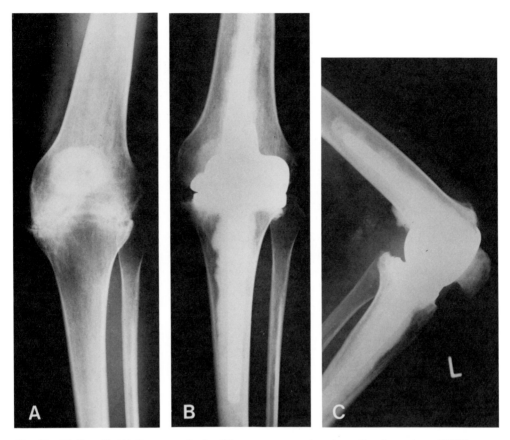

Fig. 13-10. Case 5. (A) Preoperatively; (B) antero-posterior view after insertion of GSB prosthesis in left knee; (C) lateral view, after insertion of GSB prosthesis. Note flexion to 110 degrees.

severe chronic polyarthritis that involved practically all her joints (Fig. 13-10). Despite the best possible conservative treatment at a university clinic for rheumatic diseases, there followed a rapidly progressing destruction of both knee joints and the finger joints. When her ability to walk seemed to be threatened to a high degree by pain and instability of the knee joints, and despite a successful operative correction of the painful changes in the forepart of the foot, GSB prostheses were implanted on the left and right sides when she was age 43 and 44, respectively. Figure 13-5 shows the preoperative and postoperative status on the left side. The patient has been able to walk without pain since the operation. Mobility improved from a preoperative range of 85 to 45 degrees (extension loss, 45 degrees), to 110 to 0 degrees (full extension) in the left knee joint; and from 100 to 30 degrees (extension loss 30 degrees), to 120 to 0 degrees in the right knee joint. If we evaluate all major parameters (pain, swelling, mobility, ability to walk, stability, deformity, and radiographic findings) by a point system in which 0 is the worst and 10 the best rating, the preoperative left knee score amounts to 2.5 points, and the postoperative left knee score amounts to 8.5 points; also, the preoperative right knee score amounts to 4.5 points, and the postoperative right knee score amounts to 8.5 points.

The postoperative anteroposterior radiograph shows minimal bone resection and correct axial relations. The lateral radiograph shows that, in contrast with constrained hinge-type prostheses (Walldius, Shiers, Guepar), the femur-tibia relation remains normal thanks to the rolling-sliding mechanism made possible by the traveling axis of the GSB prosthesis. The anteroposterior radiograph further shows a wide prosthesis-bearing surface on the femoral and tibial condyles. In case of complications (loosening or infection), it would be much easier to resort to another form of arthroplasty or to arthrodesis with the GBS prosthesis than with most other artificial joints.

BIOMECHANICS OF THE KNEE IN RHEUMATOID ARTHRITIS

Rheumatoid arthritis of the knee, as described in the foregoing sections of this chapter, ultimately leads to attenuation of the articular cartilage, narrowing of the articular space, collapse of the bone surfaces, disturbances in the sliding mechanism, ligament insufficiency, and severe pain. A few introductory remarks should therefore be made on the pathomechanics of the knee deformities induced by rheumatoid arthritis.

Pathomechanics of the Knee

In a normal knee, the action line of a force P resulting from the body weight minus the weight of the calf and foot of the stand leg is displaced medially from the knee (Fig. 13–11A). The knee is counterbalanced by a lateral force L, so that the resultant R lies in the center of the joint, equally loading the two condyles. A varus deformity induced by rheumatoid arthritis or some other cause shifts the resultant force R medially.

All of these displacements lead to an increase in the compressive stress on the medial condyles; this, in turn, results in an additional load-dependent destruction of the articular cartilage. However, it may also happen that the force L increases because it must counterbalance a reduction of the hip abductor muscle (gluteus medius and minimus) force. Owing to this abnormal increase in L, R is shifted laterally. With a genu valgum that comes to rest nearer to the action line P, the resultant force is consequently smaller, as shown in Fig. 13–11B. Owing to this smaller force R, a further load-dependent aggravation of rheumatoid arthritis rarely occurs, even despite a lateral displacement.

An unstable knee may also occur in rheumatoid arthritis (Fig. 13–12). A knee becomes unstable when the resultant force R comes to lie outside the centers of curvature O_1 and O_2. An unstable knee rarely leads to a varus deformity, because the lateral muscles and ligaments are much stronger than the medial ones. In cases of advanced rheumatoid arthritis in which the knee joint is destroyed to such an extent that neither physiotherapy nor medication will bring relief, and when because of ligamentous insufficiency there is severe instability, only two practical alternative procedures remain: arthrodesis or implantation of a knee joint prosthesis with an accurate motion pattern since it is only the muscles around the natural knee joint that may be regarded as being capable of a certain degree of functioning.

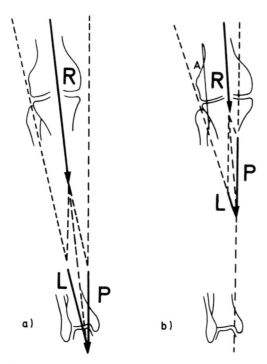

Fig. 13 – 11. Pathomechanics of the knee. (A). Normal; (B) varus.

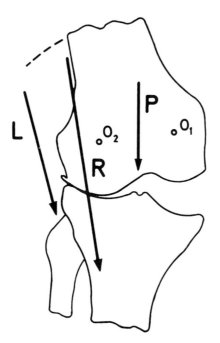

Fig. 13 – 12. Extreme varus instability in rheumatoid arthritis.

Knee Loading

In dealing with total knee placements, it is essential to understand the stresses that can occur in the knee. In normal walking, total knee loading is significant only for the first 20 degrees of flexion. For climbing stairs and walking up ramps, the loaded flexion range is about 80 degrees. In this range, femorotibial loads of 3 to 4 times the body weight are quite normal. With the menisci intact, a compression force of 2,000 to 2,500 N applied on a cadaver knee leads to a load-bearing contact area between the femur and tibia of 18 to 22 cm.² After removal of the menisci the load bearing contact area is reduced by half. In general this contact area also decreases with increasing knee flexion, and becomes approximately 9 to 11 cm² at 90 degrees flexion.

Both the varus-valgus stiffness and anterior-posterior stiffness of the knee are maintained primarily by the collateral and cruciate ligaments. To maintain equilibrium during standing, the stiffness must be a maximum as the muscles are relaxed. The forces in the ligaments can reach values of up to 350 N.

After a total knee replacement, the prosthesis must sustain all of the aforementioned knee loadings.

Position of the Axis of Rotation

A correct axis of rotation is probably the most decisive factor in the long-term success of a hinge-type knee prosthesis. The axis of rotation chiefly determines the loading of the muscles and patella. However, a fixed axis of rotation cannot be defined, since knee flexion is described by the rolling and sliding motion of the femoral and tibial condyles relative to each other.

Therefore, the concept of the instant center of rotation must be discussed. For each individual flexion angle there exists a distinct center of rotation between the

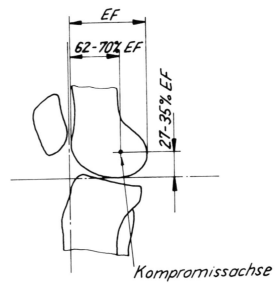

Fig. 13–13. Location of the "compromise" axis.

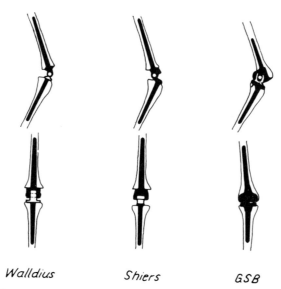

Walldius   Shiers   GSB

Fig. 13–14. Comparison of geometry of Walldius, Shiers, and GSB prostheses.

femur and tibia. Many independent studies have demonstrated that these instant centers of rotation lie within the posterior portion of the femoral condyles. A method for obtaining instant centers of rotation from radiographs is discussed in Chapter 17.

A radiographic study of 88 knee joints[10] has made it possible to establish the locus of a so-called "compromise" axis. As shown in Fig. 13–13, this compromise axis lies in the posterior region of the femoral condyles. Care must be therefore taken to ensure that the prosthesis axis is put in the same locus.

According to the position of the axis of rotation of a prosthesis, the musculature can be loaded in a considerably different manner than during physiologic motion. If the muscles are not loaded properly, this may cause premature fatigue or even pain in connection with the implanted prosthesis. The reason for our comparison of the three types of prosthesis is further evident from Fig. 13–14. In each type, the axis of rotation was at a different location. In the Walldius prosthesis, the axis of rotation lies in line with the femoral and tibial stems. During flexion, the femur is displaced dorsally relative to the tibia. The dorsal displacement increases with increasing flexion. In the case of the Shiers prosthesis, the dorsal displacement of the femur relative to the tibia is even greater, because the axis of rotation lies in line with the femoral component, but dorsally to the tibial stem. Since the axis of the GSB prosthesis relative to the anchorage stem is within the range of the compromise axis (Fig. 13–13), the femur is in the correct position with reference to the tibia. The axis is not rigid, yet the axial region is located correctly. An axis that is not rigid has the advantage of not transmitting the total force on the axial pin, but of supporting the principal loads on the condyles, as is the case with the GSB prosthesis.

The axis in the case of the GSB prosthesis is located only to guide the desired motion process. Nevertheless, in certain positions there may yet occur considerable forces that will impose severe loads on the guiding axial pin. A further advantage offered by the linkage having a non-rigid axis lies in the sparing of the prosthetic anchorage stems through some clearance between the anchorage pin in the femoral component and the fin in the tibial component. To sum up, it may be said that GSB prosthesis comes closest to having the correct position of the axis, without having a rigid linkage.

If a stem is subjected to alternating tensile and compressive loads, it will loosen much sooner than if subjected only to a compressive load. However, the type of loading may be influenced by the position of the axis of rotation. According to Schumpe and Friedrich,[13] the tibial and the femoral stem of the Walldius prosthesis are subjected to alternating loads. In the case of the Shiers prosthesis, only the femoral stem is subjected to alternating loads, but much more heavily than with the Walldius prosthesis, because the action of the forces is much greater, as will be seen later. In the case of the GSB knee prosthesis, only the posterior bearing surfaces of the femur and tibia are subjected to compressive load.

Loading of the Muscles

When considering the muscular forces acting on the knee, we must include all of the muscles of the leg, because forces acting on the knee have stabilizing effect both on the ankle and on the hip joint. In this consideration, it is convenient to idealize the

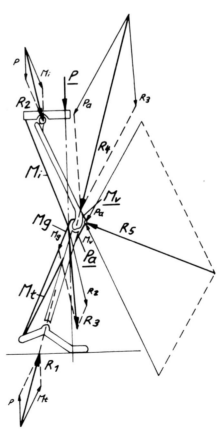

Fig. 13–15. Forces acting on the knee.

muscles. In the sagittal plane, the most important forces are applied by:

Mt: Muscles of the calf.

Mi: The hamstring group of muscles.

Mg: The gastrocnemius.

Pa: The patellar tendon.

Mv: The quadriceps.

These are shown graphically and in relative magnitude in Figure 13–15.

Muscular Force-Prosthesis Loading Relationship

As mentioned earlier, the action point of the muscles is the axis of rotation of the prosthesis. With both the Shiers and the Walldius prostheses, with the lever arms of the dorsally acting muscles become longer after surgery because of the displacement of the axis relative to the physiologic axis. This results in an increased effect on the knee joint. The anterior femoral muscles are consequently also subjected to a higher load, so as to maintain the static equilibrium.

Thus, in addition to the increased stressing of the muscles, the stress on the axial pin of these prostheses is also increased. If a GSB prosthesis has been implanted correctly, the loading of the muscles will not differ considerably from that prior to im-

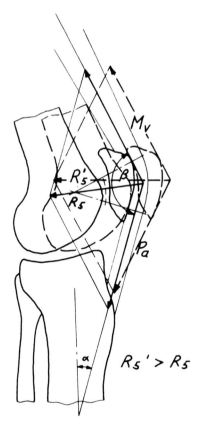

Fig. 13–16. Effect of location of rotational axis on forces on the patella.

plantation, unless—and this applies to all other prostheses as well—the prosthesis has been implanted in order to correct a varus or valgus deformity, whereby the muscles are again loaded correctly.

The Patella

The patella, connected proximally with the quadriceps and distally with the ligamentum patellae, acts chiefly as a bearing and sliding surface against the femur. Depending on the position of the patella relative to the rotational axis, the pressure between patella and femur is greater or smaller after surgery—that is, the quadriceps and the patellar tendon (ligamentum patellae) are loaded to a greater or lesser extent (see Fig. 13–16).

The Patella-Femur Compressive Force

During walking, the quadriceps works only when the gravity axis passes posteriorly to the knee. This is mainly the case during the start of the standing phase. The resultant compressive force is about three times the body weight. The increasing patello-femoral compressive force, and the associated more rapid wear of the respective cartilage surfaces, is a further disadvantage of prostheses whose axes are not within the range of the compromise axis. This is true of both the Shiers and the Walldius prosthesis.

Patellectomy

The biomechanical consequences of excision of the patella should be also considered, since patellectomy is normally performed in connection with Shiers arthroplasty of the knee. As a result of excision of the patella (Fig. 13–17A), the lever arm of the quadriceps, c, is shortened to c', without any change in the position of the femur relative to the tibia. Consequently a greater force is required from the quadriceps in order to maintain equilibrium, enhancing the risk of rupture of the extensor mechanism.

Forces about the Patellar Tendon

A simple calculation indicates the change in compressive force, R, between the femur and the tibia, and the quadriceps force in the patellar tendon M, that results from patellectomy (Fig. 13–17B). The conditions of equilibrium can be used to solve for the muscle and joint reaction forces. The body weight carried by one of the supporting limbs is taken as 400 N.

Knee with Patella

First we can write the equation for the equilibrium of the moments ($\Sigma M = 0$):

$$\Sigma M = (-400 \text{ N} \times 34 \text{ mm}) + (M \times 27 \text{ mm}) = 0$$

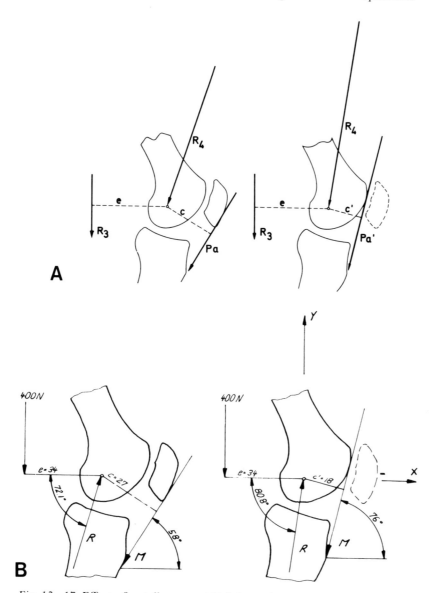

Fig. 13–17. Effect of patellectomy. (A) Schematic; (B) calculated (see text).

$$M = \frac{400 \times 34}{27} = 503.7 \text{ N} \tag{1}$$

To determine the joint reaction force, the angle of pull of the patellar tendon with the horizontal must be known. In the considered case, with the patella it is 58 degrees. We can now write the equations for the equilibium of the forces ($\Sigma F = 0$):

$$\Sigma F_x = -(M \cdot \cos 58°) + R_x = 0$$
$$R_x = (M \cdot \cos 58°) = 266.9 \text{ N} \tag{2}$$
$$\Sigma F_y = (-400 \text{ N}) + (-M \cdot \sin 58°) + R_y = 0$$

$$R_y = 400 - (-M \cdot \sin 58°) \tag{3}$$
$$R_y = 400 - (-503.7 \times 0.848) = 827.2 \text{ N}$$

The resultant joint force is:

$$R = \sqrt{R_x{}^2 + R_y{}^2} = \sqrt{266.9^2 + 827.2^2} = 869.2 \text{ N} \tag{4}$$

The angle with the horizontal is:

$$\tan \alpha = \frac{R_y}{R_x} = \frac{827.2}{266.9} = 3.099 \tag{5}$$

$$\alpha = 72.1°$$

Knee without Patella

In the same way we may write movement and force equilibrium equations:

$$\Sigma M = 0 = (-400 \text{ N} \times 34) + (M \times 18) \tag{6}$$

$$M = \frac{400 \times 34}{18} = 755.6 \text{ N}$$

$$\Sigma F_x = 0 = -(M \cdot \cos 76°) + R_x \tag{7}$$

$$R_x = M \cdot \cos 76° = 182.8 \text{ N}$$

$$\Sigma F_y = 0 = (-400 \text{ N}) + (-M \cdot \sin 76°) + R_y \tag{8}$$

$$R_y = 400 - (-755.6 \times 0.970) = 1133.2 \text{ N}$$

The resultant joint f, after patellectomy, is (an increase of 32 percent in this case):

$$R = \sqrt{R_x{}^2 + R_y{}^2} = \sqrt{182.8^2 + 1133.2^2} = 1147.9 \text{ N} \tag{9}$$

The angle with the horizontal is:

$$\tan \alpha = \frac{R_y}{R_x} = 6.199 \tag{10}$$

$$\alpha = 80.8°$$

Due to the increase of the force in the extensor mechanism, the resultant compressive force between the femur and the tibia also becomes greater. In any case, in consequence of patellectomy, the prosthesis itself, as well as the extensor mechanism, is more heavily loaded. If there are no compelling reasons for removing the patella, no prosthesis that postulates patellectomy ought to be used unless other advantages justify running such a risk.

Prosthesis Support on the Bone

Fundamentally it may be said that any external load through the prosthesis and the bone cement is transmitted to the bones adjacent to the site of anchorage.

A knee prosthesis may thus be anchored with or without bone cement. Most stemmed prostheses are currently fixed in the respective bone cavities by means of polymethyl methacrylate (PMMA) bone cement, because the prosthesis is thus firmly anchored instantly. Owing to the elastic behavior of the bone cement, part of the high stress is not transmitted by the individual components directly to the adjacent tissues. The Walldius hinged prosthesis is one of the few knee prostheses that may be anchored without cement.

The disadvantages of PMMA bone cement lie in the generation of heat and the toxicity of the monomer, with possible local tissue damage; the larger amount of cancellous bone to be removed; and last but not least in the modulus of elasticity, which is far below that of human bone.

If a prosthesis is fixed without cement, the anchorage obtained may not be fully loaded until much later, because the bone tissue must make up for the damage suffered during the operation. Even the best operative technique is not capable of adapting the bone cavities exactly to the volume of the prosthesis components. In order to shorten the waiting time prior to loading of the prosthesis, a primary fixation may be performed. Cementless anchorage of knee prostheses therefore does not impose lesser requirements on the operative technique.

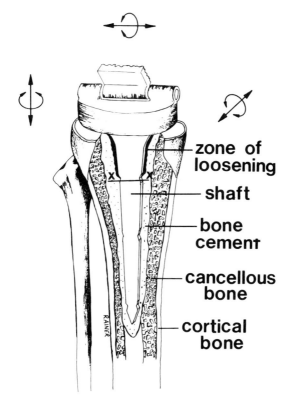

Fig. 13–18. Zone of tibial loosening (x ——— x) in cemented, hinged knee prosthesis.

With all rigid-axis knee prostheses, the prosthetic stems take up the torsional stresses between the femur and tibia in the absence of clearance. This applies to the Shiers and Walldius hinged prostheses but not to the GBS knee prosthesis, because, as mentioned earlier, the latter allows some clearance. All three types of prosthesis are supported on the resected, plane bone surfaces, and are secured against displacement by the prosthetic stems anchored in the medullary cavities.

Every prosthesis should have an additional anchorage as protection against torsion. The best support is provided by a minimally resected bone surface on which the remaining bone is loaded by the implanted prosthesis in a manner like that prior to implantation. The GSB prosthesis comes nearest to this requirement, since it largely loads the condyles and is even supported by them.

There should be no cavities below a support, for the loaded surfaces are thus subjected to additional pressure. Any increase or decrease of load from the physiologic load prior to implantation of the prosthesis will sooner or later lead to bone resorption. In this connection, only a very small permissible variance in load range is possible. The exact limits have not yet been finally established, and it is not yet known whether it is compressive stress or micromotion that is responsible for failure when the range is exceeded.

Anchorage of prosthetic stems ought to be as firm and sound as possible so as to minimize the risk of loosening over the longest possible term. If a stem has become loose, a reoperation must be performed sooner or later because of pain, bone fracture, or prosthesis breakage secondary to loosening. For this reason, removal of large amounts of bone, as required with the Shiers or the Walldius hinged prosthesis, should be avoided if possible.

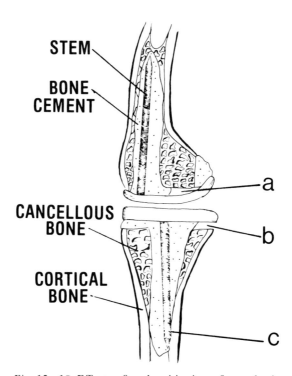

Fig. 13–19. Effects of malpositioning of prosthesis.

In constrained hinges with a rigid axis, all forces acting on the prosthesis are transmitted to the anchorage stems. At the origin of the prosthetic stem, the compressive stress is so high that the bone cement may be crushed there. The resulting depression or hollow has a very unfavorable influence on the stem. The risk of stem fracture is greatest (Fig. 13–18) in the transition zone from the loose to the firm anchorage (x − x).

This section will show by means of a few examples the effects of malpositioning of a prosthesis. The examples are chosen from observed failures due to incorrect implantation of prostheses.

Details a, b, and c in Figure 13–19 are shown on a schematic implant, without a rigid axis. However, the following remarks are equally valid for rigid prostheses. A bone cement anchorage is shown because this is the prevalent way in which to implant prostheses.

Detail a

If under a portion of the load-bearing surface of the prosthesis there is no cement or bone, the stresses on the prosthesis itself, or on the remaining supporting area, can exceed the stress limits. This can result in either a fracture of the prosthesis or bone, or owing to excess pressure, in a sinking of the prosthesis into the bone.

Detail b

This is a similar case to detail a, except that a missing support area at the boundary of the prosthesis has a more severe effect. The overhanging part will bend when loaded, and more cement or bone parts will break away. The damage will progress much more rapidly than the enclosed "hole" in detail a.

Detail c

If a stem is not centered in the bone shaft, it can very easily happen that part of the end of the stem is not surrounded by cement. In this case the bone is very susceptible to fracture. Owing to the interrupted structure of the cement anchorage (the stem is partially exposed), high local stresses can occur in the cement, the stem, and the bone. For reinforcement, the bone may react with a local thickening, or if the stress is too great it may fracture. Nevertheless, even a reinforcement of the bone may sooner or later end in a fracture, with the severe consequences of a very difficult or even impossible reoperation.

Wear

As far as wear is concerned, metal-to-metal contact is generally claimed to be unfavorable and should be avoided with prostheses. This requirement is opposed by some other mechanical requirements which make a compromise almost indispensable with regard to currently used biomaterials. From the mechanical-function viewpoint, the knee joint is more complex than the hip joint, where avoidance of metal-metal contact may be more easily fulfilled. The hip joint is a ball-and-socket joint and requires no stabilization of its own, because the remaining action of the muscles and ligaments is sufficient. With the knee joint, the situation is different. Due to external dynamic forces, movements here are possible in all six degrees of freedom.

A good prosthesis allows only the most significant degrees of freedom, so as to obtain a simple but all the more safe prosthetic device. With prostheses having a rigid axis, metal-to-metal contact can hardly be avoided. All three prosthesis types studies exhibit metal-to-metal contact at least partly. In the Walldius and Shiers prostheses, the entire load is transmitted through the axial pin. In the case of the GSB prosthesis, the bearing force is transmitted via the condyles. Owing to the polyethylene bearings incorporated in the tibial component, the GSB device has a metal-polyethylene material combination. Only the shearing forces are transmitted directly by metal against metal through the guide pin between the femoral and tibial components. Calculations have shown that these shearing forces may reach quite considerable values, and can consequently induce an appreciable amount of wear even with the GSB prosthesis. Metal wear particles produce a nasty black discoloration of the tissue, but, thanks to the currently used implant materials, they are virtually inert. Nevertheless, prosthesis designers and manufacturers ought to give priority to the elimination of metal-to-metal contact surfaces.

A further remark, although not connected directly with wear but with the material combination, concerns the annoying creaking noise associated with intermittent metal-to-metal contact.

Biomechanical Observations on the RA Patients with Implanted Knee Prostheses

The rheumatoid arthritis cases treated with knee prostheses, and described in this chapter, should also be considered from the biomechanical aspect.

The radiograph of the Walldius prosthesis (Case 1) reveals an interesting phenomenon around the anchorage stem of the tibial component. Adjacent to the stem, there is a light zone of about 0.5 mm with an adjoining darker zone of the same width. The light zone is apparently connective tissue and the darker zone is dense regenerated spongious bone, as seen with this cementlessly anchored prosthesis. Under certain circumstances, the bone is thus capable of adapting itself to the changed loading conditions, whereby the connective tissue acts as a buffer between two zones with a different modulus of elasticity (prosthetic stem-spongious bone). The connective tissue layer should not, however, be thicker than 0.5 to 1 mm, for the prosthesis will otherwise become loose.

In Case 2, the implanted Shiers prosthesis (right knee) had to be removed after a few years because of the bone that was not equal to the changed loading conditions. Although the severe joint deformity could be successfully corrected for some time, we wonder in connection with Case 2 whether a prosthesis with a non-rigid axis like that of the GSB design would not have succeeded in preventing very high loads on the bone.

The Shiers prosthesis implanted in Case 3 had to be replaced after 8 years because of a typical fatigue fracture of the stem of the tibial component. The same situation occurred as in Figure 13-19, detail b. Due to a slight yield of the PMMA bone cement on the joint side, a transitional zone developed between the firm and loose anchorage. The tibial component stem developed a crack at that location and subsequently fractured. In Case 4, the femoral component of a GSB prosthesis had to be replaced after 3 years because of fracture of one of the condyles. This case shows the problems associated with a prosthesis that is as bone-sparing as possible—that is, designed with small material thickness. A thin condyle is much more subject to the risk of fracture in case of an infrequent excessive stress concentration or insufficient underpinning of the condyles with bone cement. This fact was taken into account in the design of the new GSB knee prosthesis by providing it with fins to strengthen the condyles.

SUMMARY

The present study was deliberately based only on three fundamentally very different prosthesis types. It would be wrong to fail to mention, in the case of the same operative indications, some other well-proven knee prostheses used for the treatment of arthritic knees. Mention should therefore also be made of other constrained hinges—the Guepar, and nonconstrained hinges—the Attenborough and Sheehan knee prostheses.

As to axis relations and anchorage, the Guepar hinge prosthesis bears favorable comparison with the GSB prosthesis. A major feature deserving special mention is the patellar slide bearing, which enhances the slidability of the patella. The amount of bone resection required with this prosthesis (more than with the GSB hinge), and above all the rigid axis relations are, however, among the less favorable features of this implant.

The Attenborough prosthesis also exhibits very interesting features. Like the GSB prosthesis, it offers the advantage of an axis that is not fixed and a metal-polyethylene wear combination. On the tibial side it requires more bone resection than the GSB prosthesis. The position of the axis corresponds to that of the Walldius hinge axis and is consequently not ideal for the loading of the extensor mechanism.

The Sheehan prosthesis is also very suitable for the operative treatment of knees attacked by rheumatoid arthritis. It features an eccentric center of rotation that is favorable for the extensor mechanism, its axis is not rigid, and its femoral condyles articulate with the polyethylene of the tibial component. Less favorable, however, is the relatively massive and not so simple bone resection, and the high load of the high-density polyethylene (HDP) fin of the tibial component.

In conclusion it may be said that—in the absence of specific reasons—prostheses of a design such as that of the GSB, Attenborough, and Sheehan hinge are to be preferred to those based on the principles of the Shiers, Walldius, or Guepar prostheses.

REFERENCES

1. Deburge, A., and G.U.E.P.A.R.: Guepar hinge prosthesis. Clin. Orthop., *120*:47–53, 1976.
2. Friedrich E., Schumpe, G. and Nasseri, D.: Vergleichende Statistische Berechnungen am Kniegelenk bei der Ventralisation der Tuberositas tibiae nach Bandi. Z. Orthop., *111*:134–138, 1973.
3. Gschwend, N.: Total Knee Replacement using the GSB Knee Joint. Institution of Mechanical Engineers, London, 1974.

4. Gschwend, N.: Indikation zur Verwendung von Teilprothesen, Schlittenprosthesen und Scharnierprothesen am Beispiel des Kniegelenks. Orthop. Praxis, *12(11)*:924–928, 1975.

5. Hanslik, L., and Scholz, J.: Erfahrungen mit der Geupar-Endoprothese. Arch. Orthop. Unfallchir., *88*:347–357, 1977.

6. Jacob, H. A. C., and Huggler, A. H.: Experimentelle Spannungsanalysen im menschlichen Oberschenkelknochen-Modell mit und ohne Prosthese. Technische Rundschau Sulzer, Forschungsheft 1978, 73–83.

7. Maquet, P. G. J.: Biomechanics of the Knee. Springer-Verlag, Berlin, 1976.

8. Menschik, A.: Mechanik des Kniegelenks. Z. Orthop. 112:481–495, 1974.

9. Morrison, J. B.: The mechanics of the knee joint in relation to normal walking. J. Biomech., *3*:51–61, 1970.

10. Nietert, M.: Das Kniegelenk des Menschen als biomechanisches Problem. Biomed. Tech., *22*:13–21, 1977.

11. Schaldach, M., and Hohmann, D.: Advances in artificial hip and knee joint technology. Engineering in Medicine, *2*:90–114, Springer-Verlag, Berlin, 1976.

12. Scheier, H. J. G.: GSB-Kniegelenke, Zentralbl. Chir., *102*:1337–1340, 1977.

13. Schumpe, G., and Friedrich, E.: Statische Berechnung der Druckbelastung an den Gelenken der unteren Extremitaten unter Berucksichtigung der Muskulatur bei einer Kniebeugestellung. Orthop. Praxis, *7(10)*:415, 1970.

14. Schumpe, G., Friederich, E., Rossler, H., and Hofmann, P.: Biomechanik des Kniegelenks unter Berucksichtigung der Alloplastik. Z. Orthop., *113*:501–505, 1975.

15. Seedhom, B. B., Longton, E. B., Wright, V., and Dowson, D.: Dimensions of the knee. Ann. Rheum. Dis. *31*:54, 1972.

16. Shiers, L. G. P. Total Knee Hinge Replacement. Institution of Mechanical Engineers, London, 1974.

17. Swanson, S. A. V., and Freeman, M. A. R.: The Scientific Basis of Joint Replacement. John Wiley & Sons, New York, 1977.

18. Ungethüm, M., and Stallforth, H.: Systematisierung kunstlicher Kniegelenke unter Berucksichtigung von am naturlichen Kniegelenk abgeleiteten konstruktiven Merkmalen. Arch. Orthop. Unfallchir. *89*:227–237, 1977.

19. Walldius, B.: A Compartive Analysis of Different Methods for Arthroplasty of the Knee. Institution of Mechanical Engineers, London, 1974.

20. Watson, J. R., and Hill, R. C. J.: The Shiers arthroplasty of the knee., J. Bone Joint Surg. *58B*:300–304, 1976.

21. Wilson, F. C., and Venters, G. C.: Results of knee replacement with the Walldius prosthesis. Clin. Orthop., *120*:39–46, 1976.

14
Correction of Paralytic Foot Drop by External Orthoses

D. H. Sutherland, M.D.; J. U. Baumann, M.D.

DEFINITION OF PARALYTIC FOOT DROP

Paralytic foot drop is a gait disorder characterized by exaggerated ankle equinus and increased hip and knee flexion during the swing phase of gait. It is caused by paralysis of the anterior and lateral compartment muscles—those muscles innervated by the common peroneal nerve. Paralytic foot drop is distinguished from flail foot by the absence of active plantar flexors in the latter. Contractures of the antagonist muscles, limiting the range of ankle dorsiflexion in both stance-phase and swing-phase, may or may not be present. In spastic equinus the muscles are intact but the control system is defective. Because it is a subject requiring separate consideration, spastic equinus will not be included in this discussion.

ETIOLOGY

Trauma is responsible for the bulk of common peroneal nerve injuries.[10] Fractures of the acetabulum, femoral shaft, distal femur, and proximal tibia and fibula are frequent associated injuries. The mechanism of nerve injury may be laceration, contusion, compression, or ischemia. Early recognition and appropriate treatment will reduce the incidence of permanent, complete foot drop. Depending on the mechanism of nerve injury, early treatment measures include fracture reduction, nerve suture, elevation of the extremity, splinting, evacuation of hematoma, and fascial compartment release.

Next in order of occurrence as the cause of paralytic foot drop are peroneal nerve injuries, produced by external pressure. Circumferential bandaging for traction, or inadequate cast padding over the proximal fibula, are frequent offenders. Elderly cachectic patients are particularly susceptible to foot drop from pressure due to excessive side-lying. Improper positioning or inadequate padding of patients undergoing surgical procedures can also produce paralysis.

Herniated intervertebral discs frequently produce foot drop. Hansen's disease and poliomyelitis, both of which are significant diseases in some underdeveloped countries, can cause paralysis of the anterior and lateral compartment muscles.

Less frequent causes of foot drop are: (a) Prolonged squatting with excessive knee flexion (strawberry picker's foot drop)[8]; (b) compression of ganglion of the superior tibio-fibular joint[6]; (c) pretibial myxedema associated with Grave's disease[9]; (d) traction for knee-flexion contracture; (e) derotation osteotomy for tibial torsion (probable mechanism is ischemia due to anterior compartment syndrome)[7]; (f) Guillain-Barre's disease; (g) Charcot-Marie-Tooth muscular atrophy; (h) ECHO and Coxsackie viral infections; and (i) drug addiction (probable mechanism is ischemia due to anterior compartment syndrome).

THE ROLE OF THE ANTERIOR COMPARTMENT MUSCLES IN THE SWING PHASE OF NORMAL GAIT

The tibialis anterior, extensor digitorum longus, and extensor hallucis longus muscles act from the time of toe-off, throughout swing-phase, and into early stance-phase. Their function in swing-phase is to clear the foot. Their action in swing-phase is initially concentric (shortening) in the range covering from 62 to 80 percent of the gait cycle, then isometric in the range from 80 to 100 percent of this phase. It is extremely important to recognize that hip and knee flexion are equally important in foot clearance. The first movement to occur is knee flexion, starting at 40 percent of the cycle, followed by hip flexion at 50 percent. By the time of toe-off, when ankle dorsiflexion begins, the knee flexion angle has already progressed to 45 degrees and the hip flexion angle is 13 degrees. A patient with impaired hip or knee flexion as well as foot drop has much greater difficulty with foot clearance.

THE ROLE OF THE ANTERIOR COMPARTMENT MUSCLES IN THE STANCE PHASE OF NORMAL GAIT

From heel-strike until foot-flat the tibialis anterior, extensor digitorum longus, and extensor hallucis muscles act to resist extrinsic ankle-flexion torque. Their action is eccentric (lengthening occurs while these muscles act). The effect of this action is to allow a gradual descent of the foot to the floor after heel-strike. Another effect may be to assist the forward movement of the tibia.

When the anterior compartment muscles are no longer available to provide gradual descent of the foot to the floor, an orthosis is usually provided. If peroneal nerve function does not return by 18 months, some patients may wish to have surgical reconstruction, which can eliminate the need of an orthosis. An example of surgical treatment is the transfer of the tibialis posterior muscle through the interosseous membrane to the dorsum of the foot.[2, 3, 11]

CASE REPORTS

Case 1. This 25-year-old woman was a volunteer subject in an experimental study to determine the effects on walking of paralysis of the anterior and lateral compartment muscles by peroneal nerve block. Prior to the study, a polypropylene orthosis was constructed for an evaluation to be carried out following the block (Fig. 14–1).

The initial gait study did not disclose any departure from normal adult gait. Gait analysis carried out following peroneal nerve block revealed primary and compensatory changes. The primary change was an alteration of ankle movement (Fig. 14–2). The ankle was in plantar flexion, with a flat foot-strike following the nerve block, as compared with a neutral ankle position accompanying a normal heel-strike before

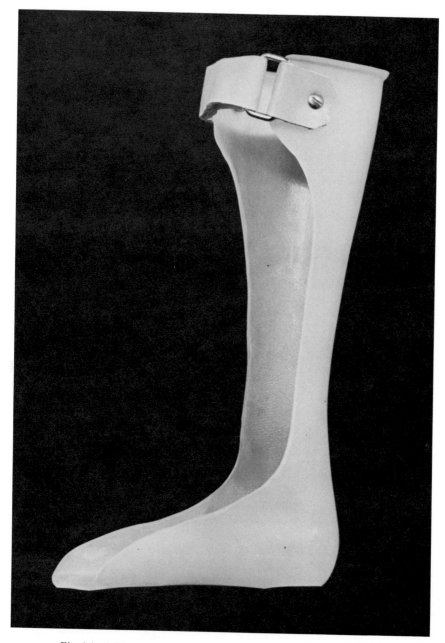

Fig. 14 – 1. Vacuum-formed polypropylene-ankle-foot orthosis.

the block. Following toe-off the foot remained in a drop-foot alignment throughout swing-phase. The compensatory changes were exaggerated swing-phase hip and knee flexion. Stride length, cadence, and walking velocity dropped slightly following the block. Work output increased by approximately 13 percent (Table 14 – 1). Gait study performed after application of the polypropylene orthosis revealed a satisfactory restoration of heel-strike and elimination of the foot drop (Fig. 14 – 2). With the wearing of the orthosis the exaggerated swing-phase hip and knee flexion was re-

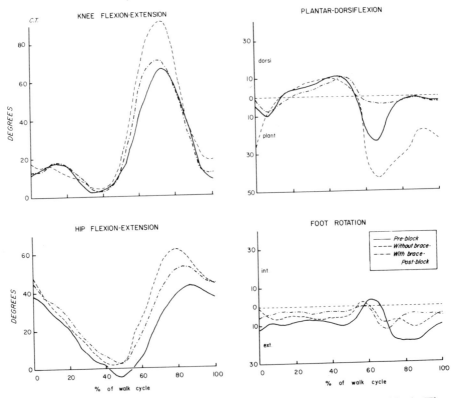

Fig. 14–2. Motion measurements of case 1 before and after peroneal nerve block. The walk cycle begins with foot-strike at the left margin of each graph. The changes in joint angle throughout one cycle with the wearing of the polypropylene ankle-foot orthosis demonstrate the correction achieved when the movement graphs are compared with the post-block joint angle rotations.

duced. Stride length, cadence, and walking velocity were restored to nearly their preparalysis levels. Work output, calculated during walking with the orthosis, returned to normal (Table 14–1).

This study illustrates the effectiveness of a polypropylene orthosis in achieving near normal gait despite total paralysis of the anterior and lateral compartment muscles. It is of interest that the patient required no training or break-in time to gain confidence and skill in the use of the orthosis.

Case 2. This 43-year-old woman had an L4 laminectomy for a ruptured intervertebral disc 2½ years before this gait analysis was performed. Her back and leg pain were relieved by the surgery, but persistent foot drop caused her to wear a polypropylene orthosis. Muscle strength by manual muscle testing was as follows: tibialis anterior = 3; extensor hallucis longus = 0; extensor digitorum longus = 3+; peroneus longus = 3+; peroneus brevis = 3+; remaining calf muscles = 5. A mild contracture of the gastrocnemius was present. The range of ankle dorsiflexion was 0 degrees with the knee extended and +10 degrees with the flexed. The patient denied any problem in walking with the orthosis.

Gait analysis was performed both with and without the polypropylene ankle-foot orthosis. Primary and compensatory movement abnormalities were demonstrated

TABLE 14–1. COMPARISON OF MEASUREMENTS
OF THE BLOCKED LEG BEFORE PERONEAL
NERVE BLOCK AND WITH AND WITHOUT A
POLYPROPYLENE SHORT LEG ORTHOSIS
AFTER THE NERVE BLOCK—CASE 1

	Before Nerve Block	*After Block: Not Braced*	*After Block: Braced*
Velocity (cm/sec)	122	100	113
Cadence (steps/min)	129	114	119
Stride length (cm)	114	106	114
Single stance (% cycle)	37	35	34
Work output (kcal/kg/km)	0.1012	0.1166	0.0791

when the orthosis was not worn. The primary abnormalities were in right ankle movement (Fig. 14–3). The ankle was plantarflexed to 15 degrees at foot-strike, and foot drop was present throughout swing-phase. To clear the foot in swing-phase a compensatory increase in hip flexion and knee flexion occurred. Walking velocity, cadence, stride length, and the duration of single-stance were reduced. Work output in walking was greater than for a normal adult (Table 14–2). When the orthosis was worn, heel-strike was restored and foot drop was eliminated (Fig. 14–3). There was a reduction in cadence, an increase in stride length, an increase in walking velocity, and a drop in work output (Table 14–2).

Case 3. This 2½-year-old male developed foot drop following an injection of penicillin into the left buttock 3 months before the gait study. Electromyogram (EMG) following the injection into the buttock demonstrated peroneal nerve involvement,

TABLE 14–2. COMPARISON OF GAIT
MEASUREMENTS WITH AND WITHOUT A
POLYPROPYLENE SHORT LEG
ORTHOSIS—CASE 2

	Not Braced	*Braced*
Velocity (cm/sec)	98	106
Cadence (steps/min)	115	110
Stride length (cm)	102	116
Single stance (involved leg) (% cycle)	31	32
Work output (kcal/kg/km)	0.1377	0.0685

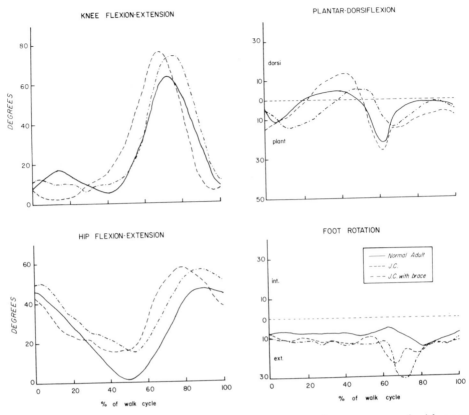

Fig. 14–3. Motion measurements of case 2 with and without braces compared with normal adult values.

including the short head of the biceps femoris. Manual muscle strength testing revealed; tibialis anterior = 0; extensor hallucis longus = 0; extensor digitorum longus = 0; peroneus longus = 1; peroneus brevis = 1.

Gait studies were performed both with and without a polypropylene orthosis.

Abnormal ankle motion was the primary movement abnormality (Fig. 14–4). The ankle position at foot-strike was 40 degrees plantar flexion. The hip and knee measurements of stance movements were relatively normal. However, after toe-off, complete foot drop was present throughout swing-phase. During swing-phase, marked internal rotation of the foot occurred. The compensatory movements for this complete foot drop consisted of exaggerated swing-phase hip flexion and knee flexion, and exaggerated hip abduction. This increased hip abduction appeared to be a circumduction movement of the hip to aid in foot clearance. This movement was not seen in the other pathological studies reviewed here, but has been seen in experimental peroneal nerve block subjects immediately following the block. The experimental subjects initially circumducted but rapidly developed the compensatory movements of steppage gait (exaggerated hip and knee flexion).

The gait changes with the use of the orthosis were striking in this child, and the walk pattern appeared to be normal. Stride-length and walking velocity were increased, and cadence and work output were reduced (Table 14–3). The primary movement abnormalities, foot drop, and foot internal rotation in swing-phase were corrected (Fig. 14–4).

TABLE 14-3. COMPARISON OF GAIT
MEASUREMENTS WITH AND WITHOUT A
POLYPROPYLENE SHORT LEG
ORTHOSIS—CASE 3

	Not Braced	Braced
Velocity (cm/sec)	68	79
Cadence (steps/min)	146	142
Stride length (cm)	56	67
Single stance (involved leg) (% cycle)	31	33
Work output (kcal/kg/km)	0.2103	0.0634

Heel-strike was restored and a return to a normal pattern of center-of-pressure progression occurred. From the gait parameters measured, it can be concluded that the polypropylene orthosis brought about a return of normal gait for this child. There is often a return of lost function after an injection into the sciatic nerve, and the plan is to maintain this orthosis as long as necessary while awaiting return of muscle function.

EVALUATION OF TREATMENT OF THE GROUP AND CONCLUSIONS CONCERNING THE ROLE OF MECHANICAL FACTORS AND THE RELATIVE OUTCOMES

The two patients with paralytic foot drop were effectively treated with a light-weight, semiflexible, polypropylene orthosis. The normal subject with peroneal nerve block was able to walk in a normal manner with a similar orthosis. The ease with which normal gait was restored by this simple orthosis suggests that the foot drop is easily controlled; however, breakage is a significant problem when the functional demands of flexibility are met. This is one example of a widely used foot drop brace of modern design, but there are a variety of brace models which provide satisfactory function for individual patients. If medial-lateral stability is required, particularly when weakness involves more than the anterior and lateral compartment muscles, some flexibility must be sacrificed.

PATHOMECHANICS OF PARALYTIC FOOT DROP

Movement Abnormalities

The primary movement abnormality in foot drop gait is an alteration in ankle joint motion. The ankle is plantarflexed at foot-strike, so that floor contact is first made with the forefoot. The ankle then moves into dorsiflexion in a normal manner through the remainder of stance-phase. After toe-off the foot fails to dorsiflex in

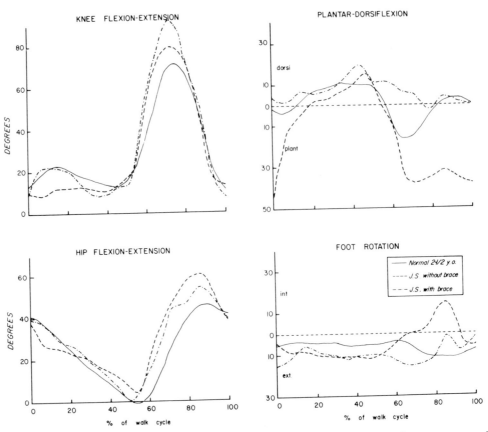

Fig. 14-4. Motion measurements of case 3 with and without a polypropylene brace, compared with 2½-year-old normal values.

swing-phase, necessitating a steppage (exaggerated hip and knee flexion in swing). In the absence of peroneal muscle function, the foot inverts sharply at the beginning of swing-phase, further increasing the risk of stumbling. Normal subjects with peroneal nerve block are most aware of the stumbling hazard when they attempt to make turns. The patient with long-standing foot drop (Case 2) had no difficulty with stumbling; however, she had some remaining power in the anterior compartment muscles as well as some function in the peroneal muscles. Peroneal muscles can be used as evertors in swing-phase to reduce the danger of stumbling. Patients with only anterior compartment muscle weakness often choose not to wear a foot drop orthosis. It appears that the peroneal muscles prevent inversion of the foot and reduce the stumbling hazard.

Floor Reaction Abnormalities

There were some overall changes in the force plate curves, but the primary abnormality appeared to be in the progression of the center of pressure. Normal subjects demonstrate progression of the center of pressure from heel to forefoot. By contrast, foot drop patients show a concentration of the center of pressure entirely in

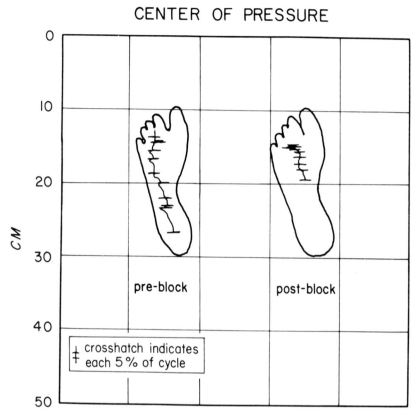

Fig. 14–5. Center of pressure recordings for case 1 before and after peroneal nerve block. The transverse marks are at 5 percent intervals in stance.

the forefoot. The contrast between the normal progression and the effect of peroneal nerve block is illustrated in the center-of-pressure recordings of the normal woman in Case 1 before and after peroneal block (Fig. 14–5).

Energy Work Output Abnormalities

When the orthosis was not worn, the three subjects demonstrated increased work output in walking as compared with normal subjects. Bracing restored nearly normal work output in each instance. (Tables 14–1, 14–2, and 14–3)

Mechanical Analysis of Treatment

Requirements for bracing can be deduced from the function of the muscles that are paralyzed. The anterior compartment muscles lift the foot in swing and decelerate plantarflexion at foot-strike. The lateral compartment muscles function in stance phase to aid in the medial-lateral balance of the foot throughout much of stance phase. The orthosis must resist the forces causing plantarflexion at heel-strike (Fig. 14–2) and return the ankle to neutral after toe-off. Mollan and James have calculated the flexion moment at the ankle following heel-strike, and have estimated the range to be from 3 to 5 N.m as the ankle moves from neutral to 15 degrees of plantarflexion.[5] They have described the manufacture of a flexible polypropylene ankle-foot orthosis. When the lateral compartment muscles are paralyzed, the orthosis must also provide some medial-lateral stability. With a polypropylene orthosis this is usually accomplished by altering the trim lines and reducing some of the flexibility.

The need for flexibility can easily be demonstrated. The reader should wear a rigid polypropylene-ankle foot orthosis, and then repeat the test with a flexible one. A rigid orthosis supplies force to the knee, interfering with the normal knee-ankle interaction. A flexible orthosis is much more comfortable because the troublesome pressures on the proximal tibia are eliminated.

While the polypropylene orthosis has been a valuable contribution to the management of paralytic foot drop, design improvements are still needed. There is no reason for the orthosis to restrict dorsiflexion, and the bending moment that is put on the orthosis when the weight-bearing ankle moves into dorsiflexion may be one factor in producing breakage. Very little resistance is required for maintaining the foot in neutral in swing-phase. The flexibility of the orthosis should be matched with the extrinsic force causing plantarflexion at heel-strike (Fig. 14-2). The proper balance should permit plantarflexion to occur at a slower rate. The orthosis should be sufficiently flexible to permit the ankle plantar flexors to function in stance phase, yet sufficiently strong to return the ankle to neutral position when the limb is unloaded.

REFERENCES

1. Atlas of Orthotics. Biomechanical Principles and Applications. Academy of Orthopaedic Surgeons. pp. 206–215. C. V. Mosby Co., St. Louis, 1975.

2. Cozen, L.: Management of foot drop in adults after permanent peroneal nerve loss. Clin. Orthop., *67*: 151–158, 1969.

3. Hall, G.: A review of drop-foot corrective surgery. Lepr. Rev., *48(3)*: 185–192, 1977.

4. Miner, K. M., Shipley, D. E., and Enna, C. D.: Rehabilitation of Paralytic Drop Foot in Hansen's Disease. Phys. Ther. *55(4)*:378–381, 1975.

5. Mollan, R. A. B., and James, W. V.: A new flexible design of drop foot orthosis. Injury, *8(4)*:310–314, 1977.

6. Muckart, R. D.: Compression of the common peroneal nerve by intramuscular ganglion from the superior tibio-fibular joint. J. Bone Joint Surg., *58B(2)*:241–244, 1976.

7. Schrock, R. D.: Peroneal nerve palsy following derotation osteotomies for tibial torsion. Clin. Orthop., *62*:172–177, 1969.

8. Seppalainen, A. M., Aho, K., and Uusitupa, M.: Strawberry pickers' foot drop. Br. Med. J., *2*:6089, 767, 1977.

9. Siegler, M., and Refettoff, S.: Pretibial myxedema: A reversible cause of foot drop due to entrapment of the peroneal nerve. N. Engl. J. Med., *294*:1383–1384, 1976.

10. Sorell, D. A., Hinterbuchner, C., Green, R. R., and Kalisky, Z.: Traumatic common peroneal nerve palsy: A retrospective study. Arch. Phys. Med. Rehabil. *57(8)*:361–365, 1976.

11. Verghese, M., Radhakrishnan, M., Chandrapal, H., and Jacob, M. V.: Phasic Conversion after Tibialis Posterior Transfer. Arch. Phys. Med. Rehabil. *56(2)*:83–85, 1975.

15

The Role of Spine Fusion in the Treatment of Problems of the Lumbar Spine

T. W. McNeill, M.D.

THE ROLE OF THE DISC IN LOW BACK PAIN

Diseases and disabilities of the lumbar spine are major challenges remaining for the orthopedic and neurosurgical communities. The understanding of the biomechanical, biochemical, and pathological states of this complex system is incomplete. Its basic functional and study unit is the "motion segment" (Fig. 15–1), consisting of two vertebrae and their common ligaments and joints. The vertebral bony elements consist of the anterior large bony cylinder called the "body," and an arch-like posterior structure that surrounds the neural elements. These are connected by several ligaments (Fig. 15–1) and paired diarthrodial joints called "facet joints." The anterior vertebral bodies are connected by a flexible linkage called the "disc joint" (Fig. 15–2).

It is not uncommon for patients to complain of very severe lower back pain several hours or even a day following a bending, twisting, or lifting accident. Frequently they give histories of having done unaccustomed heavy work, such as shoveling or mixing of concrete (Fig. 15–3), and often having felt a snap or sharp sensation in the lower back while pulling or lifting in a flexed position, frequently with a rotational component. It is not unusual for them to have continued to work for a short period of time, followed by a period of rest and progressive stiffness and pain after the slightest motion of the spine. The pain is most intense in the region of L4 and L5, lateral to the spinous processes, with radiation to the upper and middle buttocks. Such symptoms usually remain very intense for several days and then gradually diminish in about 1 to 3 weeks. The patients seldom experience paresthesias or other neurologic symptoms during this initial phase. They have little pain during rest, but even slight movement often produces excruciating back "spasm."

Two anatomic changes have been proposed to explain these initial events: (a) end plate fracture; and (b) annulus tearing (Fig. 15–4A and B). Interosseous herniations, which indicate end-plate failure, have been demonstrated at autopsy. But the earliest pathologic findings in autopsy material are fissuring and mucoid degeneration of the annulus. Furthermore, when isolated motion segments are subjected to rotational stress, the annulus characteristically fails at the same point at which the majority of posterior lumbar herniations are found.

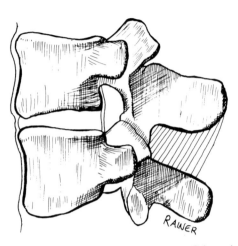

Fig. 15–1. The "motion segment" of the spine.

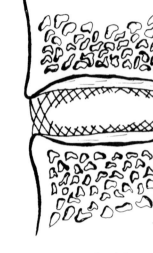

Fig. 15–2. The disc joint (schematic).

Pathologic evidence indicates that both types of failure occur throughout life.[9] End-plate fracture and annulus failure may both be implicated in explaining the etiology of the sudden "lower back syndrome." The clinical course is analogous to that in similar injuries to other joints. For example, a rotational stress applied to the ankle may result in either failure of the bone or ligament, or both, depending upon the position of the foot and leg at the time. After a partial tear of the lateral collateral ligaments of the ankle, there may be a sudden onset of pain followed by some limited function until the joint is rested. There then follows a period with excruciating pain on any motion, with involuntary contractions of the muscles bridging the joint (muscle spasm), until the initial severe reaction has subsided — usually in 1 to 3 weeks, depending on the severity. Rest during bone or ligament healing is proper for both the ankle and the lumbar spine.

After this subsidence of pain, many patients will experience no further problems with backache, but a significant minority do experience multiple severe and moder-

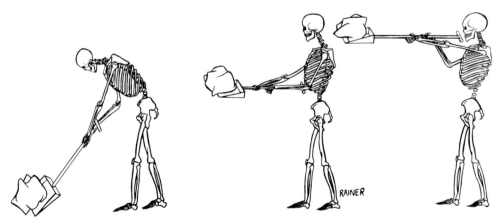

Fig. 15–3. Motions that often lead to low back pain.

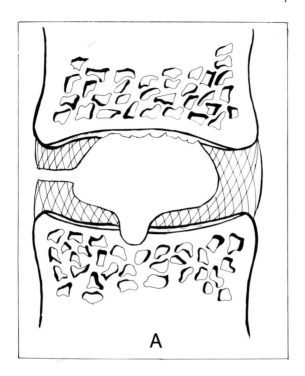

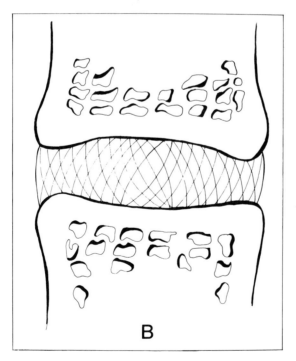

Fig. 15–4. (A) End-plate fracture and annulus tear (schematic section); (B) Normal structure (schematic section).

ately severe episodes of back pain throughout the remainder of their lives. During the first decade of this syndrome of multiple recurrent episodes of severe lower back pain following trivial trauma, x-ray study seldom discloses any significant changes.

The foregoing group of patients consists of those who have most of the acute herniations of the nucleus; and those with painful degenerative disc disease who often have histories of multiple previous attacks. The pathologic changes are: (a) progressive fissuring of the annulus and nucleus; (b) erosions of the hyaline cartilage; (c) increased collagen content of the nucleus; and (d) separation of the hyaline cartilage along with attachment of the annulus and herniations of nuclear material, usually with attached hyaline cartilage. The only treatment with consistently proven value at this stage of the disease is removal of the herniated fragment of nucleus.[14, 18] In the case of complete herniation of a fragment of the nucleus, this procedure provides relief for up to 90 percent of patients. Chemonucleolysis—partial enzymatic degradation of the nucleus—has apparently been effective in 70 percent of individuals with herniation of a fragment of the nucleus.[3] Spine fusion has failed as an initial treatment in cases with a herniated nucleus pulposus.[5]

The next stage in the development of a degenerative disc is frequently not severely painful. If symptoms are present, they are most often characterized by a backache that is more severe after a period of rest, and which is then relieved by light activity, becoming more severe after heavy activity. These symptoms usually respond to anti-inflammatory medications, physical therapy modalities, bracing, and mild limbering-up exercise in the morning. The x-rays then reveal narrowing of the intervertebral disc, marginal osteophyte formation, end plate sclerosis with, sometimes, a "vacuum sign" and similar changes in the facet joints. There may also be disordered motion of the involved segment. For surgery to be indicated, the symptoms must be severe, and they must be significantly relieved by bracing; no more than two joints should be involved; sciatica should not be present, and symptoms should be relieved temporarily by intradiscal injection of xylocaine and corticosteroids. Fusions in cases of herniation of the nucleus are not indicated except in cases of persistent backache following removal of the herniated nucleus, after the foregoing presurgical criteria have been achieved.

The end stage, perhaps four or five decades after the initial insult, is usually not disabling unless there is nerve entrapment by proliferative bone. In such patients, there may be individual root involvement (lateral recess stenosis) or generalized narrowing with cauda equina symptoms on activity (pseudoclaudication of the spinal stenosis). Here again the primary symptoms are relieved by adequate decompression, and, therefore, fusion is usually unnecessary.

Spine fusion as a treatment for lesions of the lumbar spine has a poor reputation among most orthopedic surgeons and neurosurgeons. That reputation has developed for a number of reasons, among which are:

1. The statistical results of disc surgery of herniations of the nucleus are apparently unchanged, whether or not a fusion was attempted at the initial operation.[6]
2. Fusion adds considerably to the surgical time, morbidity, and effort.
3. Solid fusion is difficult to obtain at the lumbosacral junction with the traditional Hibbs type posterior fusion.
4. Many observers have noted significant x-ray evidence of degenerative disc disease in elderly patients who have not experienced any severe back pain.

Little doubt is now expressed that the removal of a large, sequestered fragment of nucleus from between the margins of the vertebra and from under the nerve root will afford almost complete and lasting relief from the syndrome of severe back pain and sciatica in the majority of patients. Only about 15 percent of patients who have a large sequestered fragment removed will have significant residual disability from persistent back pain and sciatica.

Since there are no well-established criteria for determining which of these patients will require later fusion, it is seldom recommended at the time of the initial surgery. Froning and Frohman,[5] in an x-ray study, have demonstrated a good correlation between retained motion at the level of disc fragment excision and persistent back pain.*

Degenerative intervertebral joints need not always be a source of pain, just as degenerative interphalangeal joints or hips may not produce pain. However, degenerative intervertebral disc joints may be severely painful with motion. A solid fusion will relieve the pain in these joints, just as a fusion will relieve pain in the wrist or ankle.

The confusion in identifying back pain comes from several sources:

1. Multiple levels may be involved.
2. The source of pain, as in Case 1 below, may be from the neurologic elements per se, rather than, or as well as, from the joints per se.
3. Evaluating the solidity of the fusion is difficult.
4. Psychological factors play a large role in recovery from chronically painful conditions.

CASE DESCRIPTIONS

Case 1. This 32-year-old, 280 lb male truck driver had complained of more than 5 years of severe, intermittent lower back pain. His episodes were triggered by bend-

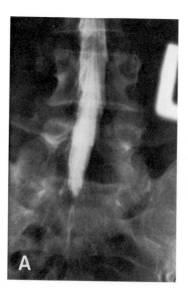
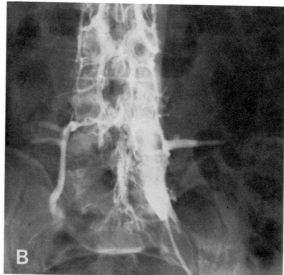

Fig. 15–5. (A) Case 1: Myelogram, soon after injection of dye. (B) Myelogram at a later time.

---

*Stauffer and Coventry[20] have stated that posterolateral intertransverse fusion has a high rate of success in properly selected patients with back pain.

ing, twisting, and lifting, and caused him to miss several weeks of work each year. These episodes were accompanied by minor complaints of sciatica, which had previously been resolved along with the back pain, after a period of conservative treatment consisting of rest and use of analgesics. However, the present episode was somewhat different than the preceding one in that sciatica was the predominant — and backache the minor — complaint. Several weeks of inpatient care, plus 6 weeks of rest at home, resulted in no relief. The patient complained of severe buttock pain on the right, and pain in the right calf and heel. Intermittently he had tingling in the right leg and lateral right foot. He was unable to sit without severe discomfort.

Physical examination revealed a markedly obese white male lying supine in bed with two pillows under his knees. Full extention of his legs caused increased pain in the right leg and lower back. Straight leg raising was positive bilaterally at 30 degrees, with pain referred to the right side in each case. There was decreased sensation to pinprick and light touch on the right foot laterally and on the posterior right calf. Deep tendon reflexes were all normal except the right Achilles tendon reflex, which was present but decreased as compared to the left reflex. The patient stood up only with help, and walked with a right-sided list and limp. He had tenderness in the right buttock and right lower lumbar spine. Even the smallest motion of the back was severely painful.

Routine x-ray films of the lumbar spine revealed only loss of lumbar lordosis; myelogram revealed an extradural defect on the right at L5/S1 (Fig. 15–5A and B); electromyogram revealed evidence of denervation in the right gastrosoleus muscle group; and surgical exploration revealed a large extruded fragment of intervertebral disc at L5/S1 on the right. This was removed.

The patient's postoperative course was uneventful, and he had immediate relief of

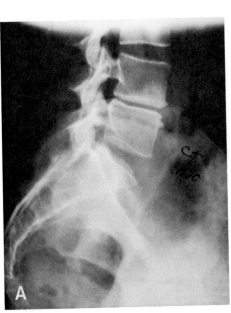
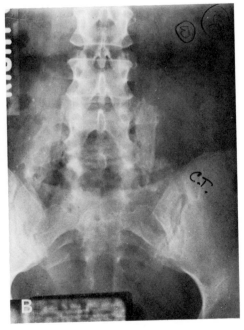

Fig. 15–6. (A) Case 2: Lateral view, at first examination. (B) Antero-posterior view.

his sciatica. He has not lost weight and is symptom-free on mild activity. He was unable to return to driving a truck because of persistent low back pain; however, he would probably have been able to undertake a more sedentary occupation.

Case 2. This 30-year-old housewife had a history of having sustained an injury to her lower back in high-school athletics. She experienced repetitive episodes of severe lower back pain and progressively severe exercise intolerance. At no point during her 15 year history did she experience sciatica, paresthesias, or weakness.

Physical examination revealed restriction of motion of the patient's lumbar spine to approximately 75 percent of normal in all planes; flexion was 60 degrees, lateral bending 20 degrees, and extension 5 degrees. A straight-leg test was 80 degrees bilaterally, without increasing the pain in either the patient's legs or back. Neurologic examination was normal, as was lumbar lordosis. Pain response was elicited on deep heavy pressure over the spinous process of L5. X-ray examination revealed moderately advanced degenerative changes at the L5/S1 level (Fig. 15–6A and B).

Trial in a Boston scoliosis brace (TLSO) was successful in partially relieving the patient's symptoms and extending her level of activity by nearly 60 percent. She was unable to eliminate the brace without recurrence of symptoms. Therefore, a bilateral posterolateral fusion was performed, using strips of autogenous iliac bone graft. The fusion became solid at 6 months postoperatively, with complete relief of symptoms 2 years later.

Case 3. This 16-year-old male athlete presented to the outpatient office with the complaint of severe lower back pain which began during the preceding football season with a rather gradual onset, and with radiation only to the upper buttocks bilaterally. No specific, severe injury was recalled. The patient became nearly symptom-

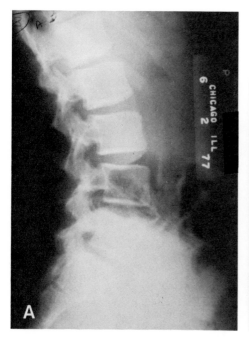
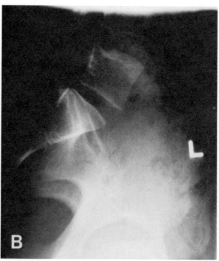

Fig. 15–7. (A) Case 3: Second x-ray examination, lateral lumbar view. (B) Second x-ray examination, lateral sacral view.

free with a few days of rest, only to have a severe recurrence with resumption of activity. The back pain was relieved by the use of a brace and was aggravated by sitting.

Physical examination revealed a 6 foot, 1 inch, 170 lb muscular male with a straight lumbar spine, antalgic gait, marked paravertebral muscle spasm, marked limitation of motion in all planes, and local tenderness over the lower lumbar spine. A straight leg-raising test was normal, as was the neurologic examination. Chest expansion was normal. Initial x-ray studies, and an HLA-B27 antigen study, were normal.

Treatment with a TLSO (thoracolumbar spinal orthosis) relieved the patient's symptoms, but they recurred with resumption of unbraced activity. Over the subsequent 6 months, the symptoms gradually became more severe and a second set of lumbar x-rays were made, which revealed a grade I spondylolisthesis, L5 on S1, with defects in the pars interarticularis of L5 (Fig. 15–7A and B). Review of the initial films did not reveal an occult lesion. Since this situation met the criteria for "instability" of the spine, and since no neurologic deficit was evident, a bilateral posterolateral fusion was performed, with solid bony fusion, with complete recovery from symptoms.

Case 4. This 19-year-old male was attempting to push a pickup truck out of a snowbank when another vehicle struck the patient from behind. He was brought to the nearest hospital, where he was treated for his multiple injuries. Physical examination was restricted to the spinal problem, and revealed a well-developed white male with bruising and tenderness of the upper lumbar spine and a mild kyphosis at the L1 and L2 levels, with a palpable gap between the spinous processes of L1 and L2. X-rays demonstrated a flexion distraction injury of L2, with comminution of the

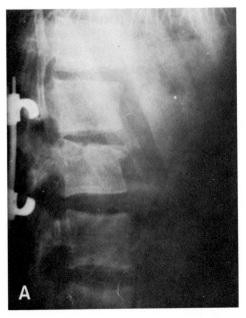
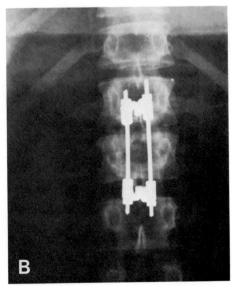

Fig. 15–8. (A) Case 4: After insertion of Harrington rods, lateral view. (B) Case 4: After insertion of Harrington rods, antero-posterior view.

body of L2 anteriorly and a fracture through the pedicles and spinous process.

Surgical treatment consisted of open reduction and internal fixation of the fracture with Harrington compression rods supplemented with spine fusion of the involved segments (Fig. 15–8A and B). The patient was able to leave the hospital in a body jacket cast after 4 weeks, and is now solidly healed without deformity or residual pain.

Case 5. This young man was involved in an altercation with a policeman, resulting in his having been shot in the abdomen with a 38-caliber revolver. The missile transected several loops of intestine, and the patient's L4/5 disc (Fig. 15–9A). He initially underwent surgical exploration of his abdomen with repair of the involved intestinal and vascular injuries. Immediately following this, the patient underwent neurosurgical exploration of his cauda equina and removal of fragments of missile, bone, and disc from his spinal canal. He made a gradual neurologic recovery following this surgery and, at the time of our examination, had full bowel and bladder control and nearly normal function of his left lower extremity. He had residual paralysis of L5/S1 and S2 on the right. He complained of severe lower back pain with any activity, and was confined to bed because of severe pain.

Physical examination revealed an emaciated male with well-healed abdominal and lumbar spine incisions. He had marked limitation of motion of the lumbar spine owing to pain, and marked tenderness over the lower lumbar spine.

A trial in a TLSO demonstrated relief of this patient's lower back pain and partial relief of his leg pain. A posterolateral spinal fusion was done, which has resulted in nearly complete relief of the patient's back pain, but only partial relief of his leg pain.

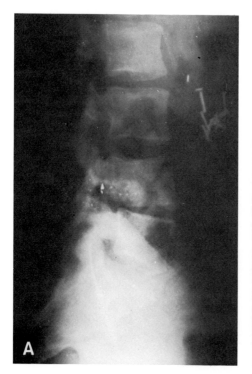
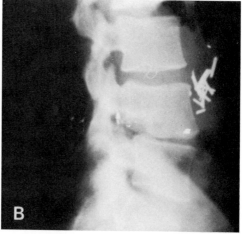

Fig. 15–9. (A) Case 5: Defect after injury. (B) Defect 3 years post injury.

Over the subsequent 3 years, he has recovered sufficient motor function on the right to abandon his leg brace, and has now returned to school. Figure 15–9B shows the x-ray appearance of the lesion at this time.

MECHANICAL ANALYSIS OF THE LOW BACK

The five cases presented seemingly represent five quite different entities. However, each of them presents a possible indication for lumbar spine fusion. The patient in Case 1 experienced a complete herniation of the nucleus pulposus, with an uncomplicated surgical removal of the nuclear fragment. This treatment is normally successful in 90 percent[18] of patients; why was it unsuccessful in this instance? Case 1, the athletic housewife, and the gunshot-injured patient all represent destruction of the integrity and function of the intervertebral disc joint. In contrast, the other two cases represent two types of "instability" of the lumbar spine.

An explanation of the problems represented in Case 1 can be made most simply with a hypothetical situation, such as shoveling clay (Fig. 15–10). In this case, a mass (the clay) is held at a distance from the body. If the mass of the clay plus the mass of the shovel were 10 kg, and its center were at a point 1 m from the instantaneous center of motion of the lumbar spine, then the movement (F × D) created would be approximately 100 Newton · meters (1 kg of mass in the earth's gravity produces a force of 9.8 N).

Furthermore, one must consider that half of the 80 kg body weight of the obese patient in Case 1 is above the lumbar motion segment. This we will assume to be centered at a point 10 cm anterior to the instantaneous center of motion, which adds another 40 Newton · meters of moment. In an equilibrium situation, this may be considered to be balanced by the force of the extensor muscles of the spine and by the intraabdominal pressure. The extensor muscles act at a very short distance (2 cm) from the instantaneous center of motion and thus, their force is almost completely compressive in nature. The maximum intraabdominal force that can be generated is approximately 600 Newtons, which acts at an estimated 10 cm for the instantaneous center of motion. This effect can only act to decrease the necessary extensor force.[13]

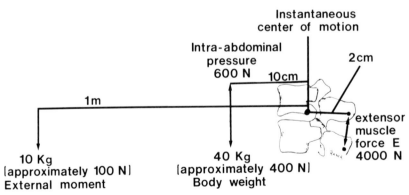

Fig. 15–10. Mechanical analysis of forces and moments on lumbar spine as a result of shoveling.

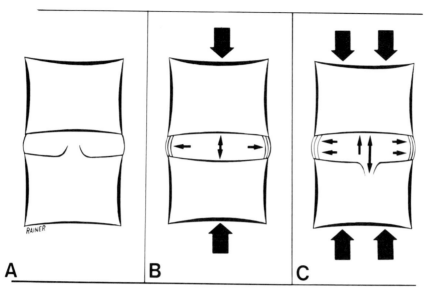

Fig. 15–11. Development of Schmorl's nodes (A) Schematic of disc with node. (B) Internal pressures with normal axial load. (C) Internal pressures, leading to defect, with twice normal axial load.

In our example, the calculated extensor force, E, of 4,000 N is well within the theoretical capability of the normal adult male's spinal extensor muscles. However, in Rolander and Blair's study of cadaver vertebra,[16] a compressive force of approximately 3,000 N was sufficient to produce a small compression fracture of the endplate of the vertebra. This compressive force of the extensor musculature must either be dissipated within other tissues, or the posterior tension may be taken up by other tissues; otherwise we have produced a situation in which an acute overload is inevitable. The weakest link in our hypothetical lifting example is the lumbar spine. Gracovetsky, Farfan and Lamy[8] have produced a theory of "optional control," in which it is suggested that the disc is a sensor and that during lifting, every neuromuscular control system will be utilized to minimize the stress within the disc. In spite of this control system, it is apparent that overload situations occur rather frequently.

In their pioneer pathological study, Schmorl and Junghanns[17] identified intraosseous disc protrusions very frequently. It is believed that these "Schmorl's nodes" represent the result of compressive overload of the spine. It is not realistic to assume that the simple two-dimensional force analysis represented in Figure 15–11 could be complete, since the calculated compressive forces are too high for the relatively light task of shoveling 10 kg weight at 1 m from the spine. The other posterior structures must also provide significant tension to resist the anterior moment. In the Lamy-Farfan model this posterior tension is generated in the posterior ligamentous structures. The theory suggests that the posterior musculature serves to adjust the geometry of the lumbar spine in response to load, and thus distributes the tension within the ligaments. This, in turn, serves to reduce the extensor muscle contractile force necessary to resist an anterior moment and, thus, also serves to reduce the intradiscal compressive forces associated with bending and lifting.

Rollander[15] subjected intact vertebral segments to compressive loading, to the

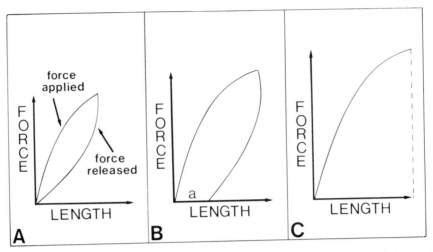

Fig. 15–12. (A) Elastic deformation of ligament. (B) Plastic (nonrecoverable) deformation of ligament. (C) Rupture of ligament.

point of failure, and found the weakest link to be the bony vertebral end-plate, which developed stellate fractures with excessive compressive loading.

Farfan[4] believes that the compression-overload type of lumbar failure is not likely to be the cause of the more common clinical syndromes of disc injury and secondary disability. He explains the initial and subsequent episodes on the basis of twisting injury with annulus tearing. With twisting injury the obliquely oriented fibers of the annulus are subjected to tensile forces that exceed the fibers' elastic limit, and permanent deformation or failure is a result of this excessive force. Evidence for this rotatory type of failure is present in many clinical instances, as deformity of the posterior bony elements as well as annular tears. This theory suggests that multiple separate instances of rotatory overload lead to secondary disc degeneration and nuclear herniation.

In underloading situations with human annular ligaments,[7] low levels of stress will result in strain that does not exceed the elastic limits of the ligament, as in Figure 15–12A. As the force is increased (Fig. 15–12B and C), there is first a permanent deformation, and eventually rupture of the ligament. This rupture usually takes place at the site of greatest stress concentration posterolaterally, and unfortunately in the lumbar spine this area is directly under the nerve root. This theory correlates well with the pathological findings at autopsy of tissues in the annulus.

Whichever mode of failure actually took place in Cases 1 and 2, the common end result was to reduce the ability of the spine to resist a flexion moment. In Case 1, two factors were operative. First, with disc rupture and the associated degenerative changes, came loss of disc height, which reduces the ability to adjust the geometry of the spine, and thus to distribute the posterior tension within the ligament structures. Second, the laminectomy disrupted both the posterior ligamentous structures and some neural control to the extensor musculature. In Case 2, the disc resorption as a late result of an adolescent athletic injury effectively shortened the intervertebral distance, rendering the extensor musculature and ligament structure ineffective. Thus, in both cases the disc became progressively less able to adjust to an anterior moment. If the disc and facet joints are pressure sensors, they may respond—with

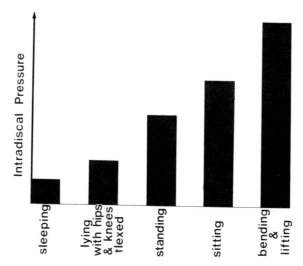

Fig. 15–13. Intradiscal pressures as a result of activity.

what would be normal activity in an otherwise healthy person — with pain as an indication of overload.

Disc excision serves to relieve neurologic injury and the secondary symptoms of sciatica, but not the back pain component secondary to moderate or heavy activity. Patients who undergo disc excision for relief of sciatica and neurologic injury often continue to have episodes of low back pain with moderate or heavy activity.

The multiple episodes of lower back pain seen in Cases 1 and 2 may be interpreted as repetitive tearing of the annulus or progressive detachment of the hyaline cartilage. Each subsequent episode requires progressively less stress than the preceding episode, until an attack may be initiated by a sneeze or leaning forward, such as in brushing the teeth. The compressive forces within the disc have been measured in volunteer subjects by Andersson, Ortengren and Nachemson.[1] They found that the force was least while sleeping and greatest while bending and lifting (Fig. 15–13).

It is most interesting to note that these activities correspond well to those best and least tolerated by back patients. They are most comfortable while lying supine with their hips and knees flexed, and least comfortable while sitting or attempting to bend and lift. Why then, was spine fusion helpful in Case 2, and why has it been suggested in Case 1? The first purpose is that of substituting for the failed posterior structures, which can no longer counteract an anterior moment; and second, to resist rotational stresses and movement within the motion segment. If the only reason for the fusion was to substitute for the lost posterior ligaments, a traditional posterior fusion would be a simple and direct solution (Fig. 15–14).

The posterior fusion would appear to be ideal in that it can be placed well posterior to the axis of motion, and can thus provide considerable leverage to resist a flexion moment. But this has not proved effective in clinical practice for several reasons. First, Rollander[15] has demonstrated that this type of fusion will still allow considerable disc joint motion with normal load levels, because of the flexibility of the intervening bone. Second, posterior fusions have been found to result in a secondary laminar bony overgrowth, leading to late spinal stenosis. Third, this procedure often has been done indiscriminately in the past for almost any complaint of back pain, result-

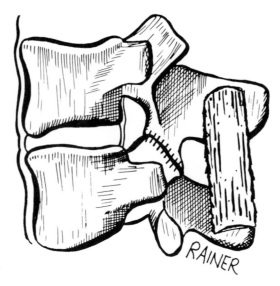

Fig. 15 – 14. Posterior fusion.

ing in the late presentation of large series of heterogenous cases with poor clinical results.

Anterior interbody fusions (Fig. 15 – 15) would have the advantages of great rigidity and providing early stability, but they have the disadvantages of a low rate of fusion and much greater risk.

Posterolateral or intertransverse fusions (Fig. 15 – 16) have the advantages of rigidity, resistance to rotatory stress, a high rate of fusion (90 percent; Stauffer and Coventry[19]), and safety. Stauffer and Coventry[19] have shown an 85 percent success rate in selected cases with degenerative disc disease.

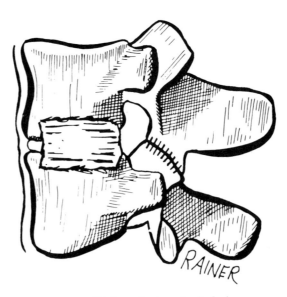

Fig. 15 – 15. Anterior interbody fusion.

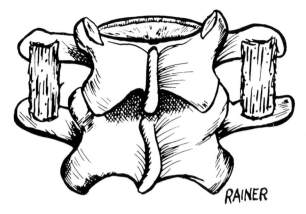

RAINER

Fig. 15 – 16. Posterolateral or intertransverse fusion.

In patients such as those described in Cases 1 and 2, the term "instability" has been used quite loosely in the past. White[21] has defined "clinical instability" as: "the loss of the ability of the spine under physiologic loads to maintain relationships between vertebrae in such a way that there is neither damage nor subsequent irritation to the spinal cord or nerve roots and, in addition, no development of deformity with excessive pain." We would add that the development of a deformity with or without pain would indicate that the spine was clinically unstable at the time the deformity developed, and that progressive deformity, whether slowly or rapidly developing, indicates existing clinical instability.

Cases 3 and 5 represent examples of clinical instability in which a deformity with excessive pain developed slowly over a period of months. The pars interarticularis defect in Case 3 probably represents a stress fracture, while in Case 5 the initial trauma plus the subsequent surgical treatment resulted in deformity and pain with physiological loads. Stress fractures are the result of the paradoxical initial weakening of

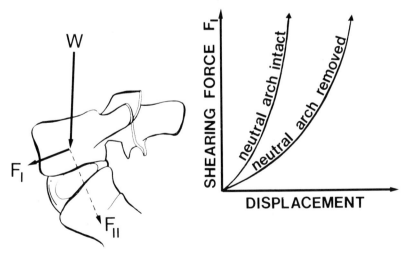

Fig. 15 – 17. Effect of lumbrosacral inclination on disc shear force.

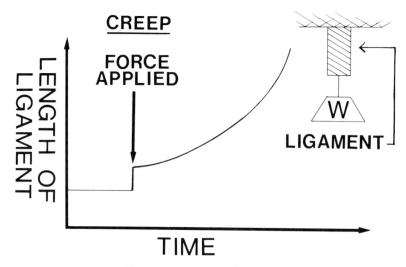

Fig. 15 – 18. Creep of ligament.

bone in response to repetitive stress. While it is true bone becomes stronger in response to stress, the initial event is resorption of the bone prior to addition of new lamellae. If excessive stress is applied to the bone in this weakened state, fractures may occur. Thus, in Case 3, with the loss of the support of the posterior element of the vertebra, a rather constant force was applied to the annulus and disc while the patient was upright. Since the lumbosacral joint is normally obliquely situated, there is a significant force vector forward (shear), which is no longer resisted by the posterior elements of the vertebra (Fig. 15 – 17). The force is now applied to the annulus, which demonstrates the viscoelastic phenomenon known as creep (Fig. 15 – 18). As a constant force is applied to a ligament it will gradually elongate with time. It is believed by some that the pain in spondylolisthesis is also primarily discogenic,

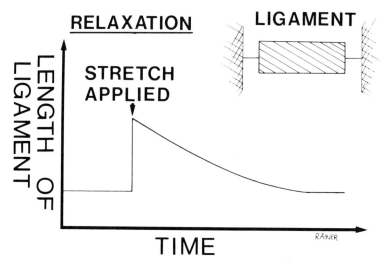

Fig. 15 – 19. Stress relaxation of ligament.

probably owing to the abnormal deformation of the joint. The converse of creep is relaxation, which is the gradual decrease of force over time under conditions of constant stretch (Fig. 15 – 19).

Case 5 represents a form of "instability" as defined by White and Panjabi.[21] In this instance, the combination of the gunshot injury to the spine and the previous wide, decompressive laminectomy resulted in progressively increasing pain with activity and motion. In this instance, although the initial injury produced the deformity and the deformity had not become progressive, the pain problem with activity had become more and more severe. For this reason, spinal fusion proved to be the treatment of choice and, in the long term follow-up, resulted in both subjective improvement of symptoms and objective improvement with reference to both activity and neurologic function.

Case 4 represents an even more obvious case of lack of spinal stability, and of the type with which most associate the term "instability." In this instance, the patient sustained fractures involving both the anterior and posterior elements of the vertebra, thus leaving no normal structural elements that could be utilized to resist deformity if the spine was loaded by gravity. In this instance, spinal fusion is again chosen as a treatment of choice; two other factors, however, intervene. The first is the need for reduction of the fracture in order to preserve space for the neural element and to prevent the late complication of neurologic involvement owing to healing of the fracture or progressive deformity. Second, the reduction must be maintained while the fracture and stabilizing spine fusion are healing. This can be achieved through the use of internal fixation devices connected to the spine, and in this instance, by providing a compressive force. The devices are called Harrington compression rods and are simply threaded, metallic rods connected to small stainless steel hooks which are attached to the normal bone of the vertebra, above and below the site of injury. Because of the large flexion component to the injury, and because of the tearing of the posterior elements, this method was chosen in Case 4 to provide reduction and temporary stability while the fracture and bone graft were healing. This is an instance in which posterior spine fusion was applied in order to provide a substitute for the torn posterior structures. Mechanically this construct is most sound in this instance because of the associated destruction of the anterior elements, resulting in an unrestrained flexion moment, until such time as the fractures have healed and posterior compressive Harrington instrumentation will provide a mechanically sound resistance to a flexion moment.

In summary, the indications for fusion in the lumbar spine are those described by White and Panjabi,[21] and grouped under the general term "instability"; or those for treatment of abnormal or excessive motion in an intervertebral joint that produces pain and functional disability. It should be noted that spine fusion is not the universal treatment for all diseases of the lumbar spine, but is a valuable tool in selected instances, such as those we have described.

REFERENCES

1. Andersson, G. B. J., Ortengren, R., and Nachemson, A.: Quantitative studies of back loads in lifting. Spine, *1(3)*:178, 1976.
2. Berkson, M., Schultz, A., Nachemson, A., and Andersson, G.: Voluntary strengths of male adults with acute low back syndromes. Clin. Orthop., *129*:84, 1977.
3. Brown, M.D.: Chemonucleolysis with discase. Spine, *1*:115, 1978.

4. Farfan, H. F., Cossette, J. W., Robertson, G. H., Wells, R. V., and, Kraus, H.: The effects of torsion on the lumbar intervertebral joints: The role of torsion in the production of disc degeneration. J. Bone Joint Surg., *52A:*468, 1970.

5. Froning, E. C., and Frohman, B.: Motion of the lumbosacral spine after laminectomy and spine fusion. J. Bone Joint Surg., *50A:*897, 1968.

6. Frymoyer, J. W., Hanley, E., Howe, J., Kuhlmann, D., and Matteri, R.: Disc excision and spine fusion in the management of lumbar disc disease. Spine, *3:*1, 1978.

7. Galante, J. O.: Tensile properties of the human lumbar annulus fibrosus. Acta Orthop. Scand. (Suppl.):*110:*1967.

8. Gracovetsky, S., Farfan, H. F., and Lamy, C.: A mathematical model of the lumbar spine using an optimized system to control muscles and ligaments. Orthop. Clin. North Am. *8:* 135, 1977.

9. Hirsch, C., and Schajowicz, F.: Studies on Structural Changes in the Lumbar Annulus Fibrosus. Acta. Orthop. Scand., *22:*184, 1951.

10. Hutton, W. C., Stott, J., and Cyron, B.: Is spondylolysis a fatigue fracture? Spine, *2,(3):* 202, 1977.

11. Kulak, R., Schultz, A. B., Belytschko, T., and Galante, J.: Biomechanical characteristics of vertebral motion segments and intervertebral discs. Orthop. Clin. North Am., *6(1):* 121, 1975.

12. Lamy, C., Bazergui, A., Kraus, H., and Farfan, H. F.: The strength of the neural arch and the etiology of spondylolysis. Orthop. Clin. North Am., *6(1):*215, 1975.

13. McNeill, T. W., Warwick, D., Andersson, G., and Schultz, A.: Trunk Strengths in Attempted Flexion, Extension and Lateral Bending in Healthy Subjects and Patients with Low-Back Disorders. Presented to the International Society for the Study of the Lumbar Spine, 1979.

14. Nachemson, A. L.: The lumbar spine, an orthopaedic challenge. spine *1(1):*59, 1976.

15. Rolander, S. D.: Motion of the lumbar spine with special reference to the stabilizing effect of posterior fusion: An experimental study on autopsy specimens. Acta. Orthop. Scand. (Suppl.), *90,* 1966.

16. Rolander, S. D., and Blair, W. E.: Deformation and Fracture of the Lumbar Vertebral End Plate. Orthop. Clin. North Am., *6(1):*75, 1975.

17. Schmorl, G., and Junghanns, H.: The Human Spine in Health and Disease, 2nd American Ed. Grune and Stratton, New York, 1971.

18. Sprangfort, E. V.: The lumbar disc herniation: A computer-aided analysis of 2,504 operations. Acta. Orthop. Scand. (Suppl):142, 1972.

19. Stauffer, R. N., and Coventry, M. B.: Anterior interbody spine fusion. J. Bone Joint Surg., *54A:*756, 1972.

20. Stauffer, R. M., and Coventry, M. B.: Posterolateral lumbar spine fusion. J. Bone Joint Surg., *54A:*1195, 1972.

21. White, A. A., III, and Panjabi, M. M.: Clinical Biomechanics of the Spine. J. B. Lippincott Co., Philadelphia, 1978.

16
The Mechanics of the Surgical Treatment Of Scoliosis

H. K. Dunn, M.D.; A. U. Daniels, Ph.D.

INTRODUCTION

In many ways the human spine behaves like a column. It is slender, flexible, loaded in compression and, with slight increases in load, is subject to sudden and dramatic changes in curvature. Alternatively, if the spine becomes deformed, it is increasingly subject to bending loads and its mechanical behavior can be described as being more like a bending beam than a column. However, unlike the typical columns or beams in buildings, the spine is not of uniform size and does not consist of a homogeneous material. Further, the ways in which it is loaded and in which it responds are not fully understood nor predictable at present. As in any biological system, these factors make the strict application of engineering formulas and mathematics impossible. However, the principles apply and, if kept in mind, can greatly enhance patient treatment.

The Disease State

The term scoliosis refers to curvature of the spine in the lateral direction; kyphosis refers to flexion or forward curvature of the spine; and lordosis refers to extension or backward curvature of the spine. It is not uncommon to see these deviations in the spine in combination, and someone who is excessively forward-flexed and laterally bent is referred to as a kyphoscoliotic patient. The discussions in this chapter will be limited to scoliosis.

When it occurs to any degree warranting treatment, scoliosis is almost invariably accompanied by some rotation of the spine. However, although this rotation is often important because of its effects on the patient's appearance and cardiopulmonary function, it is not customarily referred to or quantitated in clinical terminology. It must be kept in mind, though, that it is the rotation of the spine that principally causes the deformities seen. As a consequence of rotation, scoliosis in its severe form presents as a hunchback deformity. The hunchback is the result of rotation of the ribcage on one side posteriorly and on the other side anteriorly, carrying the scapula with it. The ribs are attached to the thoracic vertebrae, and it is the rotation of these vertebrae that causes the deformity. The resultant distortion of the thoracic cage causes the cardiopulmonary insufficiency problems seen in severe cases of scoliosis.

Scoliotic curves are measured in a standard manner, described initially by Cobb,[7] which quantitates lateral deformity. The illustrations and cases discussed here will reflect this measurement technique. The measurement is made by drawing lines across the endplates and perpendicular to the vertebral axes of the last vertebrae in a curve proximally and distally. The angle of the intersection of these two lines quantitates the degree of curvature. The vertebrae from which measurements are taken are referred to as the transitional vertebrae.

As seen in Figure 16–3A—the first roentgenograph of Case 2—the curve is 50 degrees to the right, with transitional vertebrae at T6 and T12. There is often confusion as to which is the last vertebra in the curve. For clinical purposes this is best resolved by using the vertebra that gives the greatest resultant curvature, because it is the extreme of the disease process that one should consider clinically. In clinical practice, no attempt is made to accurately measure rotation of the spine. It is roughly estimated as ranging from 1 to +4, based on the lateral asymmetry of the pedicle shadows in an anteroposterior roentgenogram.

There are many etiologies of pathologic curvature of the spine, and it is not our purpose here to discuss them completely, nor even to discuss certain aspects in great detail. However, a few points must be made regarding the resultant alterations in the spine and the forces on it owing to the various commonly recognized etiologies. Also, before proceeding further, it should be recognized that not all curvatures of the spine are abnormal. In normal anatomy there is kyphosis in the thoracic spine and lordosis in the lumbar spine.

Additionally, the spine may deviate to compensate for an abnormality elsewhere in the musculoskeletal system. In older terminology these deviations were referred to as compensatory curves. It is better, however, to refer to them as nonstructural curves to indicate that they do not represent major, permanent changes in spinal structure.

An example occurs in a patient with one leg shorter than the other and who is standing with his weight equally on both legs. In this instance the pelvis would be tilted as the result of the short leg, and there would be curvature of the spine to keep the torso and head distributed well over the base of the support. This curve in the spine would be nonstructural—that is, it would straighten as soon as the patient started ambulating and therefore became weightbearing alternately on one leg or the other, or when the patient was sitting or lying. Nonstructural curves of this nature seldom require treatment. Also, where there is curvature of the spine from a structural cause, there will initially be a nonstructural curve above and below to try to keep the head over the base of support. These nonstructural curves also correct themselves when the structural curves are corrected.

Idiopathic adolescent scoliosis is the form of spinal curvature most commonly seen in clinical practice today. It is labelled idiopathic because its exact etiology is unknown. However, a great deal is known about its character and the natural history of the disease process. Its onset is in a previously normal spine, generally at about the time the patient is starting to develop secondary sexual characteristics. The majority of its progression occurs between onset and the end of skeletal growth. Whether further changes occur in later years depends on the degree of curvature at the end of skeletal growth, and on the location of the curve.

In idiopathic scoliosis, since the paraspinous musculature is intact and can respond to stimulation, conservative treatment is often directed at selective stimulation of these muscles, so as to pull the spine straight. Casts and braces that are applied to idiopathic scoliotic patients do not shore up their spines, but irritate the patients in such a manner that they utilize their paraspinous muscles to move away from the stimulation of the brace or cast. It is an interesting analogy to note that just as guy (tension) wires will hold a relatively straight television tower upright, but will not support one that is structurally damaged, so will a brace cause muscles, which also act in tension, to hold a relatively straight spine straight. Curves of 20 to 30 degrees are easily managed in a brace, but curves in the 40 degree range and above are not. In the latter case, tension members do not have sufficient mechanical advantage.

In neurologic scoliosis, the etiology of the deformity is a deficit in the nervous system. In polio, the musculature on one side of the spine is paralyzed while the musculature on the other side remains normal, and a progressive, long, C-shaped curve develops. In cerebral palsy, the curve may result primarily from overactivity of one side as compared to the other. The most dramatic form of neurologic problem, however, is the myelomeningocele patient who has not only a deficit in the neurologic elements but also a deficit in the muscles. In these patients, the spinal column is often almost completely devoid of support, and there is a dramatic collapse.

In neurologic patients, casts and braces have very little utility; they cannot act by stimulating the paraspinous muscles to produce corrective tension. Also, casts and orthoses can only transmit to the patient as much pressure as the skin will allow. The human skin will seldom tolerate prolonged pressures greater than normal capillary blood pressure, or about 44 gmf/cm^2.[1] Surgical correction of curvature in neurologic conditions is usually directed at making the spine sufficiently structurally sound to support the eccentric loading of the weight above it.

In congenital scoliosis, the structural problem is present at the time the spine develops. There may be hemi- or triangular vertebrae that cause sharp, short angulations. The resulting structural curves invariably have nonstructural curves above them to keep the head over the torso. Since growth plates are present in the spine, the structural curves can get dramatically worse during the periods of growth. Often a bony bar will be present on one side on the spine, opposite an open growth plate on the other. As the growth plate increases the spine length, the bony bar tethers the spine at the same level, with the result being a sharp angular deformity. These structural curves will frequently become so significant that the spine above them cannot compensate, and the head and neck cannot be kept over the base of support. In these cases, the patients are referred to as being decompensated—that is, when they stand, their head and neck are not over the pelvis and they have to drop one leg or the other to get the head and neck into this position, or they have to ambulate with a crutch. The principal problem in congenital scoliosis then, is asymmetrical growth due to tethered growth plates.

Treatment

The nonsurgical treatment of scoliosis is primarily effective in patients with mild (less than 40 degree) idiopathic adolescent scoliosis. In these cases, casts and braces stimulate the paraspinous musculature, and enhance the tensioning effect of the musculature on one side of the curve as opposed to the other. In neurologic, congenital, and severe idiopathic scoliosis these small forces cannot be elicited or are not effective in the face of the deforming force. Therefore it is in all of these latter types of patients that surgical treatment is indicated.

Our discussions will concentrate on the surgical treatment of scoliosis and on the biomechanics related to the most commonly employed internal fixation device, the Harrington rod.[6] This is because this text is intended to be illustrative of biomechanical principles, rather than to be clinically comprehensive. For discussions of the clinical biomechanics of the nonsurgical treatment of scoliosis with the Milwaukee brace, the reader is referred to the work of Galante et al.[4] and Mulcahy et al.[9] Because of their indirect action, the biomechanics of such braces is complex. It should also be mentioned that there are other devices besides the Harrington distraction rod, such as Dwyer cables[3] and Weiss springs,[15] which are employed operatively in spinal stabilization and which act in tension along the spinal axis rather than in bending or compression.

In today's clinical setting, the surgical treatment of scoliosis primarily consists of two phases; first, correcting the deformity and second, fusing the spine to attempt to hold the correction. In the Harrington technique, hooks are fixed to the posterior spinal elements and a distraction rod is placed between the hooks (see Fig. 16–1). A correcting force is then transmitted by the rod and hooks to the spine.

There are severe mechanical limitations to the foregoing approach, and these account for our inability to gain more than a 40 percent correction of a given curve. (A correction to perfect straightness, or 0 degrees curvature, is a 100 percent correction.) In the structural curves encountered in idiopathic scoliosis and neurologic scoliosis, the deforming forces are in the soft tissues. As correction is applied, these soft tissues exhibit viscoelasticity; that is, an initial applied force results in a certain amount of correction, and the force is slowly dissipated as the tissues relax viscoelastically with time. There is a definite limit to the amount of soft tissue deformation possible, and past this limit, further application of force starts to tear and disrupt the tissue. However, these limits are beyond the amount of force that can be applied through a Harrington distraction rod.

One of the weakest features of the Harrington distraction system is the strength of fixation of the hooks to bone. In searching the posterior elements of the spine for a site of fixation, the strongest point has been shown to be under the lamina. The end of a hook so placed is in the spinal canal. However, it is safe to place a Harrington distraction hook in the spinal canal from L1 down. In contrast, in the thoracic spine, it is not safe to put the hook under the lamina because there is not sufficient room around the spinal cord, and one might end up with a paraplegic patient due to compromise of the cord. The area immediately adjacent to the spinal canal, just under the lamina and extending into the pedicle, is the next strongest site of attachment, and it is here that the thoracic hook should be placed. This is a difficult area in which to place the hook, and unfortunately it is often placed a bit lateral to this, on the transverse processes of the thoracic vertebrae. These processes are not nearly as strong as the other two sites, and their use as lower sites often results in loss of fixation. Finally, of all the posterior elements, the weakest is the spinous process. Unfortunately, this has been a site where individuals have earlier tried to fix other instruments (for example, Wilson plates), often resulting in failure of the fixation system.

Harrington rods themselves have definite limitations. Short rods are capable of transmitting more force than the hooks can take. Hence, in the use of short rods one will often "cut out" a hook before bending the rod. However, in 9 inch and longer Harrington distraction rods, the elastic limits of deformation of the rod are often exceeded before one exceeds the strength of fixation of the hooks. Thus permanent deformation of the rod will occur and limit the amount of correction that can be ob-

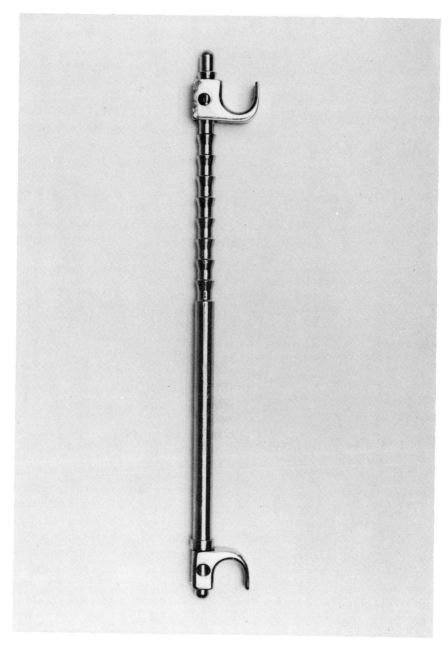

Fig. 16–1. A 3 inch (76 mm) Harrington distraction rod and hooks. Rod length is measured from the lower collar to the first notch and rods from "zero" length (actually 3.6 inches or 92 mm overall) to 12.5 inches (317.5 mm) length (16.25 inches or 413 mm overall) are available.

tained in these cases. It should also be realized that there is a definite loss of mechanical advantage of the rod system as the spine becomes straighter and straighter.

Once maximum possible correction of the spine has occurred, fusion must be obtained to maintain the correction. This is because it is not feasible with the materials and attachment systems that we have today to maintain a correction relying on the implants alone. In fusion, a portion of the spine is converted to a solid column of bone. The reaction of this column to physiologic loads is more predictable than is the case for the unfused spine. The bone in a fusion mass, like any other bone, is a living biological system useful in supporting mechanical loads. Like most materials, it is stronger in compression than in tension. Therefore a fusion mass that is primarily loaded in compression and stabilized by tension in the para-spinous muscles will stand a much better chance of maintaining its integrity through the years than a fusion mass that is on the outside of a major curve, and consequently loaded in tension and unsupported by the musculature.

The overall materials and structural performance limits of the human spine should always be kept in mind. The massive nature of the bone of the spinal column leads one to think that it could withstand virtually any force that the body might apply to it. However, the laws of biological structure dictate that there is only sufficient bone present to withstand the forces that are commonly applied. Unusual loads, as for example tetanic muscle contractions caused by electrical shocks, can cause vertebral fracture. Also, since bone is stronger in compression than in tension, the vertebral bodies are very near the center of gravity of the segment of spine in which they are located, and consequently they normally experience mainly compressive loads. Therefore, body movements leading to tension and shear loads can be expected to lead to mechanical failure. Finally, it must be remembered that in disease states that compromise the musculature, the demands put on the spinal column are even greater. To use another analogy, a spine without the support of muscular tension is endangered in the same way that the mast of a sailing vessel is when a backstay breaks.

CASE HISTORIES

Case 1. This 15-year-old girl presented with a 70 degree idiopathic scoliotic curve to the right, with transitional vertebrae at T5 and T12. The onset had been typical for this disease entity, occurring at the time the patient began exhibiting the secondary sexual characteristics of pubic and axillary hair and breast development. An attempt had been made to brace her; however, she did not cooperate with this brace program and, at the time of presentation, 4 years following the onset of her disease, she had the curve shown here. She was asymptomatic regarding her back. As noted in Figure 16–2A, her iliac crest apophysis had completed its excursion and skeletal maturity was very near.

Because of the degree of curvature and the probability of progression in adult life, it was felt that this curve should be stabilized. The patient underwent posterior spinal instrumentation and a fusion with both a Harrington distraction rod and a compression rod (Fig. 16–2B). The curve was corrected to 40 degrees and has held that position for 6 years. Postoperative management was routine and included immobilization for 6 months in a body brace, with restricted lifting and restricted strenuous physical activities for 1 year.

Case 2. This young girl presented with spinal curvature at the age of 14. The onset had been about 1½ years earlier, and was noted shortly after she began developing secondary sexual characteristics. She had just undergone her menarche.

This patient was definitely skeletally immature at the time of presentation. It was felt that her curve (50 degrees) was likely to progress if left untreated, and two treatment programs were outlined to her and her parents. It was explained to them that

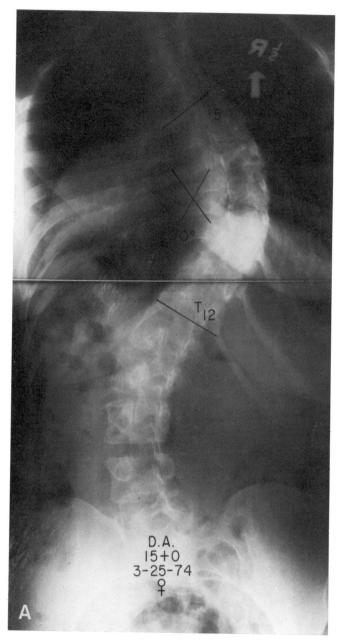

Fig. 16–2. (A) Case 1, a young girl with idiopathic scoliosis which had progressed in four years to a 70 degree curve. *(Figure continues on next page.)*

brace treatment would probably hold the curve at its present level but would not gain correction, since braces do not correct curves but only hold them at the point at which the patient is when the brace is applied. It was explained that this curve might or might not progress in adult life if it were held at 50 degrees, as shown in Figure 16–3A. It was explained that if the brace program were used, the patient would

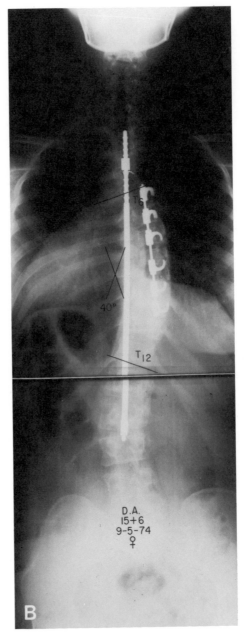

Fig. 16–2. (B) Correction to 40 degrees achieved in Case 1 with combined Harrington distraction and compression instruments.

need to wear the brace 23 out of 24 hours a day until skeletal maturity, which appeared to be a minimum of 2 to 2½ years away.

The other alternative presented was that of surgery. It was explained that surgical intervention could be expected to partially correct the spinal curvature and stabilize the spine if fusion occurred, and that curvature would not then progress in adulthood. The disadvantages of surgery, including morbidity—that is, need for a body

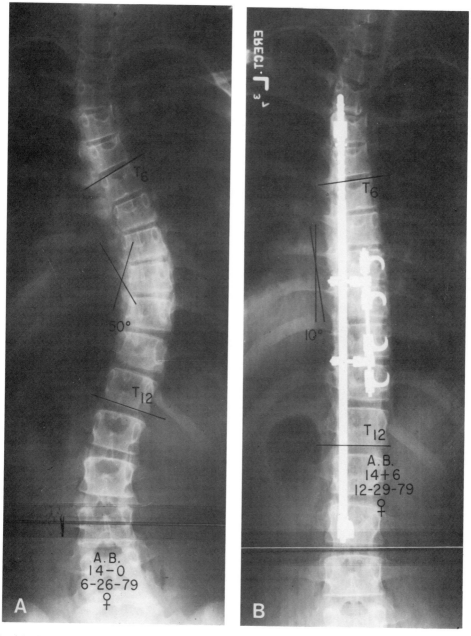

Fig. 16–3. (A) Case 2, a young girl with idiopathic scoliosis which had progressed in 18 months to 50 degrees. (B) Correction to 10 degrees achieved in Case 2 with combined Harrington distraction rod and transverse fixator.

brace for 6 months, possibilities of surgical complications such as infection and failure of fusion, failure to gain correction, and even paralysis, were discussed. It was also pointed out that the long-term effect, over 40 to 50 years, of a spinal fusion is unknown on the segments above and below the levels of fusion.

The patient and her parents elected surgery, and the patient underwent posterior

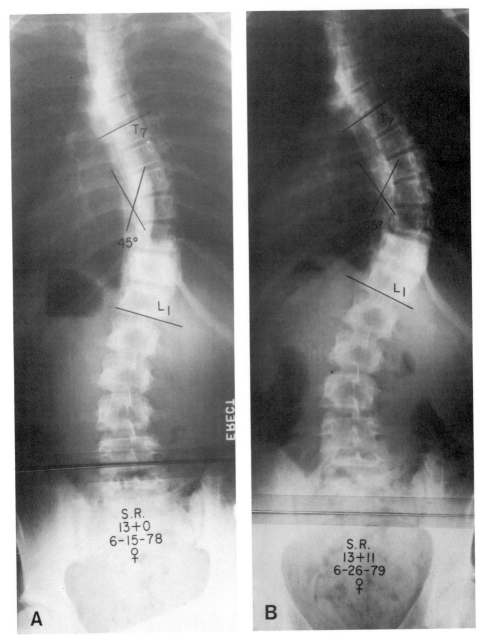

Fig. 16–4. (A) Case 3, a young girl with a 45 degree idiopathic scoliotic curve. (B) Progression of the curve in Case 3 from 45 to 65 degrees in 11 months in spite of a bracing program.

spinal instrumentation and fusion with a Harrington implant, including the recently added transverse fixator. Two of these latter devices were used. This curve was corrected to 10 degrees, as shown in Figure 16–2B.

The limitations of distraction and compression correction alone, and the added mechanical advantages of the transverse fixator in terms of gaining correction, are discussed in this chapter under applicable biomechanical principles.

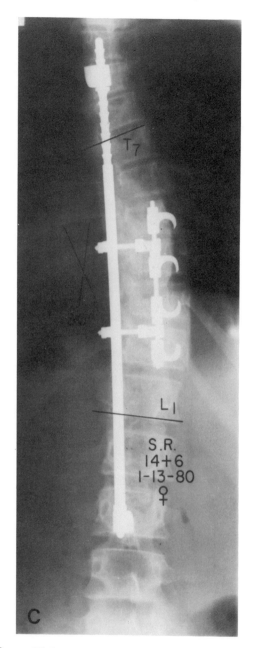

Fig. 16–4. (C) Correction to 30 degrees achieved in Case 3 using a distraction rod in combination with a transverse fixator.

Case 3. This 13-year-old girl presented with asymptomatic curvature of the spine. The curve had been detected in a school screening program a few weeks before, and the patient gave a history of having begun secondary sexual characteristic development about a year and a half previously. She had undergone her menarche about 2 months prior to presentation. She has a family history of scoliosis; her mother had a mild curve.

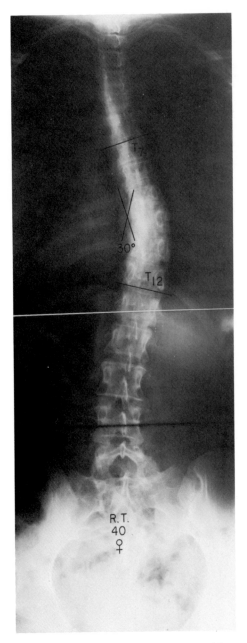

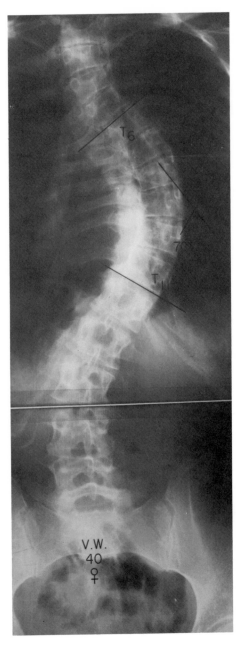

Fig. 16–5. Case 4, a mature woman with a scoliotic curve which has progressed from 25 to 30 degrees in 20 years.

Fig. 16–6. Case 5, a mature woman whose curve had progressed from 60 to 70 degrees in 18 years.

Because of the patient's skeletal immaturity, as seen in Figure 16–4A, it was thought that a brace program could probably hold the curve at 45 degrees. This would be acceptable if maintained until skeletal maturity. A brace program was outlined to the patient, and she and her mother agreed to this treatment. The patient was fitted with a Milwaukee brace and wore this for 9 months. There is some question as to how reliable the patient was as a brace wearer. She denied being out of the brace more than the allocated 1 hour a day; however, her mother felt that she often took it off when she went to school. There is no question that the brace program did cause considerable conflict between this patient and her family.

On reexamination 11 months after going into the brace, it was found that the curve measured 65 degrees (Fig. 16–4B). The rapidity of progression was considered unacceptable. The patient was nearing skeletal maturity, as evidenced by excursion of the ring apophysis, and it was further obvious that she was not a satisfactory candidate for continued brace wearing. On clinical examination in lateral bending, she was found to have a rather rigid curve.

It was elected to treat the patient surgically, and she underwent posterior instrumentation and fusion with a Harrington distraction rod system and two transverse fixators. It was possible to correct her curve to only 30 degrees, as seen in Figure 16–4C, which was made 6 months following surgical intervention. The patient was managed for 6 months in her Milwaukee brace, and was then allowed to start routine activities.

Case 4. This 40-year-old woman presented with a 30 degree scoliotic curve (see Fig. 16–5). Her history was consistent with idiopathic adolescent scoliosis. The onset of her curve was at about the time of development of secondary sexual characteristics. She had had no treatment in her teenage years and had been relatively asymptomatic all her life. She came in strictly for a checkup because she knew she had curvature of the spine and wanted it evaluated. Her curve measured 30 degrees to the right by the Cobb technique, with transitional vertebrae at T7 and T12. A chest x-ray, made at the time the patient was 20 years of age, was obtained from her family doctor, and this same curve measured 25 degrees at that point in her life.

It is obvious that the curve had not progressed significantly in the patient's adult life, nor was it causing her any discomfort. Based on our studies and those of Collis and Ponseti,[2] her scoliosis probably will not develop into a clinical problem. She also had no measurable cardiopulmonary problems and very little cosmetic deformity. Consequently, no treatment was recommended.

Case 5. This 40-year-old woman presented with painful midthoracic scoliosis. Her curve had been noticed in her middle teenage years, but she had not been treated. She had been asymptomatic until about 5 years prior to presentation, at which time there was an insidious onset of discomfort in the back, especially with standing, bending, and lifting.

On physical examination, this lady's trunk was found to be foreshortened, and she had a rigid right scoliotic curve. No neurologic deficits were found. Her curve measured 70 degrees (Fig. 16–6). Pulmonary function studies, consisting of routine spirometry, showed a vital capacity of 80 percent of that predicted, indicating a mild restrictive airway disease. Chest x-rays had been obtained on this patient from the time she was 22 years of age, and showed that the curve at that time was 60 degrees.

In contrast to Case 4, curvature in this case had progressed more significantly owing to the larger bending loads placed on the spine at this greater curvature. The patient was successfully treated surgically with Harrington rod distraction and pos-

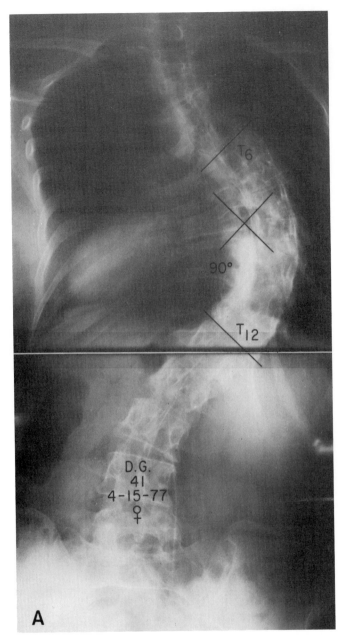

Fig. 16–7. (A) Preoperative 90 degree idiopathic scoliotic curve of Case 6, a mature woman.

terior fusion. A 40 percent correction was obtained (reduction in the curve from 70 to about 42 degrees), and further progression of the disease has been arrested.

Case 6. This 41-year-old woman presented with chronic back discomfort and a history of having lost 2 inches in height over the preceding 8 to 10 years. She was known to have developed scoliosis in her middle teenage years, but she did not receive treatment.

Because of discomfort in the spine, clinical progression of the curve, and the de-

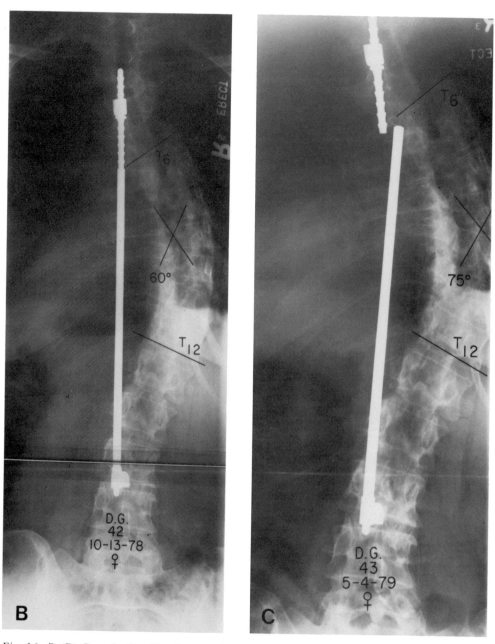

Fig. 16–7. (B) Case 6 after Harrington distraction rod instrumentation and fusion showing correction to 60 degrees. (C) Case 6 eighteen months after surgery showing loss of correction and broken rod.

gree of curvature, posterior instrumentation of this spine with a Harrington distraction rod was elected. Preoperatively, the patient's curve appeared to be that of an idiopathic scoliotic. The curve was to the right and measured 90 degrees, with transitional vertebrae at T6 and T12, as shown in Figure 16–7A. Spinal instrumentation and fusion from T4 to L3 resulted in a correction in the curve to 60 degrees, or 33 percent correction (Fig. 16–7B).

Postoperatively the patient was managed in a spinal brace for 9 months, with limited activities and restriction of lifting to no more than 5 lb. She was thought to be doing quite well until 1½ years after surgery, at which time she had a sudden sharp pain in the back during a heavy lifting episode, and then presented roentgenographically as seen in Figure 16–7C, with a broken rod and a loss of correction. She underwent surgery shortly thereafter. A pseudarthrosis was found at the T8-9 level and was repaired. A new rod was inserted, a correction of 60 degrees was again obtained, and the fusion mass was strengthened.

Mechanical factors related to the failure of this fusion are of interest. Principally, they include the fact that the fusion mass was at quite a distance from the center of gravity of the body, and was therefore under a tensile load rather than a compression load, resulting in little stimulus for bone formation. If more correction could have been obtained, it would have been possible to reduce the tensile force on this fusion mass. It might also appear that the fusion mass should have been positioned between the Harrington hooks, alongside the Harrington rod. However, this is not possible in the posterior approach because one would be encroaching on the ribs and into the pleural cavity.

APPLICABLE BIOMECHANICAL PRINCIPLES

The Spine as a Load-Carrying Device

A full treatment of the mechanical aspects of the spine as they relate to bending deformities is beyond the scope of this text. This is because the spine is an extremely complex mechanical system. By complexity it is meant that the stiffness, strength, and shape of the spine, and the loads to which it is subjected, vary along its length in manners which cannot be described by simple mathematical formulas or geometric relationships. This complexity is due to the wide variety of materials of which the spine is comprised, the large number of distinctly different mechanical components and articulations which are present, and the virtually infinite number of ways in which the spine can be loaded by its attached musculature. The reader is referred to the work of White and Panjabi[16] for a comprehensive treatment of the clinical biomechanics of the spine.

Despite the complexities described, some of the general aspects of the mechanical behavior of the spine can be understood by examining the behavior of simple mechanical components, such as columns and beams, to which the spine as a whole is grossly similar.

THE SPINE AS A COLUMN

In engineering terms, the upright, ligamentous spine, devoid of musculature, can be described grossly as a column. A column is a relatively slender, flexible structural member carrying axial compression loads (see Fig. 16–8). In a special branch of structural engineering called column mechanics,[11] equations have been developed which allow an engineer to determine, for a column of known length and bending stiffness, whether the compressive loads which the column will receive will result in buckling of the column if it is subjected to even relatively small lateral forces. These equations apply to columns made of homogeneous materials and of constant cross-

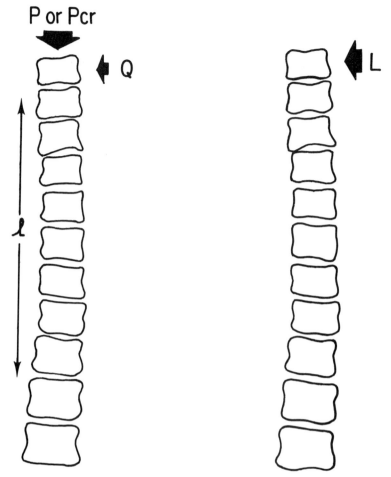

Fig. 16–8. The ligamentous spine devoid of musculature can be modeled as a column of length ℓ so as to understand its low intrinsic stability. A column is a relatively slender, flexible structural member carrying an axial compression load P. At a critical axial load, P_{cr}, columns become unstable and bend or buckle if subjected to even a minute lateral load, Q.

Fig. 16–9. If the major load on a structural member has a large transverse component, L, the member is described as a bending beam rather than as a column. For example, if a spine is sufficiently bent, a major portion of upper body weight becomes a transverse load rather than an axial load, and beam mechanics apply.

section. They do, however, grossly indicate the behavior of more complex structures, such as the spine loaded in axial compression. The original equations were derived in the eighteenth century by the mathematician Leonhard Euler, and are generally known as Euler formulas.

For various types of columns fixed in various ways at their ends, these formulas define a critical axial force, P_{cr}, at which even a minute lateral force, Q, will cause the column to bend. The general form of the formulas for columns of length ℓ is:

$$P_{cr} \simeq \frac{EI}{\ell^2}$$

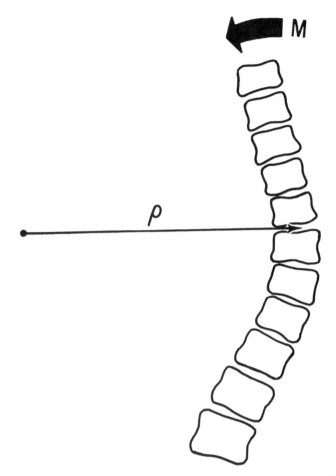

Fig. 16–10. Radius of curvature, ρ, of a segment of the spine acting as a bending beam under the application of a moment, M. Note that the radius of curvature would be *decreased* if the spine were more severely bent.

As defined in chapter 17, E is the elastic modulus of the material of construction, and I is the area moment of inertia of the cross section, so that EI is the column's bending stiffness. The Euler relationship is intuitively correct. That is, the critical load is higher for stiffer columns and lower for longer columns.

The importance of the Euler equations and column mechanics lies in the fact that below P_{cr}, no appreciable deformation of a column takes place. However, it can be shown[13] that for increased loads of as little as 1.5 percent above P_{cr}, a sudden sideways deflection equal to 22 percent of column length will occur! This is utterly unlike the situation in laterally-loaded bending beams (see Figure 16–9). In this case, deflection of the end of the beam is a linear function of the applied load L. That is, if the load L is doubled, a twofold increase in the deflection takes place. For a column before P_{cr} is reached, a doubling, quadrupling, or other multiplication of the applied force causes no appreciable effect. In the structural engineering field, the vast majority of failures have been caused by structural members acting as columns loaded at P_{cr}.

An appreciation of the precarious equilibrium of columns loaded near a P_{cr} may help us to understand why the causes of idiopathic scoliosis are so difficult to detect. It has been shown[8] that the value of P_{cr} for the ligamentous spine devoid of musculature is only 2 kgf! The low value of P_{cr} is demonstrated clinically by the spine of an unconscious person, which will not support body weight in a sitting position. In other words, the intrinsic stability of the spine is low, and most stability is provided extrinsically by the paraspinous musculature, acting as a system of "tensioning cables." It is then possible that slight, persistent, and difficult-to-detect imbalances in the lateral forces exerted by the musculature could lead to large lateral deflections of the spinal axis.

THE SPINE AS A BENDING BEAM

When a spine is grossly bent, the weight of the upper body increasingly applies lateral loads rather than axial loads. In this case, the spine acts like a bending beam, and the mechanics of bending beams can help us to understand the progression of spinal curvature. In active disease states, once spinal curvature has started, it progresses until a new equilibrium configuration is reached. This is because beams subjected to bending loads bend until the stress in the beam balances the applied load, unless the required stress in the beam exceeds the strength of the beam material.

One way of describing the curvature of a bending beam is by the radius of curvature, ρ. It should be noted, however, that the *larger* the radius of curvature the *less* the beam is bent, and vice versa (see Fig. 16-10), so that it is easier to consider the reciprocal, $1/\rho$, which increases as the beam is bent. Equations which have been derived to describe the curvature of bending beams[12] show that the reciprocal of the radius of curvature is proportional to the applied bending moment, M, and inversely proportional to the beam's stiffness, EI. That is:

$$1/\rho = M/EI$$

In a homogeneous column of constant cross-section and subject to moment M, the radius of curvature would be constant over the column length. In the spine, ρ would change as EI changes with variations in the size and material of the spinal column along its length.

In patients, curvature can be expected to progress in the inclinatory stage to a radius of curvature that is minimized by the spine's own stiffness (EI), and maximized by the net bending moment applied. This net bending moment is roughly the difference between the moment produced by the weight of the body above the curve and any counteracting moments applied by the spinal musculature. If the bending loads are high enough, or if the bending stiffness of the spine is low enough, bending may proceed beyond the inclinatory stage and into the collapse stage where, in enginerring terms, materials failure has occurred because local tensile stresses have exceeded the strength of the spine material. The clinical results are gross and irrecoverable changes in the configuration of the spinal tissue.

Bending beam principles also apply in determining the optimum configuration for the healing of a fusion. It is well known clinically that healing is stimulated, and that new bone reaches its maximum density and mass, under compressive loading without excessive localized pressure. Consequently, a fusion mass has the best chance of development if it is applied to a spine which is as straight as possible and if the mass is on the inside of the curve. When this is the case, the weight of the body will

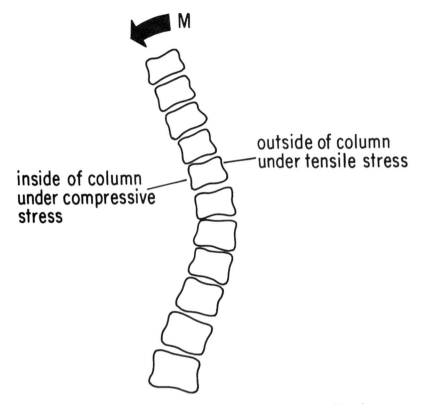

Fig. 16-11. Spinal column subjected to a bending moment M, resulting in curvature as in Figure 16-10. The material on the outside of the curve is under tension and the material on the inside is under compression.

apply compressive loads to the fusion mass. Conversely, a fusion mass located on the outside of the curve is in tension, and therefore will not be stimulated to heal as quickly or as thoroughly as is the case with compressive loads (see Fig. 16-11). This principle was discussed in Case 6.

Harrington Rod Mechanics

Two different aspects of the mechanics of the Harrington rod distraction device are important. First and most obvious is the mechanical mechanism through which the rod produces spine straightening, and the limits of this mechanism. Second and perhaps less obvious is the effect that the mechanical utilization of the rod has on the rod itself. In this latter case, it will be shown that the strength of the rod material and the length of the rod used are factors which must be considered in applying force to the rod and determining whether the rod will support the load applied, or whether it will instead be subject to plastic deformation.

EFFECT OF THE HARRINGTON ROD ON THE SPINE

It is first important to consider how the force which the Harrington rod applies to the spine is directed. As shown in Figure 16-12, the force (F_v) which is applied is

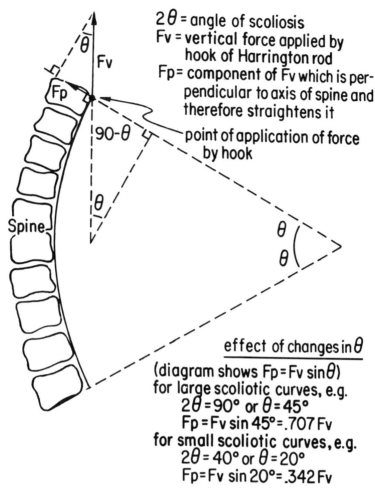

2θ = angle of scoliosis
F_v = vertical force applied by
hook of Harrington rod
F_p = component of F_v which is perpendicular to axis of spine and therefore straightens it

—point of application of force by hook

effect of changes in θ

(diagram shows $F_p = F_v \sin\theta$)
for large scoliotic curves, e.g.
$2\theta = 90°$ or $\theta = 45°$
$F_p = F_v \sin 45° = .707\,F_v$
for small scoliotic curves, e.g.
$2\theta = 40°$ or $\theta = 20°$
$F_p = F_v \sin 20° = .342\,F_v$

Fig. 16–12. Resolution of Harrington distraction rod forces and the effect of scoliotic angle. As correction is achieved, mechanical advantage is lost since $F_p = F_v \sin\theta$.

parallel to the axis of the rod and, therefore, only a component of this force acts perpendicular to the spine and exerts a straightening effect. The other component of the force is parallel to the axis of the spine and, therefore, has a stretching or distracting effect rather than a straightening effect. As described in Figure 16–12, the component of the rod force which is perpendicular to the spine axis, and thus produces straightening, depends upon the angle of bend of the spine. For large scoliotic curves, for example 90 degrees or more, as in Case 6, the straightening force is almost equal to the rod axial force. In contrast, as the spine becomes nearly straight, the force perpendicular to the spine becomes a smaller and smaller fraction of the rod axial force. In other words, as the spine becomes straight, mechanical advantage is lost and it becomes necessary to apply higher and higher rod axial loads to achieve any further correction.

For spines that are more nearly straight, devices that apply a transverse force rather than an axial force become more effective. As shown in Cases 2 and 3 (Figs. 16–3B and 16–4C) a transverse fixator may be added to the Harrington system to supply a transverse force.

With the Harrington rod, as the spine becomes straight and the rod must be subjected to higher and higher axial loads in order to achieve correction, there arises the possibility that the loading force per unit area on the bone where the hook is placed may exceed the local strength of the bony tissue. The strength of the bone in the posterior spine is extremely variable. Harrington[5] reported that the distraction forces on rods which produced local failure of this tissue varied from 9 to 136 kgf, depending on where the hook was located.

In a similar study, Waugh[14] also measured the distraction force at which the hooks caused local bone failure. In agreement with others, he also showed that the lamina represented the strongest bone available for hook placement. However, according to his measurements, the maximum force that could be exerted without breakthrough was in the range of 35 to 45 kgf. Consequently, Waugh and his co-workers[14] have recommended that these levels of force not be exceeded operatively. In this connection, Nachemson et al.[10] have reported the fabrication of a simple, force-indicating distractor for use in the operating room. To our knowledge, however, such devices have not yet become popular in clinical practice.

We are also currently involved in the measurement of intraoperative corrective forces applied both with an "outrigger," before the rod is in place, and with forces applied by the rod through the action of the distractor used to load the rod by moving the upper hook along the rod notches. We have found that forces of 60 kgf or more are occasionally applied with the distractor. When forces in this range are used and, if the bony tissue does not fail, there is some chance of damaging the rod itself and rendering it incapable of properly supporting the spine. This aspect of Harrington rod mechanics is discussed in the following section.

Effect of Loading Forces on the Harrington Rod

Since the principal component of the Harrington rod instrumentation is a rod, one's first inclination would be to assume that the rod in use is subject to simple axial loads. However, this is not true, because the hooks used to apply the loads are offset from the axis of the rod (see Fig. 16–1). This means, therefore, that the rod is subjected to a bending moment.

It is possible either to measure or to calculate the hook forces and applied bending moments which will cause rods to undergo plastic deformation and therefore compromise their ability to support spinal loads. We have calculated the maximum hook loads to which Harrington rods may be subjected without producing plastic deformation, and have verified by laboratory measurement that these calculations accurately predict the onset of plastic deformation of the rods. The calculations are based on the known yield strength of the rod material and on rod geometry, but are beyond the scope of this text. It is worth noting, however, that Harrington rods and hooks are constructed of a material that has been designated in the scientific literature and in the manufacturer's product literature as either 316L stainless steel or ASTM F138 Stainless Steel for Surgical Implants (Special Quality). These specifications are similar but not identical. As used in the Harrington rod, the steel has a 0.2 percent offset yield strength of 115,000 psi, or 8,090 kgf/cm^2. The rods are available in a range of lengths, but all have the same cross-sectional geometry.

Figure 16–13 shows the calculated maximum load which various sizes of Harrington rods can withstand without plastic deformation. It is important to note that the maximum possible load decreases markedly with rod length, whether length is de-

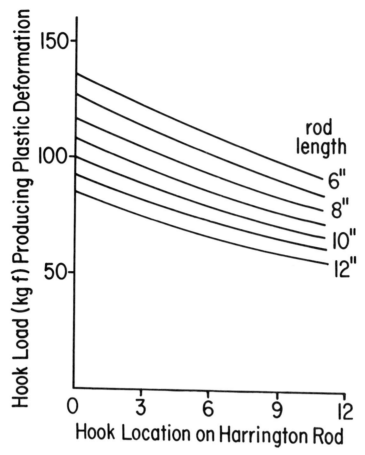

Fig. 16–13. Calculated hook loads which will produce plastic deformation of Harrington rods as a function of rod length and hook location.

termined by rod selection or hook location on the rod. Since in most surgical applications the hook is positioned between the 6th and 11th notch, the maximum permissible force range varies from as high as 110 kgf for a 6-inch rod on the 6th notch, to as low as only 55 kgf for a 12-inch rod on the 11th notch. In our experience with intraoperative Harrington rod force measurements, it is very easy for a surgeon to apply forces in this range with the rod distractor normally used. As the data show, the danger of permanently bending the rod and therefore reducing its ability to support the spine is particularly acute with the longer rods. It seems possible that new rods may be developed which increase in cross section with length, so that the load bearing ability of rods of various lengths is closer to equal.

For mechanical devices loaded to an appreciable fraction of their yield strength, the possibility of fatigue failure exists. According to Harrington,[6] this was taken into account in the design of the rods. The rods were assumed to be subject to 10,000 cyclic loads per day, and were designed to accept such loads for 1 year with a safety factor of two, or a total of 7,000,000 cycles before fatigue failure. Also, bone erosion and soft tissue movement after surgery probably lead to a gradual reduction in the forces borne by the rods postsurgically, if the fusion mass is solid and the spine thus bears the load.

In Case 6, the fusion mass did not become solid, for the biomechanical reasons described earlier. Consequently, the rod continued to bear high loads. As shown in Figure 16–7C, the rod eventually broke and the break occurred at a point where the rod cross-section suddenly changes. Under load, stress contractions occur where there are such sudden changes in the geometry of structural members. Undoubtedly, final breakage of the rod occurred under the circumstances described by the patient. However, it is likely that partial failure had already occurred due to fatigue processes initiated by cyclic loading at higher than anticipated loads.

It is apparent that the ability to make force measurements intraoperatively would be helpful in avoiding either breakdown of bony tissue at the loading points or damage to the rods. It is our expectation that such measurements will become more commonplace in Harrington procedures in the future.

REFERENCES

1. Agris, J., and Spera, M.: Pressure ulcers and treatment. CIBA Clinical Symposia, *31(5)*: 1979.

2. Collis, D. K., and Ponseti, I. V.: Long term follow up of patients with idiopathic scoliosis not treated surgically. J. Bone Joint Surg., *514*:925, 1969.

3. Dwyer, A.: An anterior approach to scoliosis: A preliminary report. Clin. Orthop., *62*: 192, 1969.

4. Galante, J., Schultz, A., DeWald, R. L., and Ray, R. D.: Forces acting in the Milwaukee brace on patients undergoing treatment for idiopathic scoliosis. J. Bone Joint Surg., *52A(3)*498:1970.

5. Harrington, P. R.: Treatment of scoliosis. J. Bone Joint Surg., *44A(4)*591, 1966.

6. Harrington, P. R.: The history and development of Harrington instrumentation. Clin. Orthop. Rel. Res., 93:110, 1973.

7. Lovell, W. W. and Winter, R. B.: The Spine. In: Pediatric Orthopaedics, vol. II, J. B. Lippincott, Philadelphia, 1978.

8. Morris, J. M.: Biomechanics of the spine. Arch. Surg. 107:418, 1973.

9. Mulcahy, T., Galante, J., DeWald, R. L., Schultz, A., and Hunter, J. C., A follow-up study of forces acting on the Milwaukee brace on patients undergoing treatment for idiopathic scoliosis. Clin. Orthop. Rel. Res., 93:53, 1973.

10. Nachemson, A. and Elfstrom, G.: "A force-indicating distractor for the Harrington rod procedure (brief note). J. Bone Joint Surg., *51A(8)*:1660, 1969.

11. Popov, E. P.: Columns. In: Mechanics of Materials. Ninth Printing. Prentice-Hall, Englewood Cliffs, N. J., 1959.

12. Popov, E. P.: Deflection of beams. In: Mechanics of Materials. Ninth Printing. Prentice-Hall, Englewood Cliffs, N. J., 1959.

13. Timoshenko, S.: Theory of Elastic Stability, p. 72. McGraw-Hill, New York, 1936.

14. Waugh, T. R.: Intravital measurements during instrumental correction of scoliosis. Acta Orthrop. Scand., (Suppl.) 92:35, 1966.

15. Weiss, M.: Dynamic spine alloplasty (spring-loading corrective devices) after fracture and spinal cord injury. Clin. Orthop. Rel. Res., *112*:150, 1975.

16. White, A. A., and Panjabi, M. M.: Clinical Biomechanics of the Spine. J. B. Lippincott, Philadelphia, 1978.

17
Principles of Mechanics

J. H. Dumbleton, Ph.D.; J. Black, Ph.D.

INTRODUCTION

The performance of such tasks as locomotion involves many different structures in the human body. The pull of muscles via tendons permits relative movement of limb segments through articulation of the joints. Thus motion occurs by the application of force or muscle pull. In addition to the muscle forces acting in the body, there are forces due to the weight of the body segments and to acceleration and deceleration during motion. Stabilization is supplied by ligaments and joint capsules, and by the action of paired muscles.

Analysis of joint forces during gait is a problem in dynamics, and is difficult to resolve owing to the number of muscles acting, and to a lack of knowledge of the direction and timing of action of each particular muscle group during the gait cycle. It is also difficult to make allowance for the constraints imposed by passive tissue structures. Despite these difficulties, progress has been made, but the solutions present difficulties in understanding owing to the highly mathematical nature of the analyses.

The analysis of joint forces during gait may be simplified by a quasi-static approach, permitting the motion to be viewed as a series of static "snapshots."

Further simplification results if static situations alone are considered. For example, the weight bearing phase of gait may be approximated as one-leg stance, a static case. The major muscles acting can then often be identified, and the action of other muscle groups neglected, reducing still further the complexity of the result. It is indeed fortunate that many problems in orthopedics can be reduced to quasi-static or simple-static cases. A two-dimensional analysis only need be considered to yield useful results to many problems.

Thus, it is possible for the orthopedist to obtain a good grasp of mechanical principles and to analyze problems involving forces on body structures, joints, and implants without having to consider highly complicated situations. Indeed, many problems may be handled graphically without the explicit use of mathematics. In this section the necessary principles are introduced, so as to allow the understanding of static and quasi-static analyses.

STATICS

Force: Definition

The definition of force rests upon observing the effect of applying a force. The results of applying a force are summed up by three laws of Newton:

1. A body (object) is stationary or has a constant velocity unless acted upon by a net force.
2. The acceleration of a body is proportional to the net force applied and inversely proportional to the mass of the body.
3. For every applied force there is an equal and opposite reaction force.

The first law defines equilibrium, since this state is achieved when the net force and the net moment each equal zero. This principle is often used in solving static problems. The third law is also useful in static cases. As an example, the reaction force of the ground upwards is equal and opposite to the resultant of forces due to body weight and muscle actions.

The second law is used in dynamic situations and may be stated as:

$$\text{Force} = \text{Mass} \times \text{Acceleration} \tag{1}$$

The mass is the amount of matter in an object. Thus the second law gives a definition of weight: Weight is the product of the mass and the acceleration due to gravity, and is therefore a force.

A force has two aspects: a magnitude (how large) and a line of action (direction of action). Thus, force is a vector quantity. Quantities requiring only a definition of magnitude, such as body temperature, are scalars.

In order to describe a force, both its magnitude and direction of action must be known. Because of this fact, the addition and subtraction of forces follow rules that differ from those governing the addition of scalars. This difference follows from the need to account for both the magnitude and direction of vectors during addition and subtraction.

Force Application Systems

Forces can be generated and transmitted in many ways. Buck's extension traction (Fig. 17–1A) is an example of force transmission by a cable. The traction force is applied through the skin by the weight acting on the cable. Since the cable must support the weight, the force in the cable is equal in magnitude to the weight.

The reaction force is the cable tension. The tension force acts along the line of the cable. In the absence of pulley friction, the magnitude of the tension force is constant along the cable. However, although the weight determines the *magnitude* of the tension force, the pulley position determines the *direction* of the cable, and hence the direction of the tension force responsible for traction (Fig. 17–1B).

A spring is an elastic structure which when stretched or compressed exerts a restoring or reaction force proportional to its displacement. The displacement is the *change* in length of the spring when a force is applied. The displacement and reaction force are related by a constant, the spring constant:

$$\text{Reaction force} = \text{Spring constant} \times \text{Displacement} \tag{2}$$

The cable in Figure 17–1A is a type of spring, and when the weight is attached stretches until the reaction force equals the weight (force) in magnitude. An ideal graph of reaction force versus displacement is a straight line—that is, the spring constant does not vary with displacement.

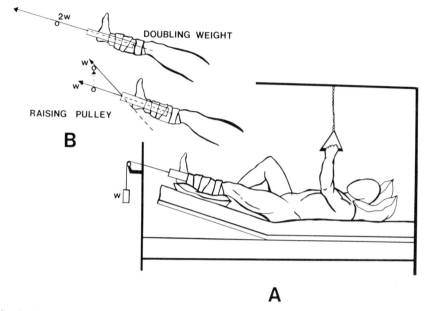

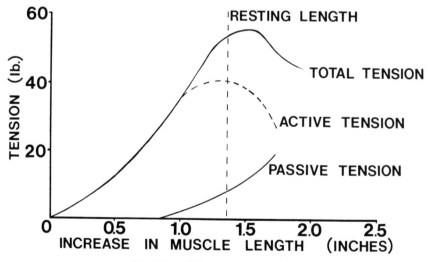

Fig. 17–1. (A) Buck's extension traction. (B) Effect of changing weight and pulley position on magnitude and direction of traction force.

Both the cable and natural fibers such as elastin are passive, since force is generated by an externally induced displacement. Furthermore, while engineering springs may be capable of extension or compression, cables and other long thin structures such as ligaments can only support tension; their compression leads to buckling.

In contrast to engineering springs, cables, or ligaments, muscle tissue displays both active and passive behavior. The total tensile reaction force is the sum of the net voluntary tension (active) and the passive reaction tension (passive), as shown in the Blix curve (Fig. 17–2). The maximum tension that any muscle fiber can develop depends

Fig. 17–2. Blix curve for muscle.

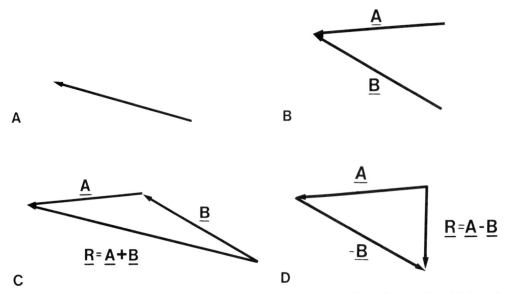

Fig. 17–3. (A) Graphic representation of a force. (B) The action of two forces. (C) Addition of two forces. (D) Subtraction of two forces.

on the fiber length in relation to the unstretched length, and is generated at a relative displacement of 1.2:1. In addition, unlike the case for a spring, the reaction force in muscle is time dependent, since the active component can be maintained only for a limited time; in general, the higher the active component, the shorter the time.

Addition and Subtraction of Forces

A force can be represented graphically by a line (Fig. 17–3A). The length of the line represents the magnitude of the force, while the direction is the line of action, with the sense of the force indicated by an arrowhead.

When two or more forces act, their effect may be represented as a net or resultant force. The simplest situation results when two forces act (Fig. 17–3B). The addition process (Fig. 17–3C) is performed by placing the head of one force vector on the tail of the other. The resultant is the vector that completes the triangle, laying head to head with one of the original force vectors.

The subtraction of forces is similar to their addition. To subtract B^* from A (Fig. 17–3D), a line is drawn of length proportional to the magnitude of B but *opposite* in direction. This vector represents $-B$, and can be added to A to give $A-B$.

The method of adding two forces can be extended to cover any number of forces. For instance, in the case of Russell's traction for femoral fractures (Fig. 17–4A), we wish to add three forces. A, B, and C are equal in magnitude but differ in direction since the constant cable tension is always directed along the cable. The resultant

*Forces and vectors in general are denoted in italics to distinguish them from scalars. However, B, a scalar, is the magnitude of B, a vector. Underscoring, $\underline{B}$, or overlining, $\bar{B}$, is also a form used for a vector.

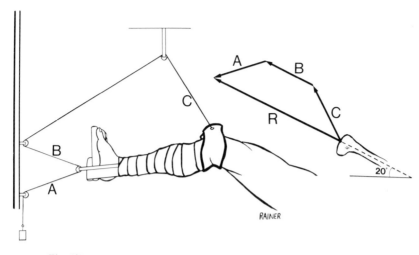

Fig. 17–4. (A) Russell's traction. (B) Determination of resultant.

(Fig. 17–4B) is nearly three times the weight in magnitude and is directed along the line of the femur. The position of the patient is important. Should the patient be moved toward the end of the bed, the directions (but not magnitudes) of *A, B,* and *C* change, and the resultant no longer lies along the femoral axis.

This treatment can be extended to cover addition (and subtraction) of forces in three dimensions. More convenient methods for three dimensional problems will be discussed later.

Resolution of Forces

Since the addition of two forces leads to a single resultant force, the effect of applying a force is equivalent to applying two seperate component forces. Any force can be resolved into components. In some cases it is easier to work with these components in solving problems.

Usually, for two dimensional problems, it is convenient to resolve a force into two perpendicular components (Fig. 17–5). A circle is drawn with *R* as the diameter. The perpendicular components are then *A* and *B,* such that *A* + *B* = *R*. Any point on the circle may be used to construct a valid set of components. It is usual to take the direction of the components along some natural axes, such as the horizontal and vertical directions.

Figure 17–6A shows a hip nail used to reduce a comminuted fracture of the femoral neck. The resolved component of the weight along the nail will tend to impact the proximal fragment on the nail and the distal fragment on the medial buttress of the intact femur; thus the movement due to the weight acts to consolidate the base. The weight force can be resolved into a component along the nail and a perpendicular (or cutting-out force) component (Fig. 17–6B). The force component causing compaction along the nail is about 1.300 N, and the cutting-out force is about 900 N. (Note that 900 N and 1,300 N does not equal 1,500 N!).

The procedure for obtaining a resultant force using the method of components is:

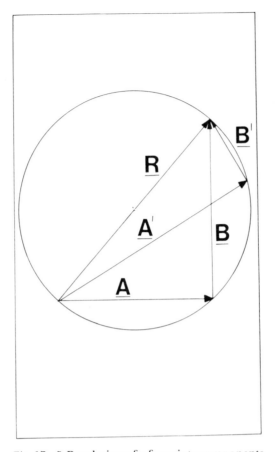

Fig. 17–5. Resolution of a force into components.

1. Resolve each force into a vertical and horizontal component.
2. Add each set of parallel components algebraically.
3. Add the net horizontal and vertical components by vector addition to yield the resultant.

Although the discussion so far has been for the two-dimensional case, the same type of treatment applies to three-dimensional problems. In these problems, forces must be resolved into three mutually perpendicular components. The three directions usually used are two in the horizontal plane and one in the vertical—such as the frontal, horizontal, and sagittal planes for the full body.

Moments

The *moment* of a force about a point is equal to the magnitude of the force multiplied by the perpendicular distance from that point to the line of action of the force. This distance is called the lever arm.

$$\text{Moment} = \text{Force} \times \text{Distance} \tag{3}$$

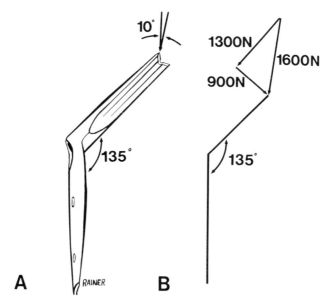

Fig. 17–6. (A) Hip nail used for comminuted fracture. (B) Determination of weight force components.

A moment is a scalar. If the unit of force is the newton and the unit of length the meter, the moment is expressed in newton-meters (N.m). Moment is also called torque.

The sign of a moment depends upon the direction of rotation that it tends to cause about the point. The usual convention is to call moments that act clockwise negative, and those that act counter-clockwise positive.

Equilibrium

The force system acting on the forearm (Fig. 17–7) consists of four forces: W, due to the weight on the hand; G, due to the weight of the forearm; B, the force of the biceps necessary to maintain equilibrium; and R, the resultant ulnar-humoral joint force. Algebraically, the biceps force magnitude, B, must equal W + R + G. However, it is intuitive that if the weight W were moved closer to the elbow, B (and R) would be less. Thus both the magnitude of W and its point of application are important in determining the overall situation. In fact the hand-held weight exerts, about the ulnar-humoral joint, a turning force (moment) that depends directly upon the distance of the weight from that joint. The conditions for equilibrium in this system, then, are:

$$B = W + R + G \text{ (Net force = 0)} \tag{4}$$

Since the forces do not act at a point, the condition for a zero resultant is not sufficient to guarantee equilibrium, because there could still be moments causing a ten-

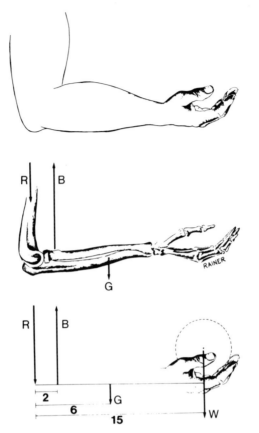

Fig. 17–7. Force system acting on the forearm.

dency to rotation. Thus, for equilibrium, the sum of the moments, as well as the sum of the forces, must be zero:

$$6\,G + 15\,W - 2\,B = 0 \text{ (Net moment = 0)} \tag{5}$$

Note that distances are calculated to the line of action of R. so that its moment at this point is equal to zero. In general, the conditions of equilibrium are as follows:

$$\text{Sum of forces} = 0$$
$$\text{Sum of moments} = 0 \tag{6}$$

or

$$\Sigma F = 0$$

$$\Sigma M = 0 \tag{7}$$

These conditions apply whether the system is analysed in two or three dimensions. Since all quantities except B are known in Equation (5), we may solve this graphi-

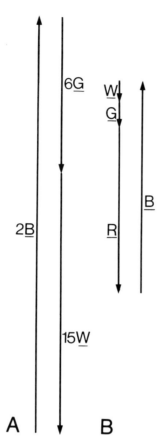

Fig. 17 – 8. (A) Solution of moments on the forearm. (B) Solution of forces on the forearm.

cally (Fig. 17 – 8A) and then use the value of *B* obtained to obtain the reaction force, *R*, from (Fig. 17 – 8B). Arbitrary values are used; it should be clear that substituting real values for *W* and *G* would yield the actual magnitude of *R*. Note however, that since B = 3 G + 7.5 W, R = 2 G + 6.5 W. Thus the typical result is obtained, that joint reaction forces tend to be large compared with loads because muscles tend to act close to joints, whereas loads act at a distance.

Center of Gravity

Experience shows that an object may balance without tipping if the point of support is chosen carefully. This point of balance is easily predicted for a symmetrical object; but for an asymmetrical object such as the human body, trial and error are required to find the point of balance. Location of the point of balance is, in fact, a location of the center of gravity (c.g.) of the object. The whole weight (vector) of the object may be assumed to act at this point. This principle is very useful and was used in the example above (Figs. 17 – 7 and 17 – 8) to locate the point of action of *W* and *G*.

Figure 17 – 9 shows the determination of the position of the c.g. of the human body with respect to the frontal and sagittal planes. The height of the center of gravi-

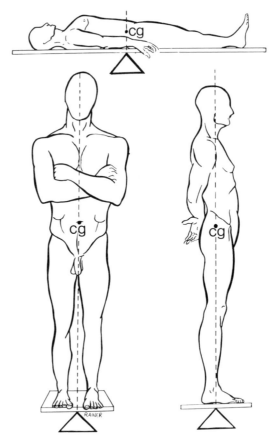

Fig. 17–9. Determination of center of gravity of the human body.

ty gives the location of the transverse plane. Similar methods can be used to obtain the c.g. of body segments.

Weights of body segments as fractions of total body weight, as well as their actual weights and c.g.'s, have been tabulated.

Free Body Diagrams

In the study of an equilibrium problem it is necessary to produce a diagram showing all of the forces that are acting. This approach has been used earlier in treating various problems; the first step is always to draw a figure and to place all the forces at the proper locations. If an object is touching the figure at a certain point and is then removed, it must be replaced by a force acting at the same point of contact to give the same effect as if the object were still present. Such a figure, showing all of the forces both known and unknown, is called a *free body diagram.*

The method of replacing an object touching the figure with a force simplifies the problem, and enables attention to be directed to the region of interest. For example, in the situation shown in Figure 17–7 involving the force system acting on the forearm, the figure drawn is actually a free body diagram. The act of holding the weight in the hand leads to forces being generated throughout the body. There will be forces acting at the shoulder, along the spine, and at other locations, and muscles will be active throughout the body in response to the act of holding the weight. Adjustment to the center of gravity of the body will also be required. However, the problem under consideration involves only the forearm; the forces generated elsewhere in the body are of no interest. In fact, the only direct influence of the body is through the humeral-ulna joint, and the body can be removed and replaced by the joint reaction force R.

It will be noted that some of the internal forces acting in a free body diagram need not be considered. In the analysis of the force system acting at the forearm, for example, it was not necessary to consider the forces acting at the wrist or between the various joints in the hand. This is because these forces are *internal*, and will cancel out by Newton's third law. The influence of the rest of the body is summed up by the joint reaction force.

DYNAMICS

Introduction

The investigation of static equilibrium situations provides an excellent basis for the determination of muscle actions and joint forces. However, such activities of everyday life as lifting objects, performing manipulative tasks, and ambulation involve dynamic considerations in which there is displacement and motion of body segments, inertial forces, and time-dependent action of muscles. Thus it is important to consider the dynamic aspects of activities.

In many cases the approximation of static equilibrium is useful, since this approach is simple and allows the basic principles behind the activity to be determined. Often, the dynamic aspects of the activity may be neglected; the static analysis will then give realistic determinations of muscle actions and joint forces. Even where the dynamic aspects are important, a static analysis is usually the first step in the solution of the problem.

Although an analysis of a dynamic situation is complicated, it is often possible to extend a static analysis to include some aspects of the dynamic situation. This is the approach which will be adopted here, and a comparison of static, quasi-static, and dynamic solutions will be presented.

A study of dynamic activities is of interest in many areas, including the analysis of joint forces, the motion of joints, and the gait. In fact, a form of dynamic analysis is used by the orthopedist for patients with conditions of the lower limbs; the gait of the patient will give a good indication of the cause and severity of the disability. In this case, experience leads to an intuitive solution to what is actually a quite complex dynamic situation.

Motion of Body Segments

Consider a gait analysis utilizing strobe photography of a subject walking, with reflective targets placed on the various body segments (Fig. 17–10). The linear motion will be considered first, and is characterized by three quantities:
1. Displacement: The linear distance through which the body segment moves.
2. Velocity: The rate of change of *displacement*, or the distance moved in unit time.

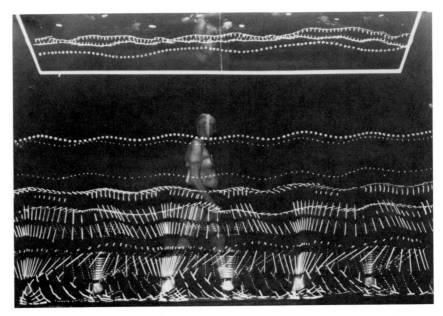

Fig. 17–10. Illustration of the relative position of body segments in the limb during motion. Both linear and angular displacements are observed. (see Fig. 9–1.)

3.   Acceleration: The rate of change of *velocity*.
The unit of linear displacement is the meter (m); the unit of velocity is m/sec; and the unit of acceleration is m/sec².
    It will be recalled that Newton's second law states that

$$\text{Force} = \text{Mass} \times \text{Acceleration} \tag{8}$$

Thus, the presence of an acceleration demands the presence of a force. The acceleration of a body segment will depend upon the magnitude of the agonist muscle forces, the magnitude of the resistive forces (antagonist muscles, soft tissue), and the mass of the body segment. During ambulation, the muscles must exert forces both to accelerate and decelerate the leg. The larger the mass of the leg the larger is the force required for acceleration and deceleration. The mass depends on the amount of matter in the object, and is related to the property of inertia, which describes the fact that the mass determines the force required to accelerate or decelerate the object.
    The phenomenon of motion will change the magnitude of the muscle actions, since the muscles must provide the accelerations and decelerations of the body segments, in addition to the forces needed for gravitational equilibrium. Since joint forces depend on the muscle actions across the joints, there will be a corresponding increase in the joint forces.
    The displacement of body segments is not limited to linear movements; there are also angular displacements, such as flexion-extension of the knee, which produces relative angular motion between the calf and thigh (Fig. 17–10). The angular motion is characterized by three quantities:
1.   Angular displacement: The change in angle between body segments.

2.   Angular velocity: The rate of change of angular *displacement*.
3.   Angular acceleration: The rate of change of angular *velocity*.
The unit of angular displacement is the radian (one radian = 57.3 degrees). The unit of angular velocity is radians/sec, and the unit of angular acceleration is radians/sec.^{2}

In angular motion, Equation (8) becomes:

$$\text{Moment} = \text{Mass moment of inertia} \times \text{Angular acceleration} \qquad (9)$$

Since angular motion implies turning or rotation, the angular acceleration derives from a force with a turning tendency—a moment. It is the sum of the moments of the muscle actions and restraint forces about a center of rotation that determines angular acceleration. Thus the moments about the center of rotation at the knee determine the angular acceleration of the lower leg.

Just as in the case of linear motion, the mass of a body segment plays a role in determining the forces (in the case of angular motion, the moments) required to accelerate the segment. However, in angular motion it is the distribution of mass in the segment, or the mass moment of inertia, that directly influences the acceleration. Figure 17–11 shows a patient with a de Lorme boot. The weight of the boot may be considered to act at the center of gravity, and the mass moment of inertia for rotation about the knee is:

$$\text{Mass moment of inertia of boot} = \text{Mass of boot} \times [\text{Distance of boot from knee}]^2 \qquad (10)$$

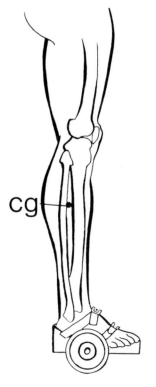

Fig. *17–11.* Patient wearing a de Lorme boot, with the leg fully extended. The mass moment of inertia of the leg plus boot is mainly influenced by the mass moment of inertia of the boot, since it is situated further from the center of rotation.

The mass of the boot does influence the mass moment of inertia about the knee, but it is seen that the distance of the boot from the center of rotation is included as the second power (square).

For an object such as a body segment, it is not sufficient to consider the mass concentrated at the center of gravity; in a limb, the weight is distributed. Thus it is necessary to divide the limb into sections, and to find the mass moment of inertia of each section. The sum of these gives the mass moment of inertia of the whole limb.

Returning again to Figure 17–10, the positions of the limb can be seen to exhibit an acceleration and deceleration during the gait cycle. Since photographs are taken at equal time intervals, the locations at which the lines are close together represent lower velocity, and the locations at which the lines are further apart represent higher velocity. Thus there is change in linear velocity, and hence in linear acceleration and deceleration. The angular acceleration of the leg may be seen in Figure 17–10 by noting that the angle that the leg makes with the ground has a cyclic variation.

Static Versus Dynamic Analysis

The effect of movement on muscle and joint forces will be examined by reference to the force system acting on the forearm with a hand-held weight. The forces acting are shown in Figure 17–7 for the forearm in 90 degrees flexion. To simplify the analysis, the weight of the forearm is neglected; that is, it is small compared to the weight in the hand. Two cases are considered: – no motion and motion.

Static case: Sum of moments about point 0, on the vector $\overline{\mathbf{R}} = 0$
or $15\mathbf{W} - 2\mathbf{B} = 0$ from (5) $(\mathbf{G} = 0)$ (11)

$$\mathbf{B} = \frac{15}{2} \cdot \mathbf{W}$$

for $\mathbf{W} = 49\,\mathrm{N}$,
$\mathbf{B} = 367.5\,\mathrm{N}$

*Dynamic case:** In this case the forearm is moving and there will be an inertia term in the equation for the sum of the moments:

The sum of the moments about a point on the Vector $\overline{\mathbf{R}}$ = Mass moment of inertia times angular acceleration:

$$0.04\mathbf{B} - 0.3\mathbf{W} = \mathbf{M} \cdot (0.3)^2 \cdot \mathbf{A}$$ (12)

where M is the mass of the weight, W, and A is the angular acceleration of W. The mass† of W is 5 kg, and A will be taken as 45 radians/sec^2:

---

*To this point arbitrary units have been used to calculate results. However, the dynamic case requires the use of a specific length measurement, the meter. We take the unit length "2" = 0.04 m, thus "15" = 0.3 m.
†In this discussion of inertia and gravity force, the quantities mass and weight were used. These quantities are related by Newton's second law as follows:
Weight = Mass × Acceleration due to gravity
Thus weight is a force and the unit is the newton. The value of the acceleration due to gravity is 9.81 m/sec^2. The unit of the mass is the kilogram and

$$\mathrm{Mass} = \frac{\mathrm{Weight}}{9.81}$$

Hence a weight of 49 N has a mass of 5 kg.

$$B = W \ 15/2 + M(15)^2 \ A/2 = W \ 0.3/0.04 + M(0.3)^2 \cdot A/0.04 \qquad (13)$$
muscle   static   dynamic
action   term   term

And for the case considered:

$$B = 367.5 + 506.2 = 873.7 \ N \qquad (14)$$

Thus the magnitude of the muscle action in the biceps must increase 138 percent to supply this acceleration to the weight W. The increase in muscle action B leads to a similar increase in joint force, since:

$$R = B - W \ [\text{from equation (4), (G = 0)}] \qquad (15)$$

The dynamic solution is valid only for the one instant of time at which the forearm is in 90 degrees flexion. But the method of solution may be used for other positions, provided the angular acceleration is known and dimensional changes — such as the change in lever arm length of the biceps — are taken into account.

There are additional differences between static and dynamic analyses beyond the inclusion of inertia terms. There is, then, a series of levels of sophistication on which a given situation may be attacked:

1. Static analysis, simple: Only major muscle(s) included with major gravitational and external forces considered.
2. Static analysis, extended: Effects of all other muscle groups as well as ligament and soft tissue contributions added to 1.
3. Quasi-static analysis — that is, inertia terms added to 1.
4. Dynamic analysis, all factors: Combination of 1, 2, and 3, as well as all other inertia terms.

It must be mentioned that the transition from 1 to 4 is by no means straightforward. The problem lies in the large amount of information required: (a) The position, velocity, and acceleration of body segments as a function of time; (b) the mass moment of inertia of the segments; the muscle insertions, line of action of muscle pulls, and muscle activity as a function of time; (d) the line of action of ligament and other constraint forces.

In all dynamic analyses the number of unknowns (muscle forces, joint forces, constraint forces) is always larger than the number of available equations for equating the sums of forces and the sums of moments. Thus from the very beginning assumptions must be made to reduce the number of unknown quantities; muscles that are not very active are eliminated, muscle groups with similar lines of action are consolidated.

Forces around the hip pose a typical problem. There are only six equations that may be written for the equilibrium of this joint: Sums of forces, including inertia terms in three planes; and sums of moments, including inertia terms, also in three planes [Equation (6)]. There are 22 muscles that act directly upon the hip joint, as well as a variety of soft tissue constraints. To some extent the information obtained on the muscle, joint, and constraint forces will depend on the initial assumptions.

Comparisons of forces between static and dynamic cases usually result in higher forces in the latter. For instance, calculations for the hip, taking various assumptions, produce a range of hip joint reaction forces from 1.5 times to 10 times body weight, depending upon accelerations involved, for typical patient body geometrics.

Instant Center of Rotation

In the analysis of rotation it has been assumed that the rotation occurs about a single point, or center of rotation. However, in many joints, the center of rotation changes position during flexion-extension. To completely specify the motion it is necessary to find the locus or path of the center of rotation. At any given time this center of rotation is known as the instant center of rotation.

Figure 17–12A shows rotation in a circle about a point 0. It will be noted that the velocity is directed tangentially to the circle and is perpendicular to the radius. In Figure 17–12B two positions of a point are shown, point A rotates about 0 to point A^1. If a line AA^1 is drawn, then the perpendicular bisector of this line passes through 0. It is then necessary to take two sets of points and to carry out the following procedure (shown in Figure 17–12C) in order to find the center of rotation, which lies along this perpendicular bisector:

1. Locate two points A, B on the object.
2. At a short time later in the motion, locate A^1 and B^1.
3. Draw AA^1 and BB^1.
4. Draw the perpendicular bisectors to AA^1 and BB^1.
5. The instant center of rotation lies at the intersection of these perpendicular bisectors, AA^1 and BB^1.

The position of the center of rotation for articulation will determine the type of motion between the two joint surfaces. This is illustrated in Figure 17–13 for a cylinder in contact with a stationary flat surface. In Figure 17–13 A, the center of rotation lies at the point of contact; the contact point will have zero velocity and the motion will be of a rolling nature. This is the type of motion that occurs when a wheel rolls freely over a surface. Sliding motion is shown in Figure 17–13 B. The rotation is about the center of the circle. The velocity at the contact point is directed along the surface; there is a velocity difference at the contact point since the lower surface is stationary, and hence slip occurs. This is the type of motion that results under accel-

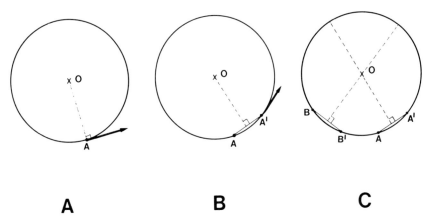

A   **B**   **C**

Fig. 17–12. (A) Rotation in a circle about a central point 0. (B) Two positions, A and A,^{1} of the rotating point. (C) Illustration of the determination of the center of rotation.

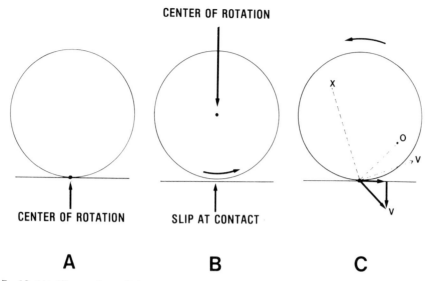

Fig. 17–13. (A) Illustration of the rotation of a cylinder in contact with a flat stationary surface: The case where the center of rotation lies at the point of contact. The motion is rolling. (B) Rotation about the center of the circle. There is sliding motion between the two surfaces. (C) Rotation about an "abnormal" point. The velocity at the contact point has a component directed into or out of the surface.

eration or in the braking of a wheel, with skidding. The motion at most joints is of the combined rolling and sliding type.

Figure 17–13 C shows a case of rotation about an "abnormal" point (o). The velocity at the contact point is directed into the surface. Since velocity is a vector, it may be resolved into its components; at the contact point there is a sliding component and a normal component. Thus the cylinder attempts to penetrate the lower surface. There could be chosen other abnormal centers that would give a velocity at the contact point tending to pull the surfaces apart (point X in Figure 17–13C). The concept of the instant center of rotation has been applied in particular to the motion at the knee. (See F & B, Chapter 5). It has been postulated that the instant center of rotation of the knee can be displaced from its normal position because of a meniscal tear, resulting in abnormal surface reaction and leading to excessive joint wear and trauma.

The path of the instant center of rotation is obtained by repeating the method given above for the joint at differing degrees of flexion. In practice, the information on the position of the joint is obtained from successive radiographs, taken at varying degrees of flexion, which are referenced to each other by super-imposing one side of the articulation—that is, at the knee the tibia may be used as the reference for each radiograph and the motion of the femur is relative to a stationary tibia.

One use of a knowledge of the position of the instant center of rotation has already been discussed—the damage caused to the natural knee by rotation about abnormal centers. However, the variation of the position of the instant center of rotation is extremely important when the design of artificial joint replacements is considered. If the design is such that rotation is constrained about a fixed point, there will be increased forces on the fixation and possible loosening. One reason why silicone rubber, finger-joint prostheses "piston" is because the center of rotation of the implant does not conincide with the natural center of rotation. For knee prostheses, an attempt to reproduce the locus of the natural instant center of rotation automatically leads to a varying radius of curvature on the femoral component, and to low conformity between the femoral and tibial components. This reduces the forces on the femoral and tibial component fixation, but increases the contact forces between the femoral and tibial components, which can lead to distortion and possibly to high wear. These are, in fact, the design constraints on a knee prosthesis, and compromises must obviously be made.

Gait Analysis

The analysis of gait can be done on several levels of sophistication.
1. Qualitative evaluation of gait using visual observation.
2. Quantitative evaluation of gait by recording the motion of the lower limb segments.
3. Quantitative evaluation of motion and quantitative evaluation of ground forces and moments during gait.

Abberations of gait, either in the borderline or pathologic case, are variations or distortions of the normal gait pattern, and can be due to several causes such as contracted limb, incoordination, sciatic radiation, muscle deficiency, joint inflammation, and so on. Since ambulation involves not only the extremities but also the spinal column, deficiencies in any of these areas can cause variation in the gait pattern. Chapter 9 provides an example of the study and analysis of hip joint motion in patients with movement limitations due to pain or loss of the passive range of motion. Similar analyses may be done for other joints as well as for different activities, such as those involved in sports or recreational pursuits.

Such application of gait analysis, and its extension to more sophisticated problems, is being increasingly made in orthopedics. The applications of gait evaluation range from diagnosis to the determination of the effects of disease and trauma on the functional mobility of joints and to the study of prosthetic joint performance. Although a complete correlation of muscle actions and segment motions is complicated, owing to the large number of muscles and joints involved in ambulation, it is possible in certain instances to directly attribute a gait pattern to a specific defect, through use of static analysis. Other gait patterns require somewhat more complete treatment, and dynamic considerations are needed. Even in those cases in which great simplification is not possible, the considerations outlined concerning gait analysis allow useful conclusions to be drawn regarding the connection between the biomechanics of the situation and the resulting gait pattern.

RESPONSE OF MATERIALS AND STRUCTURES

Introduction

Both static and dynamic analyses consider the magnitude, direction, and distribution of forces and moments. The internal effect of these forces and moments upon the object, usually a body segment, is not considered. However, changes will take place in an object whenever a force or moment acts. Thus, the imposition of too high a force will damage articular cartilage and give rise to traumatic osteoarthritis. Large forces or smaller forces applied in an inappropriate direction will lead to soft tissue damage and sometimes to broken bones or prostheses. It is convenient to first treat the response of materials in general, and then to extend the treatment to consider the response of structures.

Deformation

Forces or loads cause deformations of materials and hence of structures. A deformation is a change in dimension (size), a change in shape, or a change in both size and shape. The deformation may be recoverable, so that the size and shape changes are recovered after the force is removed and the structure returns to its original state. Thus the skin deforms during movement, but the deformation is usually recoverable. During ambulation and other activities, the skeletal structure is also subjected to forces leading to deformation of the bones; but again these deformations are recovered on removal of normal loads. It is highly desirable that the deformations in implants be of the recoverable type; otherwise failure may well occur, and in engineering generally the aim of design and materials selection is to keep the deformations in the recoverable range. Permanent deformation ensues if the load is higher than a critical value. This deformation is not fully recoverable on removal of the load.

For many materials, the deformation that results from application of a force is time dependent. This is essentially so for biological materials such as bone and soft tissue. The time-dependent effects are exhibited in different ways, depending upon the manner in which the load is applied. Under a constant load, the deformation is not attained instantaneously, but continues as time progresses. In cases in which the load changes with time, the deformation attained will depend upon the rate of change of the load with time. Figure 17–14 illustrates the deformation of wet bone at different rates of load application. If there were no time-dependent effects, a single curve of force-versus-deformation would be obtained. However, it is seen that at high rates of load application, a higher force is required for a given deformation.

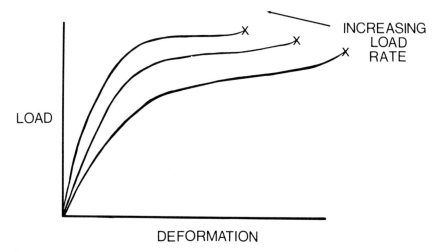

Fig. 17–14. The deformation of wet bone showing the effect of the rate of load application.

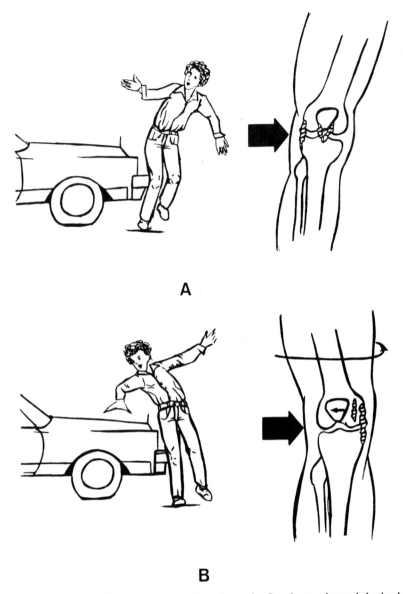

Fig. 17–15. (A) Person struck by an automobile, where the foot is caught and the body thrown violently. (B) Person struck by an automobile, where the full weight is on the struck leg and the body and thigh rotate away from the impact.

If the force is too large, the deformation will be sufficient to cause breakage. This occurs to bones and sometimes to implants. Time-dependent effects may appear. For example, bone is stronger at higher rates of loading. This has relevance in the consideration of fracture processes, which can be divided into high-energy and low-energy fractures.

There are several factors that determine whether a deformation is recoverable or permanent, whether time-dependent effects occur or whether there is failure. These factors are conveniently grouped under three headings:

1. External Factors:
   (a) The magnitude and direction of the applied forces(s).
   (b) Method of support of the structure, sometimes called "boundary conditions," since the method of support determines the deformation of the structure at its boundary.
2. Structural factors: the dimensions (size) and shape of the structure.
3. Material factors: the properties of the material from which the structure is made.

It is well known that the type of injury resulting from trauma depends on the way in which the force is applied. Figure 17−15 A gives the situation in which the foot is caught and the rest of the body thrown violently − rupture of the medial and cruciate ligaments is likely to result. Figure 17−15 B shows what happens when the force is applied to the same area with the full weight on the leg and the quadriceps muscle braced. Rotation of the body and thigh away leads to dislocation of the patella. Obviously, the situations considered are very complicated owing to the complex force system and dynamic effects. It should be noted that the mode of support plays an important role. The injury caused by the situation shown in Figure 17−15 A depends upon the foot being caught − that is, on the boundary conditions. If the foot were free, a different pattern of injury would result.

In the consideration of body injuries, the structural factors are usually not recognized owing to the complexity of the situation. However, these factors do play a part, since damage occurs preferentially to some regions whereas neighboring areas subjected to the same force will escape injury. The effect of material properties has been studied to some extent. The ease of fracture will depend on the bone-material integrity. The strength of bone decreases somewhat with increasing age. However, by far the largest decreases in bone strength will be caused by such metabolic disease states as osteomalacia and osteoporosis. Osteoporosis also influences the dimensions of the bone structure, owing to decreases in cortical thickness and is, therefore also a structural factor. The role of osteonecrosis in decreasing bone strength is less clear.

Of great importance is the consideration of the deformation of implants. This deformation is governed by the same factors given above; the loading, the degree of support (how much load is borne by the bone), the size and shape of the implant, and the material of which the implant is made. As mentioned earlier, it is highly desirable that the deformation be in the recoverable range, otherwise failure may occur.

Types of Deformation

Deformation of a structure such as a bone, ligament, or joint surface is a consequence of the application of a load (force). The type of deformation depends on the nature of the applied load. Figure 17−16 A shows the result of applying a tensile load along the axis of a cylinder: the cylinder increases in length. At the same time, there is a reduction in the lateral dimension (diameter) of the cylinder owing to the need to provide material for the increased length. It should be noted that this lateral deformation (lateral compression) is not due to an applied lateral load, since the lateral surfaces are load free. The tensile load does not lead to any shape changes since the cylindrical shape is maintained under load.

Figure 17−16 B illustrates the effect of a compressive load upon a cylinder. The cylinder decreases in length and at the same time there is an increase in its diameter.

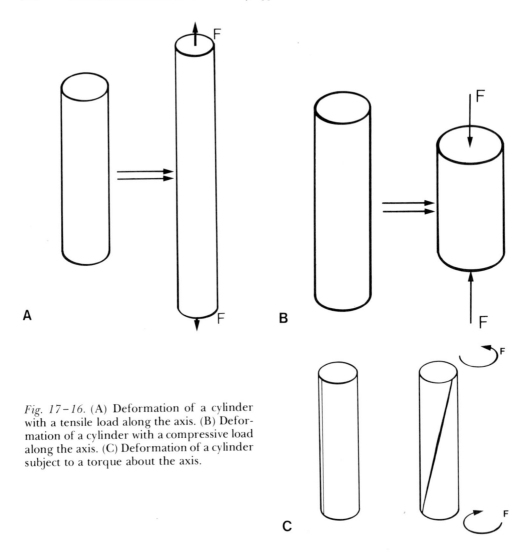

Fig. 17-16. (A) Deformation of a cylinder with a tensile load along the axis. (B) Deformation of a cylinder with a compressive load along the axis. (C) Deformation of a cylinder subject to a torque about the axis.

Ligaments and tendons are subject to tensile loads and deform in the manner described above. Bones are subject to compressive loads, as are articular joint surfaces. Figure 17-17 illustrates the effect of applying a force along the surface of a structure—this is a shear force or shear load. A shape change is produced and there are also dimensional changes. The bending of a beam is illustrated in Figure 17-18 A. The particular beam shown is known as a cantilever, since it is supported at one end only. A load may be applied at any place along the beam or may be distributed over a section of the length. Experience shows that the load is most effective in causing deformation when it is applied at the free end of the beam. Since the beam bends, there is a change in shape; there are also dimensional changes, as shown in Figure 17-18 B. It will be seen that the top surface of the beam has stretched and is therefore exhibiting a tensile elongation, while the lower surface suffers a compression.

Figure 17-16 C shows the effect of a twisting force applied to a cylinder. The cylinder will suffer a twist type of deformation, which can be shown to be shear. Circular sections within the cylinder remain circular, and there is thus no shape change. A

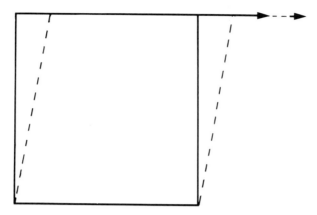

Fig. 17–17. Two-dimensional representation of shear. A force is applied along one side of the square.

line drawn on the surface, parallel to the axis of the cylinder, will take on the shape of a helix (spiral), illustrating that a twisting deformation is taking place. Torsional loads are found in the body since the muscles act at angles to the long bones in such a way as to produce twisting. In skiing accidents, high torsional loads produce the classical spiral fracture.

Although the above loading situations have been considered separately, what is usually found in practice is combined loading, in which a structure is subjected to more than one type of load. The result of combined loading can often be found by

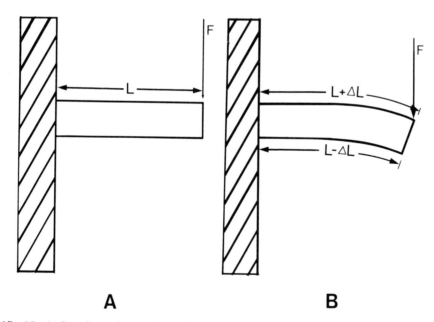

Fig. 17–18. (A) Bending of a cantilever beam. (B) Illustration of deformation in a bent cantilever.

considering each type of load as if it were applied separately — finding the deformation due to this load, and finally adding the deformations to give the result of the combined loading.

Stress and Strain

The imposition of a force or load upon an object or structure results in a deformation which is externally manifested by a change in shape, dimension, or both. However, it is clear that internal changes occur in a structure subjected to a load. These changes may be predicted by determining the internal stresses and resultant strains. Stress and strain may be illustrated simply in the tensile case (Fig. 17–19). The tensile stress in the cylinder is defined as the force divided by the cross-sectional area of the cylinder. The tensile force is perpendicular to this area. The force used is the internal force, but for the case considered, this is equal to the applied load (force), and the tensile stress is thus given by (tensile load)/(cross-sectional area of cylinder). This stress is uniform throughout the cylinder. Common experience shows that if the cross-sectional area were doubled twice the applied load would be required to give the same deformational effect, the definition shows that the tensile stress in both cases would be the same, which is consistent with experience.

The normalization of the internal movements which results in the strain can again be obtained by appeal to experience. If a tensile load is applied to a rod, there will be

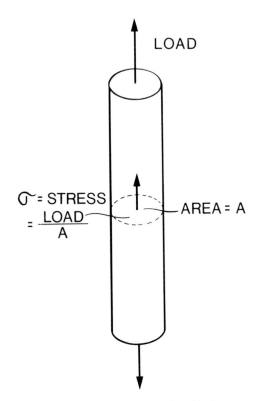

Fig. 17–19. Tensile stress in cylinder.

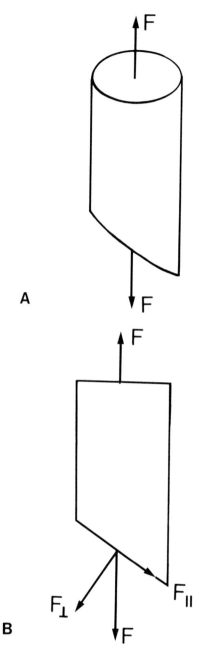

Fig. 17–20. (A) Free-body diagram for a cylinder under a tensile load, cut at an oblique angle. (B) Illustration of tensile and shear force on the oblique plane.

a resulting elongation or stretch. The value of the elongation is given by the difference between the final or stretched length and the initial length of the rod. The same load applied to a rod of twice the initial length would produce an elongation twice as large and, conversely, if the initial length were half that of the original rod, the elongation under the same load would be half as great. However, if the elongation were divided by the original length of the rod, the result would be the same for all three cases. This is, in fact, the definition of the tensile strain, which is given by (elongation) / (original length). This strain is independent of the original length of the rod, which is sensible, since the movement of the material inside the rod should not depend on the length of the rod.

Tensile stress and tensile strain have been defined. However, in an analogous manner to the existence of other loads than tensile loads, and other deformations than extension, there are different types of stress and strain. Again it is convenient to consider different cases of loading, since these have direct relevance to engineering and mechanics. However, all cases of stress may be described in terms of tensile (or compressive) stresses and shear stresses. Corresponding to these there will be tensile (or compressive) strains and shear strains.

The existence of shear stresses and strains may be seen by considering, from a different point of view, the situation of a rod subjected to a tensile load. Figure 17–20 A shows the free-body diagram, but with the cylinder cut at an oblique angle. This does not change the value of the internal force, which must balance the applied load. Figure 17–20 B illustrates how the force may be considered as the resultant of two components: $F_\perp$, which is perpendicular to the plane of the cut, and $F_\parallel$ which is parallel to the plane of the cut. The tensile stress acting on this plane of the cut is given by $F_\perp$/(area of oblique plane). The shear stress is given by $F_\parallel$/(area of oblique plane). Note that the area used to calculate the shear stress is the one containing the shear force, while the area used to calculate the tensile stress is perpendicular to the tensile load.

The term pressure refers to the force per unit area that a liquid or gas exerts against a contacting surface. Thus, pressure is a stress. Pressure is always exerted in a direction normal to a contacting surface. In a liquid, an object or, indeed, a small volume of the liquid itself is subjected to a uniform pressure acting from all directions; this is known as a hydrostatic pressure and increases with increase in depth.

Pressure must be applied to induce a liquid to flow. This is seen in the body, where a pulsating pressure is applied to produce blood flow. Pressures are generated in the synovial liquid film at articulating joint surfaces, and the film keeps the joint surfaces apart, reducing friction and wear. Pressure also exists inside the intervertebral discs of the spine. After the second decade, a comparatively minor accident may cause the nucleus pulposus to burst through the annulus fibrosus, often causing great pain.

It is important to have a knowledge of the different types of stress and strain, since deformation and failure of a structure may not result from the main stress state present. For example, in the case of a tensile load, it has been shown that both tensile and shear stresses are present. For a cylinder made from a brittle material such as a ceramic, the failure will be due to tensile stresses; and—since the maximum tensile stress occurs on the sections perpendicular to the axis of the cylinder—failure will occur on such a plane. On the other hand, metals and plastics can withstand such stresses and will show shear deformation on an oblique plane. Some materials, like wood, will withstand tensile strains but are weak in shear strength.

If a closed container contains a liquid or gas under pressure, the container will be subjected to forces that result in stresses in the container walls. The bursting of pressurized vessels is not unknown, and is usually due to the circumferential stress in the wall exceeding the tensile strength of the material. The circumferential stress is known as the hoop stress, since iron hoops were used on barrels to prevent such an occurrence. In the example of the intervertebral disc, the protrusion of the nucleus pulposus occurs partly because of damage of the annulus fibrosus and partly because the stresses generated in the weakened annulus fibrosus exceed the stress, to cause failure. The stresses generated in a cylinder and a sphere may easily be calculated if the walls are thin.

One point not so far mentioned is the distribution of stress and strain throughout a structure. The nature of the internal state of a structure under a load or loads is completely specified by giving the stress and strain at each point inside the structure. It is only rarely that uniform stress and strain states are found. Usually the stress and strain vary from point to point in the structure, but the variation is in a well defined manner which can be specified from theory or by experiment. However, on occasion, there are large changes over very small distances (in terms of stress these are known as stress concentrations) which can lead to failure, since unexpectedly high stress levels may be attained. It is the objective of good design practice to minimize stress concentrations and to arrange for as gradual as possible transitions in stress from point to point.

Stages of Deformation

When a force or load is applied to a structure, there will be a resulting deformation. The load is resisted by internal stresses. Corresponding to the external change, or deformation, in the structure, there are internal changes which, when normalized, are known as strains. Thus external loads lead to internal stresses and resultant strains.

The result of applying loads to a structure depends upon the nature of the load, the shape and dimensions of the structure, and the material from which the structure is made. In engineering calculations, materials are characterized by several parameters.

The response of a material to a load is usually determined from the application of a tensile load to a simple shape. This is the tensile case described earlier. Figure 17–21 A illustrates a typical load-extension curve for an elastic-plastic material. There are two regions to the curve. Initially the load is proportional to the extension and the response is linear. This is called the *elastic* region, since the material behaves like a spring and the deformation is fully recoverable. If the load is removed the cylinder returns to its original dimensions. If the load is great enough, the relationship between load and extension is no longer linear—this is the plastic region, and upon unloading, the deformation will not be fully recovered. The remaining deformation is known as permanent deformation. The load at which there is a transition from elastic to plastic behavior is called the yield load (on the curve this is the yield point).

The discussion has heretofore been in terms of load and extention, since these are the quantities measured during an evaluation. However, the results are better presented in a stress-strain diagram, since the stress and strain do not depend on the dimensions of the test structure, but are characteristic only of the material from which the structure is made. Figure 17–21 B shows the stress-strain curve corresponding to Figure 17–21 A. Various parameters have been marked on the graph, and are summarized below:

Stress: The ratio of force to cross-sectional area. For small changes (strains) the initial cross-sectional area is used and the ratio is called the *nominal* stress. If the actual area is used, we obtain the *true* stress.

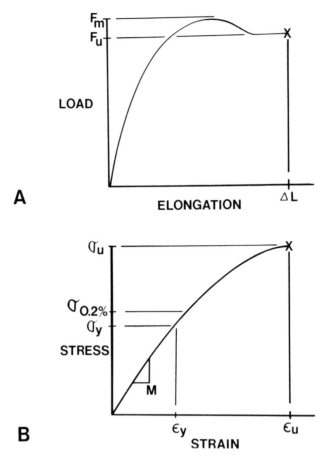

Fig. 17–21. (A) A load-extension curve for an elastic-plastic material. (B) The stress-strain curve (with nominal stresses) corresponding to the load-extension curve.

Strain: The ratio of change in length to the original length. Strain may be multiplied by 100 and expressed as a percentage.

Elastic modulus: The *slope* of the stress-strain curve in the elastic region. The modulus may be used to find the strain reached at a given stress since:

$$\text{Stress} = \text{Modulus} \times \text{Strain} \qquad (16)$$

Yield stress: Above the yield stress, the deformation is plastic; below the yield stress, it is elastic. Since the point of yielding is often hard to measure accurately, the yield stress is sometimes defined as that stress giving a certain permanent deformation of 0.2 percent. Figure 17–8 B illustrates the behavior on unloading in the plastic region, and shows both the yield stress and the 0.2 percent yield stress.

Tensile strength: The maximum stress which the material can withstand; the stress of failure.

Strain to failure: The strain at which failure occurs. As a percentage this is also called "percent elongation," or ductility.

It is general practice to design devices so that stresses within a structure under load remain within the elastic region. This is accomplished by choice of the shape and dimensions of the structure, by the material employed for the structure and, on occasion, by control of weight-bearing. Thus the only parameter required to characterize the material is the elastic modulus.

The Poisson's ratio of the material may also be required. This is the ratio of lateral strain to longitudinal strain. An elastic material may be completely characterized in the elastic region by the modulus and Poisson's ratio.

However, under some circumstances, the load may exceed the design load (the load at which the structure is expected to operate). Then the yield stress becomes important. Should the yield stress be exceeded, it is important to know whether the structure will deform or break. Thus, both the tensile strength and also the elongation to failure must be known. Design loads are exceeded in many cases during impact conditions in which the load rises very rapidly, as, for example, during a fall. Impact loads test the toughness of a material, since the behavior depends on the ability of the material to absorb the energy of impact. The amount of energy required to cause fracture is given by the area under the stress-strain curve.

The foregoing discussion has been concerned with those parameters that may be used to characterize a material on the basis of application of a tensile load. As discussed already, there are other loading situations that make use of the definitions above in a general way; thus analogous moduli, yield points, and strengths may be defined.

Structures—especially those of interest in the body—are not normally subjected to constant loads. The situation is one of repetitive loading. For example, ambulation produces a particular pattern of loading at the hip, with load maxima at heel strike and toe-off. This loading is repeated at each walking cycle. Materials behave differently under repetitive loading than under steady or steadily increasing loads. Failure of a material can occur after a certain number of load applications, even though the stress on the material is within the elastic region. This type of behavior is known as fatigue. Studies are made of the number of load applications to failure versus the applied load, to give a graph from which the lifetime of the structure is calculated. As a rule of thumb, the maximum stress which a material can withstand without fatigue failure is about 40 percent of its tensile strength.

The influence of time as well as load has been raised in connection with loads applied within a short time-span (impact), and for loads applied repetitively (fatigue). Time can enter in another context also; for some materials, especially plastics, the response to a load is not instantaneous, as it is with metals and ceramics. Plastics give a delayed response and, in addition, the magnitude of this delayed response depends upon the rate at which the load is applied. Materials that exhibit time-dependent behavior are known as viscoelastic materials.

As a group, viscoelastic materials—whether of natural or man-made origin—display the following general behavior:

1. Moduli and strength increase with the rate of load application; elongation to failure decreases.
2. Elongation beyond the yield stress, as defined by the end of the linear region on the stress-strain curve, may be fully or partially recoverable.
3. For a fixed rate of load application, the materials may be characterized in the same way as elastic-plastic materials.

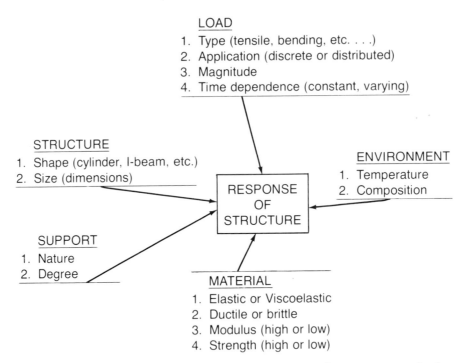

Fig. 17–22. The factors which determine the response of a structure to a load.

Earlier, it was mentioned that engineering structures are designed to operate within the elastic region, and that the only parameter required for specification of the material is the elastic modulus.

It has been tacitly assumed that material parameter(s) have been determined in air, and that a structure is loaded in air. An implant, however, is bathed in body fluid and is at 37°C. The behavior of a material in an environment other than air at room temperature may be quite different than in air, and the response should not be based on the value of parameters obtained by measurement in air. Typical chronic effects of a liquid environment include corrosion, embrittlement, and a lowering of strength and fatigue resistance.

Response of a Structure

The preceeding sections have discussed various aspects of the response of a structure. These are summarized in Figure 17–22.

For an implant, the load may not be well known, but its magnitude may be reduced by partial weight bearing. The distribution of the load may be controlled by the shape of the prosthesis and the nature of the cement distribution. For a bone plate, the distribution of the screws influences the nature of the load on the plate. The time-dependent response of viscoelastic materials, such as base and base cement may be partially controlled by controlling the rate of ambulation.

The environment of an implant can only be altered to a minimal extent. For example, a coating or layer may be put on the implant surface, so that the implant is not in contact with the environment. However, unless the layer is self-healing, such an approach may be undesirable because of the consequences of a break in the layer.

The nature and degree of support influences the load and load distribution of an implant. Fixation devices are not designed to withstand the activity of walking unless the load is shared with the bone—a sharing that is ideally about 50:50. Thus fractures must be well reduced before a fixation device will be sustained. Prostheses, on the other hand, must withstand the loads of ambulation. However, the degree of support given by the bone is also important. Disappearance of the calcar or fracture of the proximal cement reduce this support and have been implicated in fatigue fractures of femoral stems.

The shape and dimensions of a structure greatly affect its response. For an implant there are anatomic restrictions on the size; for example, skin coverage at the tibia, the matter of compromising the blood supply in the femoral neck. Despite this, the shape may often be chosen to give the desired result (usually the lowest stresses).

The choice of material used for the structure is usually of the utmost importance, especially for implants. The material chosen must have sufficient strength and ductility, but also be corrosion resistant and biocompatible.

The effect of a given load applied to a given structure made from a given material may be summarized as follows:

1. There is elastic deformation if the stresses are below the yield stress.
2. There is plastic deformation if the stresses are above the yield stress but below the breaking stress (compressive, tensile, or shear strength, as appropriate).
3. There is fracture if the stresses exceed the breaking stress.

The above conditions apply to a single load application, but it is easy to consider fatigue due to repetitive loading as an extension, time effects for viscoelastic materials, and the effect of environment.

Response of Particular Structures

BEAMS

A beam is a structure capable of sustaining a load applied in a direction normal to its axis. Most implants, such as fixation plates and screws, nails, and the femoral components of hip prostheses are essentially beams.

The response of a beam depends to a large extent on the nature of the support and the manner of loading.

Stresses in Beams

The most noticeable effect of load application to a beam is bending. Bending produces a tensile force at one surface and a compressive force on the opposing surface of the beam. The force decreases in the interior of the beam and at some position must be zero; this position is called the neutral plane, since the force changes from tensile to compressive. The magnitude of the bending stress at a point in the beam is given by the following formula:

$$\sigma = \frac{My}{I} \tag{17}$$

TABLE 17-1. AREAL MOMENT OF INERTIA

Shape	Areal Moment of Inertia
(circle, radius r)	$I = 1/4\ \pi\ r^4$
(annulus, r_o, r_i)	$I = 1/4\ \pi\ (r_o^4 - r_i^4)$
(rectangle, w × d)	$I = 1/12\ wd^3$
(triangle, side b, a)	$I = \dfrac{\sqrt{3}}{96}\ b^4$
(I-beam, w, d, a)	$I = 1/12\ [wd^3 - (w - a)(d - 2a)^3]$

where M is the value of the moment at the section where the stress is to be deter-mined, y is the distance of the point from the neutral plane, and I is the moment of inertia of the cross-section. This is *not* the mass moment of inertia discussed in the section on dynamics; the value of I depends upon the *distribution* of material across the section of the beam. Table 17-1 gives the moment of inertia for some simple shapes of section. Note that for these simple shapes the neutral plane is at the center of the section. The maximum value of the tensile or compressive stress occurs at the surface of the beam, where y is set equal to $\pm$ h for a symmetrical section.

Referring to a model of a hip nail (Fig. 17-6), the maximum moment M occurs at the fixed end of the nail. Thus:

$$M = F_\perp\ \ell \qquad\qquad (18)$$

where $F_\perp$ = component of force (= 900N) and
y is set equal to half the thickness since it is assumed that the nail has a symmetrical

section. The value of I may be obtained from Table 17-1. The maximum surface stress as a function of diameter for a circular section of nail is then:

$$\sigma = \frac{32\,M}{\pi d^3} \tag{19}$$

and doubling the diameter thus decreases the stress by a factor of 8.

Stresses in Composite Beams

Examples of composite beams include a hip nail in bone, in which the whole structure can act as a beam if the fracture is stable; a femoral component cemented into a femur; and a bone plate attached to the shaft of a bone. Of course, these also involve nonuniform cross-sections. In all three cases it is expected that part of the load will be shared with the bone, so that the stresses on the implant are reduced. However, for unstable fractures the device bears most or all the load, and in the case of cement breakdown the implant is subjected to increased loads. The response of the bone-implant combination is known as the response of a composite structure; two examples will be given.

The stresses may be calculated for an applied load F, which produces a bending moment M varying from place to place in the stem. This is a combined loading, and stresses due to both compression and bending are produced (the compressive stress was ignored in the examples given earlier for the hip nail and the femoral component). These compression and bending stresses were determined by an extension of the theory so far discussed, as follows:

$$\text{Stress due to compression } (\sigma_i)_c = \frac{E_i F}{E_b I_b + E_c A_c + E_p A_p} \tag{20}$$

$$\text{Stress due to bending } (\sigma_i)_b = \frac{E_i Y_i M}{E_b I_b + E_c I_c + E_p I_p} \tag{21}$$

where b, c, and p refer to bone, cement, and implant, respectively. E, A, and I are the elastic modulus, cross-sectional area, and moment of inertia, and Y_i is the distance from the neutral axis to the point of interest in component i. Note that these equations reduce to the case considered earlier, in which the prosthesis is supported only near the end.

$$(\sigma_i)_b \rightarrow \frac{E_p P_p M}{E_p I_p} = \frac{MY}{I} \tag{22}$$

and

$$(\sigma_i)_c \rightarrow \frac{E_p F}{E_p A} = \frac{F}{A} \tag{23}$$

the compressive stress, which was ignored.

An implant shares the imposed loading with the bone that it supports. If the total applied bending moment is M, the moments carried by each are:

$$M_p = \frac{M\ E_p I_p}{E_p I_p + E_b I_b} \tag{24}$$

$$M_b = \frac{M\ E_b I_b}{E_p I_p + E_b I_b} \tag{25}$$

where p and b refer to the implant and bone, respectively. The moments of inertia I_p and I_b are taken about the neutral axis of the composite structure; usually this does not coincide with the neutral axis of the bone or implant.

Deformation of Beams

In design, a deformation often shares an equal importance with strength. A bone plate may have sufficient strength in that the stresses are below the yield stress, but the deformation may be so great that the movement at the fracture site produces a pseudarthrosis instead of healing.

As for the case of stresses, the deformation depends on the way in which the beam is supported and the manner in which the loads are applied. Some typical results will be given.

A cantilever beam of length ℓ, with a concentrated load F applied at the end, bends in a continuous curve. The deflection, δ, of the free end, is given by:

$$\delta = \frac{F\ell^3}{3EI} \tag{26}$$

For a distributed load of w N/m, the deflection at the free end is:

$$\delta = \frac{w\ell^4}{8EI} \tag{27}$$

In the case of a simply supported beam with load F at the midpoint, the deflection of the midpoint is:

$$\delta = \frac{F\ell^3}{48EI} \tag{28}$$

For the cases considered, equations are available to give the deflection at any point along the beam. Solutions are obtainable for other types of support and other types of load. Other cases may be determined by using a superposition method. Thus, if a cantilever beam has a concentrated load F at the end and a distributed load w, the deflection at the free end is

$$\delta = \frac{F\ell^3}{3EI} + \frac{w\ell^4}{8EI} \tag{29}$$

It will be noted that the term EI appears in each of these equations. The larger the value of EI, the smaller the deflection and the stiffer the beam. Thus EI plays a large role in determining the stiffness of the beam. Again it should be noted that I is determined by the distribution of material across the section and not simply by the overall dimensions of the cross-section.

The effect of implant stiffness has been implicated in the decrease in bone mass noted some months following fracture healing when a device is left in place. It has been argued that the implant relieves the bone of some of the load, thus causing Wolff's Law-remodeling. The effect is shown to be much decreased when less stiff devices are employed.

The effect of material has been discussed only in terms of the tensile modulus, E. However, if the material of the beam is visco-elastic, there will be time effects. A calculation of the deformation may be made by using the deformation equations given, but with the understanding that the value of E is itself time dependent.

Design Considerations for Implants

The considerations involved in the designs of an implant are complex, and to date, the designs have largely been arrived at by trial-and-error. However, there are certain considerations which may be borne in mind when designing an implant.

STIFFNESS

The stiffness of the device depends on EI. For fracture fixation, an implant with a high value of EI gives a stronger support to the bone, and so gives a more stable reduction.

Change in EI may be done by choice of material; for example, titanium has an E value of 1×10^{11} N/m,^{2} compared with 2×10^{11}N/m^2 for stainless steel or cobalt-chromium-molybdenum alloy. A typical value of E for a polymer is 3.5×10^9 N/m,^{2} and for a composite material $E = 1.4 \times 10^{10}$N/m.^{2}

The value of I is changed by altering the cross-sectional shape or the overall dimensions of the section: For a rectangular section, doubling the thickness increases I by a factor of eight.

LOAD SHARING

The sharing of a load between an implant and bone depends upon the stiffness of the implant compared with the stiffness of the bone. Increasing the stiffness of the implant reduces the total deformation of the bone-implant composite, but also increases the load on the implant while decreasing the load on the bone. This could lead to stress protection and subsequent bone loss.

IMPLANT STRESSES

The maximum stress in an implant must be below the yield stress, or plastic deformation will occur. Additionally, the loads imposed on an implant vary, and fatigue failure is thus a possibility. Thus the load borne by the implant should not give rise to stresses which exceed the fatigue strength.

Increasing the stiffness increases the load on the implant. Whether the stress increases or decreases depends on whether $E_pI_p \ll E_bI_b$ or whether $E_pI_p \gg E_bI_b$. Considering the equation:

$$\sigma_{max} = \frac{M\ E_pI_p}{E_pI_p + E_bI_b} \cdot \frac{Y}{I_p} \tag{30}$$

for $E_pI_p \ll E_bI_b$

$$\sigma_{max} = \frac{M(E_p)}{(E_b)} \cdot \frac{Y}{I_b} \tag{31}$$

and an increase in stiffness of the implant leads to an increase in stress. On the other hand for $E_pI_p \gg E_bI_b$:

$$\sigma_{max} = \frac{M(Y)}{(I_p)} \tag{32}$$

and an increase in stiffness leads to a decrease in stress.

It has been assumed in these calculations that the neutral plane does not move as the stiffness is changed. The calculation is therefore appropriate for an intramedullary rod or a femoral stem.

Mechanics of Shafts

Twisting of a structure produces quite a different response than does bending. A structure that is intended to withstand twisting or to transmit a torque is called a shaft. The bones of the leg are subjected to a torque or twisting moment. For example, forcible twisting of the foot applies a twisting moment of about $20\ \mathrm{N} \cdot \mathrm{m}$ to the tibia. Torsion tests on isolated tibiae give a torque at failure between 20 and $90\ \mathrm{N} \cdot \mathrm{m}$. Spiral fractures indicate that twisting of the bone has occurred until the strength of the bone has been exceeded.

Stresses in Shafts

The internally generated shear stress τ is of most interest. This shear stress is the reaction to the applied torque, and gives equilibrium. Once the torque is applied, the shaft twists and the shear stress increases until a balance is achieved. The shear stress is zero at the center of the shaft, and at the surface of the shaft has a maximum value τ_{max}, given by:

$$\tau_{max} = \frac{Td}{2J} \tag{33}$$

In this equation, J is the polar moment of inertia, and is analogous to the moment of inertia in bending. The magnitude of J is determined by the distribution of material about the center of the section. For a solid circular section:

$$J = \frac{\pi d^4}{32} \tag{34}$$

and for a hollow circular section (an approximation to a bone) with internal diameter d_2 and external diameter d_1:

$$J = \frac{\pi(d_1{}^4 - d_2{}^4)}{32} \tag{35}$$

The efficacy of a hollow section in reducing the amount of material while still giving a high resistance to torsion is demonstrated with regard to the femur. Taking the femoral shaft as a hollow cylinder of inner diameter 16 mm and outer diameter 27 mm gives a value of J only *12 percent smaller* than if the section were a solid cylinder of diameter 27 mm.

Deformation of Shafts

The deformation of a shaft is expressed in terms of the angle Θ through which the shaft is twisted, and:

$$\Theta = \frac{T\ell}{GJ} \tag{36}$$

where ℓ is the length of the shaft and G is the modulus in shear of the material of the shaft. For many materials the value of G is about 40 percent the value of E.

The quantity GJ determines the rigidity of the structure in torsion—and is analogous to the term EI, which determines the stiffness in bending. The value of GJ may be altered by altering the value of the shear modulus, but the largest influence comes from changing J by changing the shape of the section. A cloverleaf nail design of appropriate dimensions to fit a femur has a torsional rigidity of $1.57 \text{ N} \cdot \text{m}^2$ A diamond section nail of appropriate dimensions has a torsional rigidity of 11.8 N.m.^2 Thus the solid nail is about ten times as rigid as the open-section nail; both are much less rigid than the femur. However, in order to transmit shearing stress from the bone to the nail—which is the only way in which the nail can assume load from the bone—it is necessary that the nail fit tightly in the cavity. Owing to the construction of the open-section nail, this is accomplished much more successfully than for the solid nail. Despite this statement, intramedullary nails generally have a very poor ability to transmit torsion. Under bending, however, the open-section nail performs just as well as the solid nail, and both have approximately the same stiffness.

Summary

An implant should provide sufficient stiffness so that a fracture is stabilized, but not so much that the load on the bone is reduced to the point at which bone resorption and remodeling occurs. The loads on the implant should not be so high as to cause fatigue failure, plastic deformation, or outright failure.

As was mentioned earlier, the situation is very complex and there is the added complication that different types and conditions of fracture have different degrees of stability. Ideally, the device should be matched to the fracture type with a stiffer device used for a less stable fracture. It is not certain that this is always done. However, the wide range of devices available, having very different degrees of stiffness, shows that an appreciation of stiffness as a design parameter has been realized. Nevertheless, it is hardly likely that use of these different designs in a systematic way has been attempted.

As regards prostheses, many of the same considerations apply. A prime requisite is to reduce the stresses in the implant, such as a femoral stem, so that the likelihood of fatigue failure is reduced. This is necessarily so, since the implant must be left in place, as compared with a fracture fixation device, which can be removed following fracture healing. The stresses in the implant may be reduced by increasing the stiffness, but this reduces the load on the bone, with the consequences of possible deleterious effects due to remodeling.

FRICTION AND LUBRICATION

The activities of everyday life are accompanied by frictional effects. Friction is the phenomenon of resistance to motion and, while efforts are made to reduce the frictional effects in areas such as artificial joints, friction is essential for the act of ambulation itself, since forward propulsion of the body demands that a frictional force act in the opposite direction at the moment of contact between the foot and floor.

Friction has been systematically examined on a macroscopic basis for some 200 years. The studies have been concerned with the behavior of simple systems, such as an object on a flat surface. It was noted that if a low force is applied, the object does not move. For equilibrium, an equal and opposite force—the friction force—is generated at the interface between the object and the flat surface. If the applied force is increased, the object still does not move, and the friction force must thus increase, along with the applied force. However, if the applied force exceeds a limiting value, which depends upon factors such as the weight of the object, the materials in contact, and the nature of the surfaces in contact, the object will move. It is then found that the force required to maintain motion is less than the force required to initiate motion. Note that the friction force is caused by the formation of junctions at the interface, and is manifested as a reaction force when a force is applied to cause motion. If motion does not occur, this friction force is equal and opposite to the applied force (Newton's third law—action and reaction are equal and opposite—applied to the case of friction).

The discussion of friction between two surfaces is usually carried out in terms of the friction coefficient, which is defined as:

$$\text{Friction coefficient} = \frac{\text{Applied force}}{\text{Normal force}} \tag{37}$$

For the case considered above, the normal force is the weight of the object. If the applied force is that force just giving motion, the coefficient of friction is called the static coefficient of friction. If the applied force is that needed to maintain motion, the co-efficient of friction is called the sliding coefficient of friction.

The studies of friction already mentioned lead to the "Laws of Friction."

1. The friction coefficient is independent of the normal force between the two surfaces.
2. The friction coefficient is independent of the size of the two surfaces.

These laws may be derived, with some simple assumptions, by using the microscopic view of friction. It should be noted that Law 1 means that the friction force is directly proportional to the normal force. Thus, in the example considered, if the normal force (weight of the object) is doubled, the applied force doubles.

The value of the friction coefficient depends upon several factors: (a) The materials from which the two surfaces are made: (b) the cleanliness of the surfaces; and (c) the presence or absence of lubrication.

Table 17−2 gives typical values of friction coefficients for both dry and lubricated cases of sliding. The coefficient of friction is much lower for "full film" lubrication, but even modest lubrication can substantially reduce the friction coefficient (Fig. 17−23). The presence of lubrication reduces surface interaction and hence friction, and thus less energy is required for motion. Of far greater importance is the fact that reduced surface interactions reduce wear−indeed, the reduction in wear is often far more substantial than the reduction in friction.

Frictional effects are found throughout orthopedics, but it is appropriate to cover a few cases to illustrate the calculation and importance of frictional effects.

Friction is involved at the handle of a screwdriver. The torque applied to the handle is transmitted by the interaction between the hand and the handle. The friction between the hand and the handle is equal to (the friction coefficient) × (squeezing force exerted by the hand). This friction force must equal or exceed the applied torque (the twist applied to the handle), otherwise the hand will slip.

TABLE 17−2. SLIDING CO-EFFICIENTS OF FRICTION

Material Combination	Lubricant	μ
Rubber tire/concrete	none	0.7
Rubber tire/concrete	water	0.5
Leather/wood	none	0.4
Steel/steel	none	0.5
Steel/polyethylene	none	0.1
Steel/ice	water	0.01
Cartilage/cartilage (hip)	synovial fluid	0.002
Cartilage/cartilage (hip)	Ringer's	0.01−0.005
Co-Cr/Co-Cr (hip prosthesis)*	none	0.55
	serum	0.13
	synovial fluid	0.12
Co-Cr/UHMWPE*	serum	0.08

*Weightman, B., Simon, S., Paul, I., Rose R., and Radin, E.: *J. Lubric. Tech.,* 94:131−135, 1972.

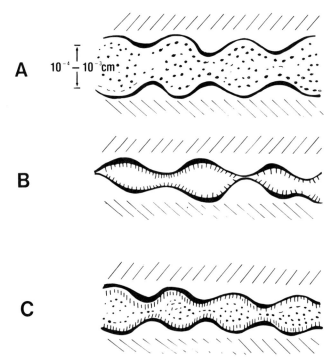

Fig. 17–23. Illustration of different types of lubrication: (A) Hydrodynamic lubrication. (B) Boundary lubrication. (C) Mixed lubrication.

A similar principle holds for the fixation of intramedullary nails. As the nail is driven in, the bone is compressed and there is a reaction force on the nail. An acting torsional force must overcome the frictional force between nail and bone, as follows:

$$\text{Applied torque} \geq \text{Frictional torque} \tag{38}$$

The frictional torque is given by the product of the reaction force, the friction coefficient, and the diameter of the nail. On the other hand, the applied torque is given by the product of a muscle force and the radius of the bone. Since the applied torque has a much larger moment arm than the frictional force, intramedullary nails do not provide a high resistance to torsion.

Figure 17–24 schematically illustrates a joint. The normal force at flat-foot is approximately 3,000 N. For a natural joint, lubrication gives a friction coefficient of 0.002. Thus, the frictional force is 6 N, acting at the joint surface in the direction opposite to motion. The frictional torque is, therefore, 0.135 N · m, where a femoral head diameter of 45 mm has been assumed. This is the torque that must be exceeded to sustain motion. However, the lubrication is so efficient, as manifested by the very low friction coefficient, that the torque is negligible compared to the forces required to move soft tissue and to accelerate body segments. For joint prostheses, the friction coefficient is higher. Thus, for highly congruent metal-on-metal joints with no lubrication, a friction coefficient of 0.55 is usual, leading to a friction force of 1,650 N and a frictional torque of 37 N · m. Lubrication with synovial fluid decreases the friction coefficient to 0.12, with a corresponding decrease in frictional torque. It should be noted that the diameters of joint prostheses are smaller than those of natural

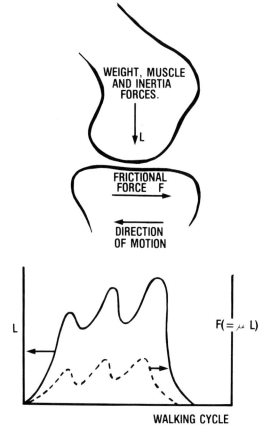

Fig. 17–24. Frictional force acting at an articulating joint.

joints, and that this will decrease the frictional torque. Metal-on-plastic joints give lower values of the friction coefficient. For stainless steel on ultrahigh-molecular-weight polyethylene, lubricated, the friction coefficient is about 0.04, and with a head diameter of 22 mm the friction torque is 1.32 N · m. The main reason for requiring a low friction coefficient is that this implies lubrication, decreased surface interaction, and greatly decreased wear.

BIBLIOGRAPHY

Statics and Dynamics

Hayden, H. W., Moffatt, W. G., and Wulff, J.: Mechanical Behavior, Vol. 3: The Structure and Properties of Materials, J. Wulff, Ed. John Wiley & Sons, New York, 1965.

Levinson, I. J.: Introduction to Mechanics. Prentice-Hall, Inc., Englewood Cliffs, New Jersey, 1968.

Marin, J.: Mechanical Behavior of Engineering Materials. Prentice-Hall, Inc., Englewood Cliffs, New Jersey, 1962.

Meriam, J. L.: Engineering Mechanics – Statics and Dynamics. John Wiley & Sons, New York, 1978.

Biomaterials

Park, J. B.: Biomaterials. Plenum Press, New York, 1979.

Dumbleton, J. H., and Black, J.: An Introduction to Orthopaedic Materials. Charles C Thomas, Springfield, Illinois, 1975.

Bloch, B., and Hastings, G. W.: Plastic Materials in Surgery. Charles C Thomas, Springfield, Illinois, 1972.

Yamada, H.: Strength of Biological Materials. In: F. G. Evans, Ed., The Williams & Wilkins Co., Baltimore, 1970.

Biomechanics

Frankel, V. H., and Burstein, A. H.: Orthopedic Biomechanics. Lea & Febiger, Philadelphia, 1970.

Frankel, V. H., and Nordin, M.: Basic Biomechanics of the Skeletal System. Lea & Febiger, Philadelphia, 1980.

Ghista, D. N., and Roaf, R.: Orthopaedic Mechanics: Procedures and Devices. Academic Press, London, 1978.

Northrip, J. W., Logan, G. A., and McKinney, W. C.: Introduction to Biomechanic Analysis of Sport. Wm. C. Brown, Dubuque, Iowa, 1973.

Radin, E. L., Simon, S. R., Rose, R. M., and Paul, I. L.: Practical Biomechanics for the Orthopaedic Surgeon. John Wiley & Sons, New York, 1979.

Wainwright, S. A., Biggs, W. D., Currey, J. D., and Gosline, J. M.: Mechanical Design in Organisms, Halsted Press (John Wiley & Sons), New York, 1976.

Williams, M., and Lissner, H. R.: Biomechanics of Human Motion. W. B. Saunders Co., Philadelphia, 1962. 2nd edition (B. LeVeau, ed.), 1977.

White, A. A., III, and Panjabi, M. M.: Clinical Biomechanics of the Spine. J. B. Lippincott Co., Philadelphia, 1978.

Implant Biomechanics

Bechtol, C. O., Ferguson, A. B., and Laing, P. G.: Metals and Engineering in Bone and Joint Surgery. The Williams & Wilkins Co., Baltimore, 1959.

Schaldach, M., and Hohmann, D.: Advances in Artificial Hip and Knee Joint Technology, Springer-Verlag, Berlin, 1976.

Swanson, S. A. V., and Freeman, M. A. R.: The Scientific Basis of Joint Replacement. John Wiley & Sons, New York, 1977.

Walker, P. S.: Human Joints and Their Artificial Replacements. Charles C Thomas, Springfield, Illinois, 1977.

Williams, D. F., and Roaf, R.: Implants in Surgery, W. B. Saunders Co., London, 1973.

Case Index

Page numbers in *italics* indicate illustrations; numbers followed by (t) indicate tables.

Subject Index

Page numbers in *italics* indicate illustrations; numbers followed by (t) indicate tables.